Pergamon Series in Analytical Chemistry
Volume 4

General Editors: R Belcher (Chairman), D Betteridge & L Meites

Chemical Methods of Rock Analysis

THIRD EDITION

Related Pergamon Titles of Interest

BOOKS

Pergamon Series in Analytical Chemistry

Volume 1
MEITES: An Introduction to Chemical Equilibrium and Kinetics

Volume 2
PATAKI & ZAPP: Basic Analytical Chemistry

Volume 3
YINON & ZITRIN: The Analysis of Explosives

Volume 5
PRIBIL: Applied Complexometry

Volume 6
BARKER: Computers in Analytical Chemistry

Other Titles

FERGUSSON: Inorganic Chemistry and the Earth

GREENWOOD & EARNSHAW: Chemistry of the Elements

HENDERSON: Inorganic Geochemistry

VAN OLPHEN & FRIPIAT: Data Handbook for Clay Materials and Other Non-Metallic Minerals

WHITTAKER: Crystallography — an Introduction for Earth Science (and other Solid State) Students

JOURNALS

Applied Geochemistry
Geochimica et Cosmochimica Acta
Ion-Selective Electrode Reviews
Organic Geochemistry
Progress in Analytical Spectroscopy
Talanta

Full details of all the above publications/free specimen copy of any Pergamon journal available on request from your nearest Pergamon office.

Chemical Methods of Rock Analysis

by

P G JEFFERY, *Deputy Director (Resources)*
Laboratory of the Government Chemist, London, UK

and

D HUTCHISON, *Geochemistry and Petrology Division, Institute of Geological Sciences (N.E.R.C.), London, UK*

THIRD EDITION

PERGAMON PRESS

OXFORD · NEW YORK · TORONTO · SYDNEY · FRANKFURT

U.K.	Pergamon Press Ltd., Headington Hill Hall, Oxford OX3 0BW, England
U.S.A.	Pergamon Press Inc., Maxwell House, Fairview Park, Elmsford, New York 10523, U.S.A.
CANADA	Pergamon Press Canada Ltd., Suite 104, 150 Consumers Road, Willowdale, Ontario M2J 1P9, Canada
AUSTRALIA	Pergamon Press (Aust.) Pty. Ltd., P.O. Box 544, Potts Point, N.S.W. 2011, Australia
FEDERAL REPUBLIC OF GERMANY	Pergamon Press GmbH, Hammerweg 6, D-6242 Kronberg, Federal Republic of Germany
JAPAN	Pergamon Press Ltd., 8th Floor, Matsuoka Central Building, 1-7-1 Nishishinjuku, Shinjuku-ku, Tokyo 160, Japan
BRAZIL	Pergamon Editora Ltda., Rua Eça de Queiros, 346, CEP 04011, São Paulo, Brazil
PEOPLE'S REPUBLIC OF CHINA	Pergamon Press, Qianmen Hotel, Beijing, People's Republic of China

First edition 1970
Second edition 1975
Reprinted (with corrections and additions) 1978
Third edition 1981
Reprinted (with corrections) 1983
Reprinted 1986

British Library Cataloguing in Publication Data

Jeffery, Paul Geoffrey
Chemical methods of rock analysis.—3rd ed.—
(Pergamon series in analytical chemistry)
1. Rocks—Analysis—Laboratory manuals
I. Title II. Hutchison, D.
552'.06 QE438

ISBN 0-08-023806-8

Library of Congress Catalog Card no.: 81-81234

Printed in Great Britain by A. Wheaton & Co. Ltd., Exeter

Preface to the Third Edition

There have been many changes in the third edition of this book quite apart from the more obvious ones dictated by the changes in printing and the need for economy. We have, for example, endeavoured to be a great deal more selective in the material presented, preferring to excise details of those older methods that have failed to keep their place in the laboratory in favour of methods based on newer ideas and techniques. This has inevitably and inexorably continued the swing from 'classical' (largely gravimetric and titrimetric) methods towards instrumental (spectrophotometric and atomic absorption) methods. Encouraged by the reception of the earlier editions, we have listened carefully to our reviewers and made significant changes to certain chapters (the noble metals in particular) in line with their recommendations. It is a matter of considerable regret that we have not been able to deal with a number of physical methods of analysis that are directly applicable to geological materials, particularly plasma source emission spectrography, x-ray fluorescence, - and micro-probe analysis.

The first and second editions incorporated a chapter on statistical methods. Its disappearance in this edition does not indicate any change of view of the importance of such methods, rather that such techniques are now a part of the stock-in-trade of analysts, and that therefore there is (or at least there should be) no need for their inclusion in a specialist text of this kind. It is assumed that a rock analyst using this book will be familiar with measures of central tendency, assessment of errors, dispersion of results, variance, correlation and tests of significance. Likewise simple curve fitting, design of experiments and elementary computer programming and operation can all now be considered as essential or near essential tools for the analyst of the eighties. There is no shortage of books on this topic ranging from student texts to advanced treatise. Recent volumes include Fundamentals of Mathematics and Statistics for Students of Chemistry and Allied Subjects by C J Brookes, I G Betteley and S M Loxston, Wiley 1979, and Statistical Methods in Trace Analysis by C Liteanu and I Rica, Ellis Horwood Ltd (Wiley) 1980.

Acknowledgements

The authors thanks are extended to the Director, Warren Spring Laboratory and the Director, Institute of Geological Sciences (N.E.R.C.) for permission to publish this book, and to the various authors and editors for permission to reproduce material published elsewhere. In addition, the authors gratefully acknowledge help and assistance from numerous colleagues, extending over many years.

Contents

Chapter 1. The Composition of Rock Material page 1

- Geochemical reference material 3
- Elements determined 9
- Reporting an analysis 12
- The selection of material for analysis 13
- Crushing and grinding 14
- Contamination 16

Chapter 2. Sample Decomposition

- Decomposition with mineral acids 18
  - Decomposition with hydrochloric acid 18
  - Decomposition with nitric acid 18
  - Decomposition with hydrofluoric acid 19
- Fusion procedures 21
  - Fusion with alkali fluoride 21
  - Fusion with potassium pyrosulphate 21
  - Fusion with sodium carbonate 22
  - Fusion with alkali hydroxide 24
  - Sinter or fusion with sodium peroxide 25
  - Fusion with boric oxide and alkali borates 26

Chapter 3. Classical Scheme for the Analysis of Silicate Rocks 29

- Decomposition of the sample 30
- Determination of silica 32
  - Separation and collection of silica 32
  - Ignition and volatilisation of silica 34
- Determination of "mixed oxides" 35
  - Precipitation of the "mixed oxides" 35
  - Recovery of the "aluminium traces" 37

Ignition of the "mixed oxides" 37
Recovery of "silica traces" 38
Determination of aluminium by difference 38
Determination of calcium 39
Determination of magnesium 40
Determination of manganese in the magnesium residue 42

Chapter 4. The Rapid Analysis of Silicate Rocks 43

Individual schemes of rapid rock analysis 48
Schemes of Shapiro and Brannock 48
Scheme of Riley 48
Scheme of Langmyhr and Graff 48
Scheme of Ingamells 49
Scheme of Shapiro 49
Scheme of Belt 50
Scheme of Bernas 50
Scheme of Langmyhr and Paus 51
Schemes of Abbey 51

Chapter 5. The Alkali Metals (Lithium, Sodium, Potassium, Rubidium, Caesium) 59

Determination of sodium and potassium in silicate rocks by flame spectrophotometry 61
Determination of potassium in micas 64
Determination of lithium 64
Determination of rubidium and caesium 65

Chapter 6. Aluminium 70

Gravimetric determination with 8-hydroxyquinoline 73
Titrimetric determination with CyDTA 75
Photometric determination with 8-hydroxyquinoline 77
Photometric determination with pyrocatechol violet 78

Chapter 7. Antimony 81

Spectrophotometric determination of antimony in silicate rocks 82
Determination of antimony by atomic absorption spectroscopy 84

Chapter 8. Arsenic 86

Spectrophotometric determination of arsenic 86
A molybdenum blue method using acid digestion 88
A molybdenum blue method using a fusion 90
An arsine-molybdenum blue method 91
Method using silver diethyldithiocarbamate 93
Determination of arsenic by atomic absorption spectroscopy 94

Chapter 9. Barium 96

Gravimetric determination of barium in silicate rocks 96
Determination of barium by atomic absorption spectroscopy 99

Chapter 10. Beryllium 102

Spectrophotometric determination of beryllium with beryllon II 102
Determination of beryllium by atomic absorption spectroscopy 105

Chapter 11. Bismuth 107

Chapter 12. Boron 109

Spectrophotometric determination of boron in silicate rocks 109
Titrimetric determination of boron in tourmaline 112

Chapter 13. Cadmium 115

Chapter 14. Calcium 117

Gravimetric determination of calcium 117
Titrimetric determination of calcium with EDTA 118
Determination of calcium in carbonate rocks (low in magnesium) 120

Photometric determination of calcium 121

Determination of calcium by flame photometry 121

Determination of calcium by atomic absorption spectroscopy 122

Chapter 15. Carbon 123

Determination of carbon dioxide 124

Method of Shapiro and Brannock 125

Method using Schrotter's apparatus 126

Gravimetric method using absorption tubes and acid decomposition 127

Gravimetric method using absorption tubes and thermal decomposition 130

Titrimetric determination of carbon dioxide 133

Determination of carbon dioxide by gas chromatography 133

Determination of carbon ("non-carbonate carbon") 134

Combustion tube methods 134

Determination of organic matter 135

Chapter 16. Chlorine, Bromine and Iodine 138

Determination of water-soluble chlorine 138

Determination of acid-soluble chlorine 138

Determination of total chlorine 139

Spectrophotometric determination of chlorine 140

Determination of bromine 142

Determination of iodine 143

Chapter 17. Chromium 145

Determination of chromium with EDTA 146

Determination of chromium as alkali chromate 148

Determination of chromium using sym-diphenylcarbazide 151

Determination of chromium by atomic absorption spectroscopy 153

Chapter 18. Cobalt 155

Spectrophotometric determination of cobalt using nitroso-R-salt 156

Determination of cobalt by atomic absorption spectroscopy 159

Chapter 19. Copper 160

Spectrophotometric determination of copper with 2,2'-diquinolyl 161

Determination by atomic absorption spectroscopy 163

Chapter 20. Fluorine 165

Spectrophotometric determination of fluorine 166

Spectrophotometric determination without distillation 169

Pyrohydrolytic determination of fluorine 169

Determination of fluorine using an ion-selective electrode 173

Chapter 21. Gallium 175

Spectrophotometric determination of gallium 175

Determination of gallium by atomic absorption 177

Chapter 22. Germanium 179

Spectrophotometric determination of germanium 179

Determination of germanium by atomic absorption spectroscopy 181

Chapter 23. Hydrogen 182

Water evolved at 105° 182

Method of Dean and Stark 185

Total water 186

Courville-Penfield method 186

Shapiro and Brannock-Penfield method 186

Harvey-Penfield method 187

Gooch-Sergeant method 189

Chapter 24. Indium 191

Chapter 25. Iron 192

Determination of metallic iron 192
Determination of ferrous iron 194
"Pratt method" 196
"Wilson method" 197
Spectrophotometric determination of ferrous iron 198
Determination of ferrous iron in carbonaceous shales 200
Determination of ferric iron 201
Determination of total iron 203
Spectrophotometric determination of total iron 204
Titrimetric determination of total iron 207

Chapter 26. Lead 213

Spectrophotometric determination of lead 214
Determination of lead by atomic absorption spectroscopy 216
Determination by sublimation procedures 218

Chapter 27. Magnesium 221

Gravimetric determination with 8-hydroxyquinoline 225
Titrimetric determination of calcium and magnesium 226
Determination of calcium and magnesium by atomic absorption spectroscopy 228

Chapter 28. Manganese 231

Spectrophotometric determination of manganese as permanganate 233
Oxidation with potassium periodate 234
Oxidation with ammonium persulphate 235
Spectrophotometric determination with formaldoxime 236
Determination of manganese by atomic absorption spectroscopy 237

Chapter 29. Mercury 238

Determination of mercury in silicate rocks 238
Vapour absorption for the determination of mercury 239
Method using acid decomposition 240
Method using thermal decomposition 242

Chapter 30. Molybdenum and Tungsten 245

Spectrophotometric determination of molybdenum and tungsten in silicate rocks 246
Determination of molybdenum in silicate rocks by atomic absorption spectroscopy 250

Chapter 31. Nickel 252

Gravimetric determination with dimethylglyoxime 253
Spectrophotometric determination with dimethylglyoxime 255
Determination of nickel by atomic absorption spectroscopy 258

Chapter 32. Niobium and Tantalum 260

Spectrophotometric determination of niobium 262
Determination of tantalum 264

Chapter 33. Nitrogen 265

Determination of nitrogen 265

Chapter 34. Phosphorus 267

Determination of phosphorus using a vanadomolybdate reagent 269
Determination of phosphorus as a molybdenum blue 271

Chapter 35. Scandium, Yttrium and the Lanthanide Rare Earths 274

Determination of scandium in silicate rocks 274
Determination of yttrium 275

Determination of the lanthanide rare earths 275

Spectrophotometric determination of rare earths in silicate rocks 277

Determination of yttrium and certain rare earths by atomic absorption spectroscopy 280

Chapter 36. Selenium and Tellurium 284

Determination of selenium 284

Photometric determination using 2,2'-diaminobenzidine 285

Determination of tellurium 288

Chapter 37. Silicon 290

Gravimetric determination of silica in quartzites and sandstones 292

Gravimetric determination in silicate and carbonate rocks 293

Determination in carbonate rocks 294

Gravimetric determination without dehydration 295

Spectrophotometric determination of silica 296

Combined gravimetric and photometric determination of silica 300

Determination of silica by atomic absorption spectroscopy 302

Chapter 38. Silver, Gold and the Platinum Metals 304

Determination of silver in silicate rocks 304

Determination of gold in silicate rocks 305

Determination of platinum metals in silicate rocks 308

Chapter 39. Strontium 311

Determination of strontium by flame photometry 312

Determination of strontium in silicate rocks by atomic absorption spectroscopy 314

Chapter 40 Sulphur 317

Determination of free sulphur 317
Determination of sulphide-sulphur 317
Water-soluble sulphur 319
Determination of acid-soluble sulphur 319
Determination of total sulphur 320
Gravimetric determination 321
Titrimetric determination 322

Chapter 41. Thallium 325

Spectrophotometric determination with brilliant green 325
Fluorimetric determination in silicate rocks 327

Chapter 42. Thorium 330

Spectrophotometric determination using fluoride precipitation 332
Spectrophotometric determination using anion-exchange separation 335

Chapter 43. Tin 338

Spectrophotometric determination with catechol violet 340
Spectrophotometric determination using salicylidene-amino-2-thiophenol 343
Determination of tin by atomic absorption spectroscopy 344

Chapter 44. Titanium 345

Spectrophotometric method using hydrogen peroxide 346
Spectrophotometric method using diantipyrylmethane 348
Determination of titanium by atomic absorption spectroscopy 350

Chapter 45. Uranium 352

Spectrophotometric determination following anion exchange separation 353

Chapter 46. Vanadium 357

Spectrophotometric determination of vanadium as phosphotungstate 357

Spectrophotometric determination using N-benzoyl-o-tolylhydroxylamine 359

Determination of vanadium by atomic absorption spectroscopy 361

Chapter 47. Zinc 363

Spectrophotometric determination following ion-exchange separation 363

Determination of zinc by atomic absorption spectroscopy 366

Chapter 48. Zirconium and Hafnium 368

Spectrophotometric determination of total zirconium and hafnium 368

Determination of hafnium 370

AUTHOR INDEX 371

SUBJECT INDEX 377

CHAPTER 1

The Composition of Rock Material

Since early times man has speculated upon the origin and composition of the earth and the great variety of rocks and minerals of which it is composed. For many of the eminent chemists of the eighteenth and nineteenth centuries, the uncharacterised minerals provided the challenge that led to the identification and subsequent isolation of the elements missing from the periodic table. By the end of the nineteenth century, Berzelius, Lothar Meyer, Lawrence Smith and others had laid the foundations of the classical scheme of silicate rock analysis as we know it today and, by the end of the century methods for the determination of all elements present in major amounts had been proposed and evaluated. By 1920, when Washington had issued the third edition of his book, "Manual of the Chemical Analysis of Rocks"(1) and Hillebrand his "The Analysis of Silicate and Carbonate Rocks"(2) (itself a revised and enlarged version of earlier texts), interest in silicate rock analysis had spread to those elements present in only minor amounts. Barium, zirconium, sulphur and chlorine - elements that could all be determined gravimetrically by well-established procedures - were soon added to the list of major components required for a "complete analysis". Elements such as titanium, vanadium and chromium were recognised as essential components of certain silicates, and new procedures were devised for their determination.

The interest in the minor components of silicate rocks has continued almost without a break to the present day, extending to elements at lower and lower concentration as more and more sensitive techniques have become available.

As with other well-defined applications of classical analytical chemistry, the ability to undertake a good analysis depended upon the skill of the analyst in making his separations and in completing his determinations gravimetrically or titrimetrically, although for manganese a visual comparison of colours provided an early example of the use of a colorimetric method. The general sensitivity of photometric methods, coupled with the improvements in the design of instruments available from about 1950 onwards has resulted in a considerable extension in the use of such methods. At first this extension was limited to the minor and trace components such as titanium, phosphorus and fluorine, but this was later extended also to those elements present in major amounts - silicon, iron and aluminium.

Some considerable effort by a number of analysts has been devoted to devising new schemes of rock analysis based upon spectrophotometric methods, with complexometric titration for the determination of calcium and magnesium. Most of the early schemes suffered from some disadvantage - some of the procedures were analytically unsound, some required the services of an exceptionally skilled analyst, and most if not all were too inflexible to be applied to a wide range of rocks without modification.

Although many chemists regarded these early schemes for "complete analysis" of silicate rocks by spectrophotometry with suspicion, the prospect of obtaining large numbers of such analyses cheaply and rapidly has been welcomed by many geologists. Unfortunately this enthusiasm has not always been accompanied by an understanding of the chemistry (and the errors!) of the processes involved, or of the difficulties in making precise spectrophotometric measurement. The ease with which agreement between duplicate results can be obtained is often taken as an indication of the accuracy of the determination. What is all too often forgotten is that the "rapid" (sometime approximate) analyses, valuable in a series of similar analyses for comparative studies, may later be used by other workers and then given equal weight with analyses obtained by more rigorous methods.

The extensive introduction of spectrophotometric methods to silicate rock analysis was followed by the use of other instrumental methods. Emission (optical) spectrography, became a valuable additional technique in many rock analysis laboratories. In some of these it became the practice to make a qualitative examination of all silicate rocks prior to chemical analysis. This served to identify elements of interest that might subsequently warrant determination by other means. It also gave the analyst a guide to the approximate values that he could expect to find. Emission spectrography has provided the geologist with his dream of large numbers of rapid, cheap analyses - at least for the minor and trace components of silicates. Attempts to use it for obtaining "complete analyses"(3) have not been widely followed.

More recent introductions to the rock analysis laboratory include β-probe analysis, x-ray fluorescence, inductively coupled plasma emission, direct reading emission, atomic absorption and atomic fluorescence spectroscopy.

One of the most tedious of the determinations in the classical scheme for the complete analysis of silicate rocks is that of the alkali metals, involving a difficult decomposition procedure and a number of subsequent separation stages. It is therefore easy to see why the use of flame photometry was widely adopted, even before the difficulties associated with its use were properly understood and defined.

Gravimetric methods and the separation of the alkali metals soon became unnecessary. The determination of the rarer alkali metals, previously seldom attempted and even more rarely successfully achieved, was now possible on a routine basis. Calcium, strontium and barium, elements with characteristic flame emission, were also determined by this technique, although rather less readily than sodium and potassium, also with less enthusiasm on the part of the rock analyst with the availability of other techniques.

Schemes of rapid rock analyses usually included titrimetric procedures for calcium and magnesium, although difficulties were sometimes encountered in the presence of much manganese. In recent years atomic absorption spectroscopy has provided an acceptable alternative technique for both calcium and magnesium, as well as for manganese, iron and many other elements at major, minor and trace levels - now rivaling spectrophotometry in the extent of its application.

The difficulties inherent in collecting and determining all the silica by the classical method can be avoided by using a combined gravimetric and photometric method.[(4)] The major part of the silica is recovered following a single dehydration with hydrochloric acid, and is then determined by volatilisation with hydrofluoric acid in the usual way. The minor fraction that escapes collection is determined in the filtrate by a photometric molybdenum-blue method. Atomic absorption spectroscopy may also be used to determine the minor fraction of silicon.

Geochemical Reference Material

Geochemical reference material in the form of distributed samples has been available for so long that it is now difficult to see how rock analysts can manage without them. The need for such material has grown with the availability of it. The number is now so large (Table 1), the compositions so variable and the compositional information so detailed, that no book of this kind can do justice to any kind of evaluation of the data relating to them.

The first materials to be available as reference samples were those prepared primarily for industrial and commercial use. Both the National Bureau of Standards (USA) and the Bureau of Analysed Samples (UK) had prepared a number of sample materials of prime interest to the ceramic industry which were also of use to rock analysts and geochemists. These included alkali felspars, clays and refractories. Such samples are still available and are widely used. As befits sample materials prepared primarily for industrial and commercial use, the major interest was in their major constituents and those minor constituents of importance in the use of large tonnages of these materials.

The widespread adoption of instrumental analysis in industry introduced the widespread need for "standard" or "reference" samples by which such methods could be calibrated. Analysed samples of metals for the ferrous and non-ferrous metal industries were used not only for the rapidly developing optical emission spectrographic and x-ray fluorescence techniques, but also by the 'wet chemists' confronted by problems of tighter product specifications, the introduction of rarer elements in increasing proportions, and a requirement to complete analyses within ever decreasing timescales. This need included reference ores, minerals and related products of interest to the geochemist.

It is difficult to compare the results of one laboratory with those of another unless an adequate series of standards are available covering the range of determinations currently being performed in the laboratories concerned.(5) The "Co-operative Investigation of the Precision and Accuracy of Chemical, Spectrochemical and Modal Analysis of Silicate Rocks" reported in 1951 in the United States Geological Survey Bulletin No 980(6) showed that such comparisons were long overdue. This investigation involved the distribution of two ground silicate rock samples, a granite G-1 and a diabase W-1, to a number of laboratories regularly making rock analyses. A detailed comparison was then made of the large number of results subsequently reported. One of the more important points to emerge from this investigation was that the agreement between analysts and between laboratories was not of the order that could be expected from individual estimates of the accuracy and precision of the procedures used.

In USGS Bulletin No 980, Fairbairn(6) noted that "whatever the outcome of the present investigation, possession of a large store of such standard samples would be of immense future value to analysts of all kinds as a means of both intralaboratory and interlaboratory control." It was clear that at that time the need for geochemical standards had been recognised, and that G-1 and W-1, although not originally intended as such, had become the first of such geochemical reference materials.

With the gradual acceptance of the idea that all rock material contains all elements and that this could be demonstrated if sufficiently sensitive methods could be devised for their detection, began what is now seen as a challenge to rock analysts to devise ever more sensitive techniques for those elements in G-1 and W-1 then not yet reported. This impetus for revised and new methods of analysis came also from the rapid development of geochemistry as a clearly defined branch of science, and an appreciation of its importance in our understanding of the rock forming processes, and the origins of the elements themselves. The more refined and esoteric techniques became, the more they demonstrated the need for analysed geological samples. The more important abundance data became, the greater the need to ensure the validity of comparison between one worker and another.

The experience gained in selecting geological material, preparing the samples, distributing and in evaluating the results has been invaluable in what may reasonably be called the second and third generation of reference materials. The criticisms, made in the earlier editions of this book, concerning the "rash" of new standards is now no longer fully justified, although still relevant to some of the work in this area. It relates to those materials that have not been prepared with the care and attention to detail required of international standards. It did not appear to have been realised that the preparation, including selection, collection, crushing, grinding and sampling of a large bulk of material, in a state of homogeneity and free from contamination, was a task of considerable magnitude. Reference material produced in varying amounts, under differing conditions, in laboratories often isolated from each other inevitably produced an inadequate selection of material, with a great emphasis on certain rock types (esp. granites) to the detriment of sedimentary rocks and rock-forming minerals.

From even a brief perusal of the extensive literature that now exists on geological reference material, it is abundantly clear that the preparation and dissemination of new material is not a task to be lightly undertaken. The supposition that because there is need for such reference materials to be available, there is merit in proposing or preparing additions to the list, may now be seen as somewhat naive. The first and paramount consideration is the justification for the enormous amount of energy, time and resources that are needed in terms of the objectives that may be set; for example see Engels and Ingamells[7], Valcha[8] and Steele[9].

There is also a considerable literature related to the results of the determination of elements in standard reference materials. Such phrases as 'preferred value', 'recommended value', 'best value' indicate that some selection process has been used to discard or give minimum weight to results that differ markedly from mean values. It has long been recognised that occasional 'outlier' results can have a disproportionate effect on the calculation of mean values, and for this reason modal values have been preferred.[10,11] There is, of course, no guarantee that such modal values will give 'true' or accurate values for the rock in question, only that they will give values that are acceptable to the majority of analysts doing the work. The term 'concensus value' is probably the most appropriate to choose. Reservations must always be made[12] in respect of any attempt to derive 'best values' for particular constituents in reference material.

Table 1, listing geochemical reference material, is intended only to be illustrative and in no way exhaustive. The materials listed are those that are believed to be both commonly available from the sources listed and of more than unique use to the rock analyst. Abbey[13] has commented on many of these materials with observations on many of the reported results. His tables of 'usable values' for major, minor and trace components are particularly useful in selecting calibration standards. The value of these materials tends to fall as the stock become depleted, and any measure that extends the life of individual standards is therefore to be recommended. Flanagan[14] for example, suggests that any recipient of these standards should obtain about 10 kg of two or three rocks from his area, process these as "in-house standards", and calibrate them against the international standards.

TABLE 1

STANDARD GEOCHEMICAL REFERENCE MATERIALS

No.	Rock Name	Contact
G-2	granite	
AGV-1	andesite	
GSP-1	granodiorite	
PCC-1	peridotite	F J Flanagan
BCR-1	basalt	U S Geological Survey
MAG-1	marine mud	National Center 972
BHVO-1	basalt	Reston Va 22092
QLO-1	quartz latite	U S A
RGM-1	rhyolite	
SCo-1	Cody shale	
SDC-1	mica schist	
SGR-1	oil shale	
STM-1	nepheline syenite	
W-2	diabase	
BIR-1	basalt	
DNC-1	diabase	
GA	granite	K Govindaraju
GH	granite	Centre de Recherche Petrographic et Geochimique
BR	basalt	
Mica-Mg	phlogopite	Case Officielle No 1
Mica-Fe	biolite	54500 Vandeouvre-les-Nancy
		FRANCE

TABLE 1 Cont'd

AN-G	anorthosite	
GS-N	granite	Group International de Travail
DR-N	diorite	Association Nationale de la Recherche Technique
BX-N	bauxite	
FK-N	felspar	
DT-N	disthene (kyanite)	contact K Govindaraju
UB-N	serpentine	
GL-O	glauconite	
VS-N	synthetic glass	
MA-N	granite	
BCS 309	sillimanite	
BCS 368	dolomite	P D Ridsdale
BCS 319	magnesite	Bureau of Analysed Samples
BCS 389	magnesite, high purity	Newham Hall
BCS 395	bauxite	Middlesbrough
BCS 375	soda felspar	ENGLAND TS8 9EA
BCS 376	potash felspar	
BCS 392	fluorspar	
BCS 372	Portland cement	
183	lithium ore (lepidolite)	
182	lithium ore (petalite)	
181	lithium ore (spodumene)	
79a	fluorspar	Office of Standard Reference Materials
180	fluorspar, high grade	National Bureau of Standards
120b	phosphate rock	Washington DC 20234
88a	dolomitic limestone	U S A
70a	potash felspar	
1c	argillaceous limestone	
97a	flint clay	
98a	plastic clay	
99a	soda felspar	
69b	bauxite	
696-8	bauxite	
JB-1	basalt	A Ando
JG-1	granodiorite	Geological Survey of Japan
		135 Hisamoto-cho
		Kawasaki-shi
		JAPAN
ST-1A(2001)	trap	L V Tauson
SGD-1A(2003)	gabbro	Institute of Geochemistry
SG-1A(2005)	albitised granite	PB 701
		Irkutsk 33
		USSR
NIM-D	dunite	
NIM-G	granite	H P Beyers
NIM-L	lujavrite	South African Bureau of Standards
NIM-N	norite	Private Bag 191
NIM-P	pyroxenite	Pretoria
NIM-S	syenite	SOUTH AFRICA

TABLE 1 continued

SY-2	syenite	Canada Centre for Mineral and Energy Technology
SY-3	syenite	555 Booth Street
MGR-1	gabbro	Ottawa, Ontario
SU-1	sulphide	CANADA K1A OG1
ASK-1	larvikite	O H J Christie
ASK-2	schist	University of Oslo PO Box 1048 Oslo 3 NORWAY
LLL-1	limestone	E Schroll
ADT-1	dolomite	Geotechnische Institute BUFA Arsenal A-1030 Vienna AUSTRIA
Qr-1	adularia	N H Suhr
M-1	muscovite	Pennsylvania State University
Ab-1	albite	University Park
Px-1	pyroxene	Pennsylvania 16802 U S A
KH-2	limestone	K Schmidt
TB-2	slate	Zentrales Geologisches Institut
GM	granite	Invalidenstrasse 44
GM	basalt	104 Berlin
TS	shale	G D R
LO1-1	blast furnace slag	G Jecko
MO8-1	ferriferrous marl	Institute de Recherches de la Siderurgic Station d'Ersais Maizieres-les-Metz 57 FRANCE

It is uncertain whether the following materials are still available:

NS-1	nepheline syenite	A A Kukharenko Leningrad State University Leningrad V-164 USSR
T-1	tonalite	Commissioner, Geological Survey PO Box 903, Dodoma TANZANIA
I1	aplitic granite	A B Poole
I2	dolerite	Department of Geology
M2	pelitic schist	Queen Mary College
M3	calcsilicate	Mile End Road LONDON E1 4NS England

Elements Determined

In his examination of silicate rocks the petrologist is primarily concerned with the mineralogical composition, and his interest in the chemical analysis is largely directed towards the major components of the rock forming minerals, that is towards those elements present in major proportion. There is a small group of elements which, calculated as oxides, account for 99 per cent or more by weight of a large number of silicate rocks. All analyses of igneous rocks that claim to be complete must include values for these thirteen constituents:

silicon
aluminium
iron (ferrous and ferric)
magnesium
calcium
manganese
titanium
phosphorus
sodium
potassium
water (water evolved above and below 105°)

for many rock analysts, a complete analysis will include not only these thirteen components, but also a number of other elements that are occasionally present in rock specimens in amounts of up to several per cent. Those frequently reported include:

sulphur (sulphide and sulphate)
carbon (carbonate and non-carbonate)
chlorine
fluorine
chromium
vanadium
barium
nickel
cobalt

The elements reported in sedimentary rocks are essentially the same as those in igneous silicate rocks. In sandstones and quartzites, silica is the dominant and sometimes only major component, all other elements being present in minor or trace proportion only. Shales, muds and slates resemble the igneous silicates, with the same group of major elements present in somewhat similar proportions, although carbon dioxide, organic matter and pyritic sulphur are likely to be present in increased amounts. Some limestones are little more than calcium carbonate, but others contain major amounts of magnesium and iron. Those limestones with an arenaceous fraction may contain appreciable amounts of silica, aluminium, iron and other elements.

Some of the most difficult rocks to analyse are the carbonatites. These igneous carbonates vary considerably in mineralogical composition, but often include appreciable amounts of certain silicates particularly pyroxenes and micas, oxide minerals such as magnetite, phosphates such as apatite and monazite, and sulphides. Many of the carbonatite occurrences are of economic importance as sources of niobium (pyrochlore), iron ore (magnetite), phosphate (apatite), copper (sulphide minerals) or vermiculite. The difference between the known deposits is so great that it is not possible to draw up a list of elements that should be determined in a "complete analysis".

With the introduction of more sensitive methods of analysis, it is now clear that the list of elements that can be reported from igneous silicate rocks could, if methods were available, be extended to include all the naturally occurring elements of the periodic table.

In the earlier editions of this book, information was provided in the form of brief notes on the occurrence of the element or group of elements described in each chapter. The purpose of these notes was to indicate to the practising analyst the range of values that he might reasonably expect in his analyses.

The information available for the individual elements varied considerably in quantity and quality and this was reflected in these earlier notes. Attempts to achieve a uniformity of presentation for this edition gave rise to much repetition. For this reason these notes on occurrence in silicate rocks have been replaced by Table 2 in which they have been summarised.

This table is a guide only, care must be taken in using it. The classification of rocks into ultrabasic, basic, intermediate, granitic and alkalic groups is a gross simplification - a more detailed classification would be inappropriate in a book of this kind. It should also be remembered that occasional high or low values can be

TABLE 2 GENERAL GUIDE TO ELEMENT VALUES (PPM) IN SILICATE ROCKS

Element	Ultrabasic rocks	Basic rocks	Intermediate rocks	Granitic rocks	Alkalic rocks
Li	2-10	10-20	3-30	5-50	20-40
Be	0.05-0.5	0.3-2.0	1-4	1-10	2-10
B	1-10	4-20	15-40	2-40	3-20
F	1-20	300-500	500-1000	200-2000	500-2000
P	10-100	1000-2000	1000-2000	1000-2000	1000-2000
S	300-5000	300-20000	100-500	50-500	100-500
Cl	100-400	100-400	100-400	100-400	200-2000
Sc	1-10	20-50	2-15	2-10	1-10
Ti	2000-5000	5000-20000	3000-5000	1500-5000	5000-20000
V	20-300	10-1000	10-300	3-300	100-300
Cr	500-15000	100-1000	20-100	3-30	50-200
Mn	600-1500	1000-1500	300-800	200-600	1000-1500
Co	50-150	30-60	10-20	1-5	40-60
Ni	1000-1500	50-300	10-30	10-20	
Cu	10-50	10-100	30-100	5-50	5-30
Zn	25-100	50-200	50-100	10-100	50-200
Ga	1-3	10-30	10-30	10-30	
Ge	0.5-1	1-2	1-2	1-2	
As	0.5-5	0.5-5	0.5-5	0.5-5	
Sc	0.01-0.1	0.01-0.1	0.01-0.1	0.01-0.1	
Br	0.1-2	0.1-2	0.2-2	0.5-2	
Rb	0.1-3	10-50	20-100	100-1000	200-1000
Sr	0.2-100	20-1000	50-2000	2-1000	10-2000
Y	1-10	10-50	10-50	10-50	10-50
Zr	1-50	50-500	50-500	25-500	100-2000
Nb	0.2-2	5-50	10-25	10-25	50-1000
Mo	0-1	0.5-4	0.1-2	0.1-2	
Ag	0.03-0.08	0.02-0.2	0.02-0.2	0.02-0.05	
Cd	0.01-0.1	0.1-2		0.1-2	
In	0.01-0.1	0.01-0.1	0.01-0.1	0.01-0.1	
Sn	0.1-2	1-10	1-10	1-10	
Sb	0.1	0.1-2	0.1-1	0.1-1	
Te		ca 0.001(?)	ca 0.001(?)	ca 0.001(?)	
I	0.1-0.4	0.1-0.4	0.1-0.4	0.1-0.4	
Cs	0.01-0.1	0.2-2	0.2-2	2-10	1-10
Ba	1-50	10-500	50-1000	50-2000	1000-3000
La	1-5	1-20	5-50	20-100	50-500
Hf	0.1-1	1-5	1-5	2-25	5-100
Ta			0.5-2	0.5-2	10-50
W	0.1-1	0.1-2	0.5-5	0.5-5	1-5
Au	0.0004-0.02	0.0004-0.02	0.0002-0.01	0.0002-0.01	
Hg		0.005-0.2	0.005-0.2	0.005-0.2	
Tl	0.05-0.5	0.05-1	0.1-2	0.5-5	0.05-5
Pb		1-5	2-20	5-100	2-50
Bi		.01-0.2	.01-0.2	.01-0.2	
Th	0.1-5	0.1-5	1-10	10-50	0.5-50
U	0.01-0.2	0.1-1	0.5-5	1-10	

encountered in particular rocks and occasionally in what appears to be a normal, otherwise unexceptional rock specimen. This, although particularly true for chalcophilic elements such as copper, zinc, cobalt and nickel, is true also for very many other elements.

Reporting an Analysis

Published chemical analyses of igneous rocks are often put to uses that were not considered by the analyst. The potential value of an analysis is therefore greater than the sum of the determinations, and this should be increased by including with the analysis full details of the origin of the specimen and notes on the petrographic examination. The name of the analyst, address of the laboratory where the analysis was made and the date of the analysis should also be included. These notes will give the means of recovering more detailed information, such as the procedures used, if these are wanted at a later date. Hamilton(15) making a plea for more information of this sort gives the following example from Shaw(16) of the type of petrographic information that should accompany the analysis:

> L 62 Staurolite schist
> A silvery crumpled schist. Porphyroblasts of staurolite, with quartz-graphite inclusions, and of occasional garnets, in a matrix of quartz, muscovite and biotite, with crumpled schistosity planes, Staurolite crystals have been fragmented by shearing along micaceous folia. Principal opaque mineral is graphite, but iron oxides are also present. Minor felspar and sphene. Grade: staurolite zone.

In many cases it will not be possible to give such a detailed description of the specimen, but what information is available should be recorded in such a form as to leave no doubt as to what was analysed.

The conventional way of reporting the detailed analysis of a silicate rock is to express each element in the form of its oxide, and to give the results to the second decimal place. This can lead to certain difficulties, as for example in reporting the ferrous iron content of rocks containing much pyrite or carbonaceous matter. Attempts have been made to depart from these traditions by giving results to only the first decimal place and by expressing the constituents as elements in place of oxides. Neither of these suggestions has so far been widely adopted. The summation of the oxides is widely regarded as a test of the skill of the analyst, and for this reason alone is unlikely to be discarded. The analyst should, however, be aware of the possibility of compensating errors occurring in his analysis, giving a fortuitously good total. Chalmers and Page(17) have suggested that where the results are made the basis for comparisons, each complete chemical analysis should include an estimate of the precision and accuracy of the results. It will usually be found that not more than three significant figures can be justified. Claims to accuracies comparable with, or better than, those of the accepted atomic weights of the elements should be resisted.

Constituents that are present in only trace amounts are usually reported as parts per million of the element, rather than of the oxide. At lower concentrations "parts per billion" (1 in 10^9) is sometimes preferred.

The Selection of Material for Analysis

Problems involved in the selection of the material are largely the concern of the field geologist, but the geochemist and rock analyst should appreciate the difficulties of ensuring that material which arrives in the laboratory is representative of the exposure from which it has been taken. Great care must be taken in the choice of material, in the collection of a suitable bulk and in the proper labelling and storage of the material before despatch to the laboratory. The importance of keeping full detailed field notes concerning the rock exposure and in the proper indexing of all specimens cannot be over-emphasised.

In general, where there is no shortage of material, it is easier to collect too much at the first visit, than to return later to collect more. It will be required for petrographic and possible mineralogical studies as well as for chemical and spectrographic analysis. Reserve specimens should also be retained for further study and also for future reference.

At this stage it is important to understand fully what the sample is intended to represent. An outcrop of granite has sometimes been represented by a single specimen. Neither this nor a series of chips taken over the exposed surface is likely to be typical of the granite in depth - each specimen collected represents the granite only at the place from which it was taken. The practice of combining chips from as much as possible of the exposed area to give a composite sample has only one merit - it reduces the analytical effort required. The results for composite samples tend to reflect the way in which the composing was done and may suggest an overall composition that occurs nowhere in the outcrop. Wherever possible such outcrops should be sampled over the whole of the exposed area, but the specimens taken should be kept separate and if possible analysed separately. The results for constituents of interest are of greater value if they indicate both a mean value and the extent of variation from it.

It will be appreciated that the size of the sample necessary to give a representative specimen will vary with the mineral grain size. A far smaller quantity will be required of a fine-grained rock with no phenocrysts, such as dolerite, than of a coarse-grained or porphyritic rock. For this reason it is difficult to lay down any rule as to the weight of rock material that should be taken. In general, it is only for the very coarsely crystalline rocks, such as pegmatites, that a sample size of greater than 20 kg is necessary. For dolerites and other fine grained rocks a minimum of 2½ kg should be collected.

If necessary chemical analysis can be made on a very much smaller sample weight - amounting to no more than a gram or two. But in such instances the task of the analyst is rendered more difficult by limiting his choice of methods and by leaving no margin for repeat determinations. It is highly undesirable to use all the material leaving none for reference or future work.

Due care should be taken to see that the rock is as fresh as possible and that no skin of altered material is included. Likewise, fragments with obvious mineral veins and inclusions should be excluded (or preferably be collected and analysed separately). Paint should not be used to label a specimen, as this can give rise to contamination of one or more trace constituents. If it is necessary to use paint - as, for example, in tropical areas where paper labels or containers are likely to be eaten by ants - it should be removed in the laboratory before the specimen is crushed. Fresh bags should be used to contain the rock material - previous use of sample containers is a frequent source of contamination.

Crushing and Grinding

The first step in preparing the sample is to examine the total bulk, reject any contaminated or suspect material and select the portion required for analysis. This latter may conveniently be done by selecting the portions for petrographic examination and for the reserve collection and crushing what is left. In the case of coarse-grained and porphyritic rocks, a sample of not less than 10 kg should be available for crushing, and proportionately less of the fine-, even-grained rocks. At this and all subsequent stages in the sampling, crushing and grinding, an intelligent approach is necessary to ensure that the introduction of extraneous matter is kept to a minimum. Only then can the results of the chemical analysis of the prepared sample be taken to represent the chemical composition of the material collected.

The following notes are based upon the procedure used by the senior author to prepare igneous silicate and carbonate rocks for analysis. A simplified procedure is used for friable rocks such as unconsolidated sediments which can usually be fed directly to a mechanical agate mortar and pestle. All other samples are fed first to a small jaw-crusher. The product is screened and any oversize material returned to the jaw-crusher, now set with the jaws giving a slightly smaller gap. The whole of the sample can be reduced in this way to pass a No 5 mesh sieve. This rock material is then riffled to give about 500 g which is sieved on a No 10 mesh sieve, the over-sized material then being cram-fed to the jaw-crusher on its finest setting. The product is riffled once more to give 75 to 100 g of material, all of which is subsequently ground to give the sample for analysis. The grinding is done by feeding small quantities at a time to a mechanical agate mortar and pestle. The grinding is stopped from time to time to remove the 100-mesh material by sieving through bolting cloth.

Once the grinding is complete, the sample material is transferred to a large bottle - an 8 oz bottle is a convenient size - and is thoroughly mixed by shaking and rolling. After this the material is transferred to a smaller bottle and labelled with details of the specimen, locality from which it was taken, serial or catalogue number and the notebook reference. The details recorded in the notebook should include notes on the sampling, crushing and grinding procedures, sieving operations, weight of the prepared material and the date. The sample is then ready for analysis.

The practice of coning and quartering is not recommended for the sampling of the small amounts of material collected for rock analysis. A series of riffles of varying sizes can be kept for this purpose, or alternatively a rotary sampling machine can be used.

Samples produced by grinding in a mechanical agate mortar and pestle are similar to those ground by hand in that they contain a great deal of fine material.

Anyone who has attempted to reduce specimens of mica minerals or rock samples containing large amounts of mica, will have experienced the difficulty of reducing platy minerals to an impalpable powder. Where mechanical mortars are used, the harder, more brittle minerals are preferentially ground, leaving the mica minerals to enrich the latter fractions. Care must be taken to ensure that none of the mica fraction is lost or discarded, and the powdered sample is thoroughly mixed before portions are taken from it. Abbey and Maxwell[18] have reported that pre-ignition of mica samples makes the grinding stage easier, but that the ignited product slowly gains weight, making accurate weighing virtually impossible. These authors recommend the use of a "blender" with blades rotating at 15,000 rpm for size reduction of mica.

Contamination

Agate mortars and pestles are a frequent source of contamination, but introduction of extraneous material can occur at all stages in the preparation of the sample. Contamination from painted labels has already been noted. If such labels have been used, they should be removed by chipping or grinding before the sample is crushed. Jaw-crushers must be cleaned particularly carefully and thoroughly if cross contamination is to be avoided. If the crusher is fitted with jaws of mild steel, small fragments may be shorn from the faces giving appreciable errors in the determination of ferrous iron. The amounts of tramp iron introduced in this way can be considerably reduced by fitting jaws of hardened manganese steel, but this may in turn introduce small amounts of other elements such as chromium into the sample.

The practice of using iron jaws to crush the sample, followed by the removal of the introduced iron fragments with a magnet, is not to be recommended, as any magnetite present in the rock will be similarly removed together with smaller quantities of other iron minerals such as pyrrhotite and ilmenite. This would materially affect the composition of some samples - carbonatites for example.

The use of nylon bolting cloth supported in a ring of plastic material cut from an acrylic pipe, can eliminate metallic contamination at the sieving stages, but care must be taken that loss by dusting is kept to the minimum.

If many silicate analyses are to be made it is preferable to reserve a special agate mortar for grinding these samples. If ore minerals are ground in it, traces of these minerals can usually be found in subsequent samples no matter how carefully the cleaning is done.

The introduction of extraneous material is not the only change occurring in rock samples during the grinding. Water may be lost, or in some cases gained, whilst both ferrous iron and sulphur may undergo partial oxidation. These effects are enhanced by excessive grinding. For this reason Hillebrand[2] recommends that silicate rocks should be reduced in size only to pass a 70-mesh sieve. This may reduce oxidation changes, but at the same time increases the difficulty of decomposing the rock. Oxidation during grinding was clearly shown by French and Adams,[19] who reported that a sample of diabase (dolerite), I_3, containing approximately 10 per cent FeO, gave a progressively lower FeO content as the grinding period was prolonged. After 20 minutes the FeO content had decreased by more than 0.5 per cent and this rate of oxidation was maintained throughout the grinding period. Grinding of the rock material in the same agate mortar for 10 minutes but with continuous moistening with acetone produced a sufficiently fine material with no detectable oxidation.

References

1. WASHINGTON H S., Manual of the Chemical Analysis of Rocks, Wiley, New York, 3rd edition, 1918.
2. HILLEBRAND W F., The Analysis of Silicate and Carbonate Rocks, U S Geol. Surv. Bull. 700, Washington, 1919.
3. AHRENS L H., Quantitative Spectrochemical Analysis of Silicates, Pergamon, Oxford, 1954.
4. JEFFERY P G and WILSON A D., Analyst (1960) 85, 478
5. INGERSON E., Geochim. Cosmochim. Acta (1958) 14, 188
6. FAIRBAIRN H W. and others, U S Geol. Surv. Bull. 980, 1951
7. ENGELS J C and INGAMELS C O., Geostds. Newsl. 1977, 1, 51
8. VALCHA Z., Geostds. Newsl. 1977, 1, 111
9. STEELE T W., Geostds Newsl. 1977, 1, 21
10. CHRISTIE O H J and ALFSEN K H., Geostds. Newsl. 1977, 1, 47
11. ELLIS P J, COPELOWITZ I and STEEL T W., Geostds. Newsl. 1977, 1, 123
12. ABBEY S., Geostds. Newsl. 1978, 2, 141
13. ABBEY S, X-Ray Spectrometry (1978) 7, 99 (also Geol. Surv. Canada Paper No 77-34)
14. FLANAGAN FJ., Geochim. Cosmochim. Acta (1973) 37, 1189
15. HAMILTON W B., Geochim. Cosmochim. Acta (1958) 14, 253
16. SHAW D M., Bull. Geol. Soc. Amer. (1956) 67, 919
17. CHALMERS R A and PAGE E S., Geochim. Cosmochim. Acta (1957) 11, 247
18. ABBEY S and MAXWELL J A., Chem. in Canada (1960) 12, 37
19. FRENCH W J and ADAMS S J., Analyst (1972) 97, 828

CHAPTER 2

Sample Decomposition

The process of sample decomposition - the first step of all analyses - consists of the destruction of some or all of the original minerals as part of or prior to the dissolution of the constituent of interest. The processes of decomposition vary considerably - from extraction with water, organic solvents or mineral acids to the more elaborate techniques of sintering or fusion. Few of these techniques will decompose completely all types of rock material, nor is this always desirable. Many of the decomposition procedures serve to dissolve the major part of the constituent minerals but leave a minor fraction as a residue that can be separated from the solution by filtration. Whether or not this residue will require separate decomposition will depend upon the amount of residue and more particularly whether it can be expected to contain the elements of interest.

Decomposition with Mineral Acids

DECOMPOSITION WITH HYDROCHLORIC ACID

With the exception of certain minerals of the scapolite group, carbon dioxide containing minerals are decomposed, either in the cold or on digestion at an elevated temperature, with dilute hydrochloric acid. This method of decomposition is therefore of particular use for carbonate and carbonatite rocks, where it serves to dissolve the carbonate fraction. However, except for rocks that mineralogically are rather simple in composition, the separation is unlikely to be perfect. Silicates such as wollastonite or fayalite, sulphides such as sphalerite, phosphates such as apatite and oxide minerals such as magnetite and haematite may be wholly or partially decomposed by heating with hydrochloric acid. Nevertheless useful separations can sometimes be made - examples include the separation of pyrite from carbonate rocks and pyrochlore (frequently with other accessory minerals) from carbonatites.

DECOMPOSITION WITH NITRIC ACID

Concentrated nitric acid serves to decompose not only carbonate minerals, but also any sulphide minerals present. This is probably its most important application in rock analysis, leading to one method for the determination of sulphide sulphur. Other applications include the extraction and subsequent determination of heavy metals occurring as sulphide minerals in a silicate matrix, particularly those of copper, cobalt, lead and zinc. Such determinations are of considerable economic significance in the exploitation of sulphide mineral deposits, but clearly any heavy metal present in the silicate matrix will be largely unrecovered.

Platinum metal is appreciably attacked by mixtures of hydrochloric and nitric acid. For this reason platinum apparatus should be avoided whenever such mixtures are used. Glass vessels are suitable for most applications, and basins of PTFE can be used if hydrofluoric acid is to be added in a subsequent stage of the analysis.

DECOMPOSITION WITH HYDROFLUORIC ACID

Hydrofluoric acid has long been used for the decomposition of silicate rocks, usually in combination with nitric, perchloric or sulphuric acid and in platinum apparatus. This combination enables substantially all the fluorine as well as all the silica to be removed, leaving a residue that can be dissolved in dilute acid and used for the determination of the alkali metals, alkaline earths, iron, aluminium, titanium, manganese, and phosphorus. With many rocks a small residue consisting of acid-resistant minerals such as zircon, topaz, corundum, sillimanite, tourmaline, chromite and rutile may remain, together with barium sulphate, particularly if the sample material contains much barium and sulphuric acid is used for the decomposition. This procedure is still widely used, although the determination of titanium is not always satisfactory, possibly due to residual traces of fluorine in the solution. With the introduction of highly sensitive atomic absorption spectrometers, the rock solution obtained in this way can be used also for the determination of other metals, present in trace amounts. PTFE vessels are now widely used in place of platinum.

The use of hydrofluoric acid without the addition of other mineral acid has been recommended by May and Rowe[(1)]. A platinum-lined bomb was used at a temperature of 400-450° and a pressure of 6000 psi. Langmyhr and Sveen[(2)] also recommend hydrofluoric acid but at temperatures of up to 250° in a PTFE lined bomb. The results given which indicate complete decomposition are somewhat meagre. The advantages claimed for high temperature - high pressure decomposition with hydrofluoric acid are that the procedure is more effective than when sulphuric acid is included in decomposing refractory minerals. In addition, because silicon is not volatilised in the closed system, it can be determined by spectrophotometry.

Disadvantages obviously include the need for expensive specialised apparatus and a requirement to remove fluorine from the solution and any residue before proceeding with other determinations.

A somewhat simpler procedure using a polyethylene or other similar vessel with a close fitting lid was described by Antweiler[3]. Large rock fragments (up to 50g) were decomposed by digestion at a temperature of 85° for 24 hours.

Most authors have, however, preferred to use hydrofluoric acid in the presence of some other mineral acid. This serves to moderate the initial reaction between hydrofluoric acid and finely powdered silicate material (for this reason it is recommended that all powdered rock material should be moistened with water prior to adding hydrofluoric acid; failure to do this can result in overheating and consequent loss of material by spitting). Nitric acid is often added to decompose any traces of carbonate minerals, to oxidise sulphides and organic matter and to convert iron and other elements into their higher valency states.

Evaporation with perchloric-hydrofluoric acid mixtures has frequently been recommended for the decomposition of silicates. This evaporation is a great deal more easy to carry out than the similar evaporation with sulphuric acid, there being less tendency for the solution to spit, as the perchlorate salts crystallise more cleanly than the corresponding sulphates. The perchlorate residue, unlike the sulphate residue, is readily soluble in dilute acid - aluminium and ferric sulphates in particular, once dehydrated, can only be dissolved with difficulty. In addition the perchlorate ion, unlike the sulphate ion, does not have a depressant effect upon the flame emission of the alkali metals.

The evaporation with a mineral acid addition to the hydrofluoric acid serves also to remove much of the fluoride ion which otherwise interferes with the determination of aluminium, titanium, potassium and certain other elements. The order of effectiveness in removing residual amounts of fluorine increases in the order nitric-perchloric-sulphuric acid. Langmyhr[4] has shown that a double evaporation with perchloric acid at a temperature of 180° reduces the fluorine level to a value that can be reached in a single fuming with sulphuric acid at a temperature of 250°, and that only microgram amounts can then be recovered from the residue.

Work in the laboratory of one of the authors has broadly confirmed these observations, except that larger amounts of fluorine were recovered in each case, and that the only really effective way of removing these traces of fluorine was to add potassium pyrosulphate to the residue obtained from evaporation of the excess sulphuric acid and to convert the evaporation into a fusion. This further stage has an additional advantage in that the pyrosulphate melt is readily soluble in hot dilute hydrochloric acid, in contrast to the sulphate residue, which is soluble only with difficulty. Such solutions cannot of course be used for the determination of potassium.

Certain authors have recommended that the silicate rock material should be allowed to stand overnight with hydrofluoric acid, either at room temperature or at the temperature of a steam bath. The addition of perchloric or other mineral acid and subsequent evaporation is then undertaken on the following day. This procedure is particularly effective for decomposing those rocks that are rich in magnesium and/or quartz, and is recommended as applicable to most silicate rocks.

Fusion Procedures

FUSION WITH ALKALI FLUORIDE

Fusions with ammonium fluride have been recommended for the decomposition of beryl and other silicate minerals.(5) Not all silicates are attacked and attempts to decompose sillimanite, kyanite and zircon are ineffective.(6) In most instances where alkali fluoride is used the fluoride melt is converted to a pyrosulphate melt by heating with sulphuric acid. This serves to decompose complex fluorides, to convert all metal fluorides to sulphates and to remove most of the fluorine from the melt.

FUSION WITH POTASSIUM PYROSULPHATE

Potassium bisulphate has been advocated for certain purposes; it is converted to pyrosulphate in the earlier stages of the fusion and its use is not recommended as considerable spitting can occur in the conversion stage. Moreover, very little attack of oxide minerals can occur until this removal of water has taken place. (Potassium bisulphate is readily converted to pyrosulphate by heating in platinum until a quiescent melt is obtained. Care must be taken to avoid excessive or prolonged heating with consequent loss of sulphur trioxide. The melt may be cooled and the solid material broken up for use as described below. Alternatively, in some cases it may be possible to weigh the sample material directly onto the solidified pyrosulphate).

Silicate minerals are not decomposed by direct fusion with potassium pyrosulphate, which should be used only for the decomposition of the residue remaining after an evaporation with hydrofluoric acid. This can be done immediately after the evaporation as described above, or after the major part of the metallic constituents present as perchlorates or sulphates have been removed by dissolution in dilute acid. The residue then obtained is often quite small, but it frequently contains a variety of minerals, some silicate (zircon, tourmaline, andalusite, etc), some oxide (rutile, ilmenite, cassiterite, chromite etc) and some phosphate (monazite).

With most silicate rocks this assemblage is best decomposed by fusion with anhydrous sodium carbonate, but if certain oxide minerals preponderate (ilmenite or rutile, for example), then potassium pyrosulphate can be used. Chromite, cassiterite and zircon, some of the commonest accessory minerals, are not appreciably attacked in a pyrosulphate fusion.

In addition to its use in dissolving the residue remaining after decomposition of the rock material with hydrofluoric acid referred to above, fusion with pyrosulphate is also widely used to dissolve the residue remaining from the silica evaporation, and the residue from a sodium carbonate fusion.

Platinum crucibles and dishes, although used and recommended for pyrosulphate fusions, are not the first choice of vessels for this purpose. Sulphur trioxide is readily lost from the melt leaving potassium sulphate which is not effective in the decomposition of oxide minerals. The loss of sulphur dioxide is very much less when silica crucibles are used. Platinum is appreciably attacked in the course of pyrosulphate fusions, introducing platinum into the rock solution. This can interfere with subsequent determinations, as for example that of vanadium. For this determination and also that of total iron, the rock material should be decomposed by evaporation with hydrofluoric acid in a PTFE vessel and the residue transferred to a silica crucible for the pyrosulphate fusion.

FUSION WITH SODIUM CARBONATE

All silicate rocks are decomposed more or less completely by prolonged fusion with anhydrous sodium carbonate, normally in a platinum crucible. Crucibles of a platinum-iridium alloy, which has a much higher mechanical strength and larger resistance to deformation, have been used. Palau crucibles (a gold-palladium alloy) are also suitable being not only more rigid than pure platinum, but also much cheaper. The amounts of platinum or other noble metal introduced into the melt are very small and can usually be ignored.

Platinum crucibles usually become iron-stained after a few fusions of rock material with sodium carbonate. This indicates that some reduction of iron to the metallic state has occurred, which has then become alloyed with the platinum. This is often difficult to see, but is usually visible as a purple coloured stain when the apparently clean crucible is heated in an electric furnace. This stain can be removed by alternate roasting in the furnace and leaching with 6M hydrochloric acid. Some small amount of platinum is inevitably taken into solution. It is essential to remove this iron from the crucible before reusing it, and also if iron is to be determined later in the analysis.

In the analyses of acidic and intermediate rocks, this small amount of alloying by iron can sometimes be ignored, but with basic rocks a small amount of potassium nitrate or chlorate can be added to maintain the melt in an oxidised condition. This addition increases slightly the extent to which platinum is attacked and removed from the crucible.

Reducing melts can be obtained from rock samples containing much sulphide or carbonaceous matter; these elements should be removed by roasting prior to adding the sodium carbonate, although small amounts of sulphide minerals or organic matter can be tolerated as these will be oxidised by the added potassium nitrate or chlorate.

Complete fusion of 1 g of most silicate rock is obtained by using 5 g of sodium carbonate. Larger quantities are not justified, even for basic rocks, whilst as little as 3 g will give a fluid melt with acidic rock materials. After fusion for 1 hour at a temperature of about 1000°, the silicate rock matrix and most of the accessory minerals will be completely decomposed, although further heating at 1200° for an additional period of about 10 minutes is recommended for the decomposition of the small amounts of zircon, rutile and chromite that are sometimes present.

Although fusions with sodium carbonate are usually preferred certain authors have noted that sintering will often suffice. Finn and Klekotka,[7] for example, sintered 0.5 g of silicate rock material with 0.6 g of anhydrous sodium carbonate. This method of decomposition has the advantage of reducing the volumes of acids and other reagents added in subsequent stages of the analysis, of reducing considerably the amount of sodium salts to be washed from later precipitates, of reducing the contamination from introduced platinum from the crucible and any impurities present in the sodium carbonate (perhaps no longer as important as it may at one time have been!), and more particularly of reducing the time necessary for the complete analysis.

Hoffman[8] used 0.5 g of sodium carbonate with 0.5 g of rock material and sintered in a 75-ml platinum dish at a temperature of 1200°. The addition of hydrochloric acid to the sinter gave an insoluble silica residue that could be dehydrated in the same 75-ml dish, in place of the clear solution usually obtained by treating the fused melt, and which requires evaporation and dehydration in a much larger basin. The silica residue obtained in this way appears to contain somewhat larger amounts of other elements from the rock material.

This 1:1 ratio of sample weight to weight of anhydrous sodium carbonate is not recommended for the decomposition of kyanite, sillimanite, andalusite or silicate rocks containing large amounts of these alumino-silicates. These minerals tend to fuse and form glassy-melts with a well-ordered structure not readily broken down by the addition of hydrochloric acid. This difficulty does not arise if larger amounts of sodium carbonate are used, not less than 4 g of flux should be used for a 1 g portion of these silicates.

The melts or sinters obtained with sodium carbonate are usually extracted with hot water prior to acidification with hydrochloric acid. An alternative procedure by Flaschka and Myers[9] is to make use of isothermal diffusion of hydrochloric acid vapour. The melt or cake is covered with a little water and placed with a beaker of hydrochloric acid in a vacuum desiccator which is then evacuated to insipient bubbling of the acid. It may be necessary to repeat the evacuation to remove liberated carbon dioxide. The dissolution takes place so slowly that no effervescence occurs and danger of splattering is avoided.

FUSION WITH ALKALI HYDROXIDE

Sodium and potassium hydroxides are extremely efficient fluxes for the decomposition of silicate minerals. This decomposition occurs rapidly at temperatures very much less than those required for fusions with sodium carbonate. The ease with which silicate minerals dissolve in molten alkali is deceptive in that the accessory mineral fraction is likely to remain unattacked unless the fusion is prolonged. Although 5 minutes fusion is more than sufficient for felspars and other silicate minerals, a full hour is recommended for silicate rocks.

As molten alkalis are particularly corrosive, this fusion should be carried out at as low a temperature as possible, with the bottom of the crucible at only a faint red heat. Earlier workers recommended a spirit lamp for this decomposition, and positioned the crucible in a hole cut in asbestos board. This served to keep the upper parts of the crucible cold, preventing "alkali-creep" over the edge of the crucible.

Platinum crucibles are subject to considerable attack from molten alkalis and should not be used. Silver and gold crucibles have been suggested, as the attack by molten alkali is very much less. However, some attack of metal does occur and the silver or gold introduced into the analysis in this way should be removed from the solution at a later stage. It must also be remembered that silver and gold have somewhat lower melting points (960° and 1063° respectively) than platinum, and crucibles can easily be damaged by overheating.

For many purposes iron or nickel crucibles can be used for these fusions. Although there is an appreciable attack of the metal, most crucibles will stand up to at least a dozen fusions before becoming porous. They cannot be used for determinations where the introduced iron or nickel would interfere with the subsequent analysis, but have long been used for the determinations of such elements as chromium and vanadium that form anions in their higher valency states. These crucibles have also been used for the determination of silica by a photometric method, where a rapid, effective decomposition of the silicate fraction is adequate. Both sodium and potassium hydroxides may contain traces of absorbed water and should in the first place be fused in the crucible without sample material.

Zirconium crucibles are excellent for this decomposition. They are much more resistant than either nickel or iron. Very little zirconium is introduced into the analysis.

Nickel crucibles have been preferred for the determination of silica, but they should not be used if the determination of iron is required, as inevitably some loss of iron to the crucible occurs when silicates are fused with sodium hydroxide.[10]

SINTER OR FUSION WITH SODIUM PEROXIDE

Sodium peroxide is particularly useful in mineral analysis, as it is the only flux that can be easily and readily used for the complete decomposition of cassiterite and chromite. Earlier authors have tended to avoid its more general use, partly because of the uncertain quality of the reagent then available and partly because of the corrosive action of sodium peroxide on the materials used for crucibles - that is platinum, gold, silver, nickel and iron. Where the obvious advantages of using sodium peroxide could not be overlooked, as for example in the analysis of silicates containing appreciable amounts of chromite, then iron or nickel crucibles were used and discarded after a few determinations. In more recent years these difficulties have been largely overcome and the use of sodium peroxide is now more generally possible. Certain batches of reagent have been found to contain calcium, and these should be avoided if complete analyses are to be made.

One method of avoiding excessive attack of platinum is to line the crucible with a thick layer of fused anhydrous sodium carbonate before adding and mixing the sodium peroxide flux with the sample material. This technique is successful only if the subsequent fusion is not unduly prolonged. Nickel crucibles can be protected from excessive corrosion by a similar lining of the base of the crucible with fused sodium hydroxide.

Zirconium crucibles have been shown[11] to have superior resistance to molten sodium peroxide, although old crucibles may contribute appreciable amounts of zirconium to the melt, particularly if fusions have been conducted at temperatures in excess of 700°. Rafter and Seelye[12] have shown that most minerals occurring in silicate rocks are rapidly decomposed by sintering with sodium peroxide at a temperature of $480^\circ \pm 20^\circ$. This operation can be conducted at temperatures of up to 540° in platinum crucibles without introducing platinum into the rock sinter or solution. Rafter[13] has recommended that samples for decomposition in this way should be ground to pass a 240-mesh sieve, but this fine grinding is probably not necessary for most silicate rocks which are readily attacked at 100-mesh size by sintering.

Sodium peroxide melts or sinters are readily disintegrated by reaction with water, giving a highly alkaline solution containing much of the silica and aluminium, and a residue containing iron, titanium and other metals as hydroxides. If silica is to be determined then, as with alkali hydroxide fusions, the use of glass beakers must be avoided. Beakers of stainless steel or polypropylene should be used. The reaction of sodium peroxide with water can be violent and on no account must water be added directly to the melt in the crucible, as this may give rise to local overheating and the splattering of caustic alkali, as well as partial loss of sample material.

FUSION WITH BORIC OXIDE AND ALKALI BORATES

Boric oxide and boric acid, although apparently attractive fluxes for the decomposition of silicate rocks, have never been widely used. This may be due in part to the extremely viscous nature of the melts which makes then difficult to use, and in part to the necessity of removing boron at a later stage of the analysis.

The use of borax (sodium tetraborate), or combinations of boric oxide, boric acid or borax with sodium carbonate has achieved some prominence in the analysis of materials rich in alumina, and has been recommended[14] for the decomposition of refractory minerals such as corundum, and chromium- and zirconium-bearing materials. It can be used with advantage for the analysis of kyanite, sillimanite and other aluminosilicates.

Borate-carbonate melts disintegrate readily in dilute hydrochloric acid giving solutions that can be evaporated for the determination of silica. Methyl alcohol is added to the solution before commencing the evaporation, in order to remove boron as the volatile methyl borate. Failure to remove the boron at this stage will give high values for silicon, as some boron will be trapped with the silica on dehydration and subsequently be lost in the evaporation of the weighed silica with hydrofluoric and sulphuric acids. This evaporation with methyl alcohol is not necessary if silica is to be determined photometrically, as boron does not interfere with either the silicomolybdate or the molybdenum blue methods. Bennett and Hawley[14] have noted that is difficult to remove boron from materials with greater than 50% silica, indeed it is doubtful if all the boron can be removed with methyl alcohol in this way.

Biskupsky[6] has suggested using a flux composed of boric acid and lithium fluoride for the decomposition of silicate rocks and minerals. Lithium tetraborate is formed in the fusion, whilst silica is removed as the volatile tetrafluoride. Both boron and excess fluoride are removed by heating the melt with concentrated sulphuric acid. Advantages claimed are that only 12 to 13 minutes fusion time is required and that zircon, sillimanite, topaz, spinel, corundum, rutile, kyanite and other refractory minerals are decomposed without difficulty.

Lithium metaborate ($LiBO_2$) has been suggested by Ingamells[15,16] as a suitable flux for the decomposition of silicate rocks preparatory to determining silicon, phosphorus, iron, titanium, manganese, nickel and chromium by spectrophotometry. Sodium and potassium can be determined by flame photometry[17] and other elements by an emission spectrographic solution technique giving an essentially complete analysis (less FeO, CO_2, H_2O and certain minor components) from one sample portion. Lithium metaborate is now widely used in a variety of schemes for rock analysis. Graphite crucibles are generally preferred as the fusion beads can be easily poured from the crucible into the solvent.

This chapter contains only a summary of techniques available and the problems encountered in rock analysis. More exhaustive reviews of the methods available for sample decomposition have been given by Dolezal et al[18] and by Bock.[19]

References

1. MAY I and ROWE J J., Anal. Chim. Acta (1965) 33, 648
2. LANGMYHR F J and SVEEN S., Anal. Chim. Acta (1965) 32, 1
3. ANTWEILER J C., U S Geol. Surv. Prof. Paper 424-B (1961), p.322
4. LANGMYHR F J., Anal. Chim. Acta (1967) 39, 616
5. CHEAD A C and SMITH G F., J. Amer. Chem. Soc. (1931) 53, 483
6. BISKUPSKY V S., Anal. Chim. Acta (1965) 33, 333
7. FINN A N and KLEKOTKA J F., Bur. Std. J. Res. (1930) 4, 813
8. HOFFMAN J I., J. Res. Nat. Bur. Std. (1940) 25, 379
9. FLASCHKA H and MYERS G., Z. Anal. Chem. (1975) 274, 279
10. BENNETT H., EARDLEY R P and THWAITES I., Analyst (1961) 86, 135
11. BELCHER C B., Talanta (1963) 10, 75
12. RAFTER T A and SEELYE F T., Nature (1950) 165, 317
13. RAFTER T A., Analyst (1950) 75, 485
14. BENNETT H and HAWLEY W G., Methods of Silicate Analysis, Academic Press 1965 (2nd ed.) p.41
15. INGAMELLS C O., Talanta (1964) 11, 665
16. INGAMELLS C O., Analyt. Chem. (1966) 38, 1228
17. SUHR N H and INGAMELLS C O., Analyt. Chem. (1966) 38, 730
18. DOLEZAL J., POVONDRA P and SULCEK Z., Decomposition Techniques in Inorganic Analysis, Iliffe (English edition), London, 1968
19. BOCK R., A Handbook of Decomposition Methods in Analytical Chemistry, International Textbook Co., 1979

CHAPTER 3

Classical Scheme for the Analysis of Silicate Rocks

In the classical scheme for the analysis of silicate rocks, provision was usually made for the determination of a total of the thirteen most commonly occurring constituents. Of these, the alkali metals were determined in a separate portion of rock material, as were moisture, total water and ferrous iron. Most rock analysts preferred also to determine manganese, titanium, phosphorus and total iron in separate portions, leaving only silica, "mixed oxides", calcium and magnesium to be determined in what was known as the "main portion". Where the silicate rock sample was available in only small amounts, the sample portion used for the determination of moisture was used also for the elements in the main portion, as well as for total iron and sometimes also for titanium. Strontium, when present in more than trace amount, was precipitated with calcium as oxalate, and then separated and determined gravimetrically.

One of the most serious criticisms of the classical scheme is that any error in the determination of some of the constituents - iron, titanium or phosphorus, for example - was reflected in a similar error in the aluminium content, which was always obtained by difference.

This chapter is concerned with the analysis of the main portion, that is with the determination of silica, the total of elements precipitated with ammonia and known collectively as the "mixed oxides", "ammonia group" or by certain analysts as the "R_2O_3 precipitate", together with calcium and magnesium. In the original classical scheme of analysis manganese appeared partly with magnesium in the phosphate precipitate and partly with iron and other elements in the ammonia precipitate.(1) Procedures have been devised to collect all the manganese in one fraction, but these are not entirely successful. Chromium, vanadium, zirconium and other elements are also precipitated with ammonia and, when present in more than trace amounts, can introduce errors into the reported aluminium content.

The scheme of analysis is given in outline form in Fig 1. It is based upon the use of a 1 g portion of ground silicate rock material. The broad outline of the scheme and some of the methods used were devised in the nineteenth century, but continuous development has occurred since then, largely by way of refining procedures in the light of subsequent knowledge and experience.

It should also be noted that gravimetric procedures for the determination of calcium and magnesium have now been almost entirely supplanted by procedures based upon titration with EDTA or atomic absorption spectrometry. They are included here only for the sake of completeness. The alternative, now widely used procedures are described in the relevant chapters.

Similarly it should be noted that in the classical scheme of silicate rock analysis, separate portions were always used for the determination of alkali metals, ferrous iron, 'moisture', total water, and usually for total iron, titanium, total manganese and phosphorus. These determinations also are dealt with in their respective chapters.

Decomposition of the Sample

Procedure. Ignite a clean platinum crucible of 25 to 30-ml capacity, together with its lid over the full heat of a Meker burner for a few minutes. Allow to cool for a few seconds, transfer to a desiccator and weigh after 30 minutes. Accurately weigh approximately 1 g of the finely powdered silicate rock material into the **crucible and with the lid displaced slightly to allow water vapour etc to escape,** heat over the Meker burner, gently at first, gradually increasing the flame to give full heat for about an hour. Allow the crucible to cool and re-weigh after 30 minutes. Record the loss in weight. This gives a useful check for total water plus carbon dioxide after making allowances for the gain in weight due to oxidation of ferrous iron.

Add to the ignited sample 3-5 g of anhydrous sodium carbonate and mix with a platinum or glass rod. Brush any particles of rock material or flux from the rod back into the crucible, cover with a platinum lid and heat over a Bunsen burner or in an electric furnace to a dull red heat (furnace set at about 700°), and maintain at this temperature for about 30 minutes. Slowly raise the temperature to about 1000° and maintain at this temperature for a further 30 minutes, finally transfer the crucible to a Meker burner (not a blast Meker) or to an electric furnace set at 1200°, and heat for a further 10 minutes. Allow to cool, rotating the crucible held in a pair of platinum-tipped tongs, so as to allow the melt to solidify in a layer around the walls of the crucible.

The fusion cake may be dissolved from the crucible as described below. Alternatively with the crucible held upside down over the platinum dish, it may be gently flexed until the fusion cake falls into the dish. The crucible can then be half filled with 6 M hydrochloric acid and warmed until all solid matter has dissolved. This solution is then added to the main portion after evolution of carbon dioxide has ceased.

Add water to the fusion cake in the platinum dish, or platinum basin if it has been dislodged from the crucible, together with 2 or 3 drops of ethanol and allow to stand overnight. On the following day rinse the solution and residue from the crucible into a large (6-inch) platinum dish, wash the crucible with water and set it aside. In the presence of much manganese the melt is tinged green with alkali manganate, but this is reduced by the ethanol on standing overnight.

This procedure serves to decompose completely all the minerals present in silicate rocks. As noted in Chapter 2 the quantity of sodium carbonate used is now regarded as excessive, and can be reduced, as for example by using the sintering technique of Hoffman,[2] as follows:

Procedure. Accurately weigh approximately 0.5 g of the finely powdered silicate rock material and 0.5 g of anhydrous sodium carbonate into a 75-ml platinum dish and mix together with a glass rod. If the rock material contains much ferrous iron add also 0.05g of potassium nitrate. Brush the mixture into the centre of the dish and then spread out in the form of a disc of about 3 cm diameter. Cover the mixture as evenly as possible with a further 0.5 g of anhydrous sodium carbonate. Transfer the dish to an electric muffle furnace and heat, slowly at first then more strongly until a temperature of 1200° is reached. Maintain the dish at this tempeature for 15 minutes, then allow to cool with a cover over the dish to prevent loss of material by spitting in the cooling stage.

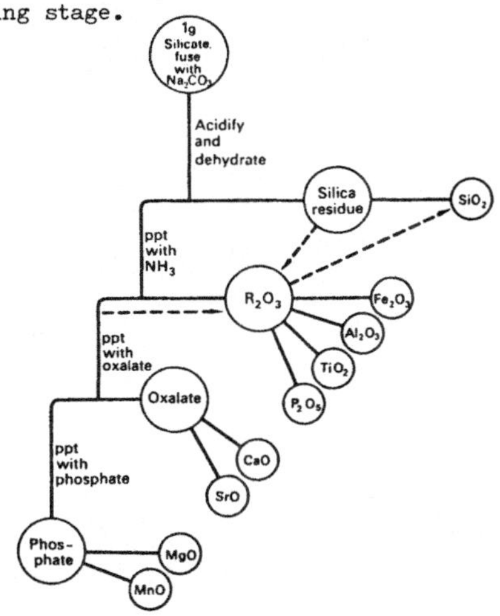

FIG 1. Classical scheme for the analysis of the main portion.

The advantages of this method of decomposition have been listed in Chapter 2. It has found application in the field of glass technology, where it has been strongly recommended by Chirnside.(3) A sodium peroxide sinter in a platinum crucible can also be used for the decomposition of the main portion. A grade of sodium peroxide free from calcium is required for this. It should be noted that this is not readily available.

Determination of Silica

SEPARATION AND COLLECTION OF SILICA

The rock material, decomposed as described above, is acidified with hydrochloric acid, and the chloride solution evaporated to dryness. Most silicate rock analysts prefer to use platinum apparatus for this evaporation, but porcelain dishes can also be used. The major part of the silica present in the solution is recovered by dehydration and filtration, leaving aluminium, iron, alkali and alkaline earth elements together with the minor part of the silica in the filtrate. In the classical scheme of analysis, the filtrate is returned to the platinum basin for a second evaporation and dehydration to recover an additional silica fraction. Only a few milligrams of silica remain in solution after this second evaporation and these cannot be recovered by a third evaporation. These traces will be precipitated together with iron, aluminium, titanium and other elements by adding ammonia.

In the procedure described by Hoffman,(2) the evaporation and dehydration of silica are conducted in the 75-ml platinum dish used for the sample decomposition, and the silica fractions recovered by filtration are returned to this dish for ignition and subsequent treatment.

Procedure. Cover the platinum dish with a large clock glass, and by displacing it slightly, carefully add 15 ml of concentrated hydrochloric acid. Replace the cover and allow to stand for a few minutes until all vigorous action has ceased.

Add 5 ml of concentrated hydrochloric acid to the platinum crucible used for the decomposition of the sample, cover with a small clock glass and transfer to a steam bath for 10 minutes. Allow to cool and then rinse the contents into the large platinum basin with a jet of water. Carefully wipe the crucible out with damp filter paper (or alternatively use a rubber tipped glass rod, "policeman") to remove all traces of silica adhering to it, add these pieces of paper to the solution in the dish. If the same crucible is to be used for the ignition of the silica precipitates, ignite the crucible over a Meker burner, allow to cool, weigh after exactly 30 minutes, and then set it aside until required. (Note: platinum-iridium crucibles although suitable for sodium carbonate fusions should not be used for ignitions as they tend to lose weight at high temperatures.)

Remove the cover, rinse down with water and then replace it over the dish. Transfer the dish to a steam bath and heat until no further effervescence is apparent, then rinse down and remove the clock glass, and evaporate the solution to dryness on the steam bath. As the last traces of water and acid are removed, the deep yellow colour of the residue is replaced by a paler tint. When this stage is complete, test for complete expulsion of hydrogen chloride with the stopper of an ammonia bottle. Leave the dish on the steam bath for 30 minutes after fumes of ammonium chloride can no longer be detected. More rapid expulsion of hydrogen chloride can be obtained by drying the residue at a temperature of 120 to 150° in an electric oven.

Remove the dish from the steam bath, allow to cool and add 10 ml of concentrated hydrochloric acid, tilting the dish to ensure that all the residue is wetted with acid. Rinse down the clock glass and the sides of the dish, adding sufficient water to give a total volume of about 100 ml. Stir the solution with a stout glass rod and warm on a steam bath until all soluble salts have dissolved, leaving only a gelatinous residue of silica.

Collect the silica residue on a small medium-textured filter paper and wash at least six times with cold water and twice with hot water, to remove all soluble chlorides from the residue. Rinse the filtrate and washings back into the large platinum dish, transfer to the steam bath and again evaporate to dryness. As the evaporation proceeds, break up all crystals of sodium chloride with the flattened end of a glass rod. When all moisture and hydrochloric acid have been removed, transfer the dish to an electric oven and dry at a temperature of 105° to 110° for 1 hour.

Moisten the residue with 10 ml of concentrated hydrochloric acid and dissolve the chloride salts in about 100 ml of water as before. Collect the small residue, consisting largely of silica on a small close-textured filter paper and wash first with cold, then hot water, as described above. Carefully wipe the large platinum dish with wet filter paper to collect any silica adhering to the dish, and add this paper to the residue in the filter funnel before washing. Reserve the combined filtrate and washings for the subsequent analyses.

IGNITION AND VOLATILISATION OF SILICA

The silica residues contain small amounts of iron, aluminium and even smaller amounts of other elements of the ammonia group - titanium, zirconium and phosphorus. Calcium, magnesium and strontium are not likely to be present, and if the washing has been correctly and adequately performed, the alkali elements are also unlikely to be present.

The total weight of the silica residue is determined after ignition in platinum. Silica is then removed by evaporation with hydrofluoric and sulphuric acids:

$$SiO_2 + 4HF = SiF_4 + 2H_2O$$

Iron, aluminium and other elements present in small amounts are converted to sulphates, but on strong ignition these are again converted to oxides. The difference in weight corresponds to the silica lost in the evaporation with hydrofluoric acid. A small correction arising from the presence of a trace of silica in the filter papers, and from involatile residue in the hydrofluoric acid, should be determined. With present grades of hydrofluoric acid, this correction should be very small, amounting to no more than about 1 mg. This correction must also be made to the "mixed oxides". Before calculating the silica content of the sample material, the "silica traces" must be added. These traces are recovered from the ammonia precipitate at a later stage of the analysis.

Procedure. Clean, ignite in an electric furnace and weigh a platinum crucible of about 25 to 30-ml capacity. Transfer to it the moist filter papers containing the silica residues and dry carefully over a small flame or in an electric oven. Allow to cool, moisten with 4 or 5 drops of 20 N sulphuric acid, and continue the heating over a low flame, burning the paper away and giving a white residue. Transfer the crucible to an electric furnace, set at a temperature of 1050°, cover with a platinum lid - slightly displaced - and heat strongly for 40 minutes. Allow to cool in a desiccator and weigh after exactly 30 minutes. Repeat the ignition for a period of 10 minutes, cooling and weighing as before; repeat the ignition as necessary to obtain constant weight.

Moisten the residue with 1 ml of water and add 5 drops of 20 N sulphuric acid and 10 ml of concentrated hydrofluoric acid. Transfer the crucible to a hot plate and evaporate the silica and excess hydrofluoric acid. Raise the temperature towards the end of the evaporation to remove free sulphuric acid. Allow to cool. Transfer the crucible to a silica triangle and heat over a Bunsen burner to decompose sulphates and then over a Meker burner until constant weight is obtained. The loss in weight is the uncorrected main fraction of the silica. Set the crucible aside for the ignition of the ammonia precipitate.

To determine the correction, transfer filter papers equal in number to those used in the silica determination, to a clean, weighed platinum crucible, burn off the carbon and ignite over a Meker burner. Allow the crucible to cool and then weigh the residue obtained. This gives the weight of filter paper ash. Now add 5 drops of 20 N sulphuric acid and 10 ml of concentrated hydrofluoric acid, transfer the crucible to a hot plate and evaporate the hydrofluoric and sulphuric acids as with the silica evaporation. Finally ignite over a Meker burner, cool and weigh. There is usually a small increase in weight after the ignition, corresponding to the arithmetic total of a small loss by volatilisation of silica from the filter paper ash, combined with a gain in weight from the non-volatile residue in the hydrofluoric acid. The overall increase in weight must be added to the silica value previously obtained.

Determination of "Mixed Oxides"

PRECIPITATION OF THE "MIXED OXIDES"

The mixed oxides are precipitated in the filtrate from the silica determination by adding ammonia to the hot solution until it is just alkaline to methyl red or bromocresol purple indicator, ie at a pH of about 7. Iron, aluminium, phosphorus, zirconium, vanadium and chromium are precipitated together with a number of other elements present in only minor or trace amounts including beryllium, gallium, indium, thorium, scandium and the rare earths. The very small amounts of nickel, cobalt and zinc present in most silicate rocks are not precipitated, but accompany calcium, strontium and magnesium into the filtrate. If nickel is present in more than trace amounts, some will be caught in the ammonia precipitate.[4] Some small amount of calcium and magnesium will be entrained in the ammonia precipitate, but these amounts are recovered by dissolving the precipitate in dilute hydrochloric acid and reprecipitating with ammonia.

Although the bulk of the aluminium is precipitated with ammonia, some small amount is found in the filtrates ("aluminium traces"), from which it can be recovered and added to the ammonia precipitate.

Small amounts of manganese are not usually precipitated with ammonia, but pass into the alkaline filtrate and are subsequently precipitated as phosphate with the magnesium. Larger amounts of manganese are divided between the two fractions, the major part with the magnesium. It has been noted that manganese will be precipitated with the elements of the ammonia group if oxidising agents are added. Bromine water has been used for this by Holt and Harwood[5] and ammonium persulphate by Hillebrand.[6] Neither technique gives complete precipitation of manganese.[1]

The use of oxidising agents such as persulphate converts chromium into a higher valency state which is then not precipitated with ammonia. Moreover, the addition of oxidising agents increases the difficulty of adjusting the pH of the solution. In the authors' opinion, the simplest and best procedure is not to attempt to precipitate manganese with the ammonia group, but to collect most of it with the magnesium and then determine it photometrically in the phosphate residue. This, together with a determination of total manganese in a separate portion, enables the minor fraction in the ammonia precipitate to be calculated. The amount of manganese collected in the oxalate precipitate can generally be ignored. There is little point in removing manganese from the filtrate after the precipitation of the ammonia group by treatment with bromine water, or by co-precipitation with zirconium, as described by Peck and Smith[7] as the small amount of manganese incorporated in the ammonia precipitate would still remain to be determined.

Procedure. Add 5 ml of concentrated hydrochloric acid to the combined filtrate and washings from the removal of silica (for rocks rich in magnesium, add 10 ml), and concentrated ammonia solution until the formation of a precipitate that only just dissolves on stirring. Heat the solution just to boiling, add a few drops of bromocresol purple indicator solution and continue adding ammonia until complete precipitation is obtained, and the supernatant liquid is purple in colour. Avoid adding an excess of ammonia.

With rocks containing much iron, the end point will be obscured, and the addition of ammonia is best continued until the precipitation is essentially complete before adding single drops of indicator solution. The colour of the indicator can then be observed as it mixes with the solution. Bring the solution to the boil again and allow the precipitate to settle somewhat. If the supernatant liquid is not purple in colour, add more ammonia, drop by drop, until the purple colour is restored. Stir in a macerated filter paper and allow to stand for 1 minute.

Collect the residue on a large open-textured filter paper and wash six times with a wash solution containing 20 g of ammonium nitrate per litre and made just alkaline with ammonia to bromocresol purple indicator. Reserve the filtrate and washings.

Transfer the filter paper and residue back to the beaker in which the precipitation was made, add 50 ml of water and 15 ml of concentrated hydrochloric acid. Cover the beaker with a clock glass and digest on a steam bath until complete dissolution is obtained, then dilute to 250-300 ml with water. Again precipitate with ammonia as described above, collect the ammonia precipitate on a large open-textured paper and wash as before. Combine the filtrate and washings with those obtained from the first precipitation with ammonia, and reserve for the subsequent stages of the analysis. Transfer the filter to the platinum crucible previously used for the volatilisation of silica.

RECOVERY OF THE "ALUMINIUM TRACES"

The small amount of aluminium present in the filtrate after the precipitation with ammonia can be recovered by evaporating the solution to small volume and re-precipitating with ammonia. If ammonium persulphate has been added, chromium will also have passed into the filtrate, and will be collected with the aluminium traces.

Procedure. Add concentrated hydrochloric acid drop by drop to the combined filtrates and washings until the solution is just acid, then transfer the beaker to a steam bath or hot plate and evaporate to a volume of about 200 ml. Now add concentrated ammonia drop by drop until the solution is just alkaline to bromocresol purple, cover the beaker with a clock glass and digest on the steam bath for 10 minutes. The traces of aluminium form a small gelatinous precipitate. Collect this precipitate on a small open-textured filter paper, wash four or five times with the ammonium nitrate wash solution, dissolve in dilute hydrochloric acid and re-precipitate with ammonia. Collect the precipitate as before, wash well and add to the ammonia precipitate in the platinum crucible. Reserve the combined filtrates and washings for the determination of calcium and magnesium.

IGNITION OF THE "MIXED OXIDES"

The ammonia precipitate is ignited in the crucible used for the volatilisation of silica, and still containing the small amounts of iron, aluminium and other elements co-precipitated with the silica. Opinions differ as to the best temperature for the ignition. The conversion of ferric oxide to magnetite at temperatures in excess of 1100°, have led some authors to suggest that 1100° is the maximum that should be used. However it is known that alumina is not completely dehydrated at this temperature, and for this reason other authors have recommended 1200°. For ammonia precipitates consisting largely of alumina, this temperature of 1200° can safely be used, but for precipitates rich in iron, this temperature should be used only if an oxidising atmosphere can be ensured.

Procedure. Dry the filters in the platinum crucible, and using a Bunsen burner with a low flame, burn off the carbon at a low temperature. Increase the flame and ignite in the uncovered crucible over the full flame of the burner for 30 minutes. Transfer the crucible to an oxidising muffle furnace and ignite to constant weight at a temperature of 1200°. If an oxidising muffle is not available, use a blast Meker burner, but as far as possible keeping the flame away from the upper parts of the partly covered crucible.

RECOVERY OF "SILICA TRACES"

The ignited mixed oxides are particularly refractory after ignition at a temperature of 1200°. They can however be brought into solution following a fusion with potassium pyrosulphate. A small amount of silica known as the "silica traces", is usually recovered from the mixed oxides and can be filtered off and determined as before. If required, iron, titanium, vanadium and phosphorus can be determined photometrically in the sulphate solution.

Procedure. After weighing the mixed oxides, add 6-7 g of potassium pyrosulphate, cover the crucible with a close-fitting platinum lid and fuse gently over a small Bunsen flame. Too high a flame should not be used, as this results in the rapid loss of sulphur trioxide from the melt. Finally heat the crucible over a full Bunsen flame for 5-10 minutes and allow to cool. Place the crucible on its side in a 250-ml beaker and add 100 ml of 4 N sulphuric acid. Cover the beaker with a clock glass, transfer to a steam bath and digest until the solid melt has completely disintegrated. Rinse and remove the crucible and lid and also the clock glass.

Transfer the beaker to a hot plate, evaporate to fumes of sulphuric acid and allow to fume copiously for 10 minutes. Allow to cool. Cautiously dilute with about 100 ml of water and digest on a steam bath until all soluble material has passed into solution. At this stage a clear solution should be obtained, in which a few milligrams of silica are visible as a slight residue. Collect this residue on a small medium-textured filter paper, wash with cold water and determine the silica by volatilisation with hydrofluoric and sulphuric acids as described previously. Combine the filtrate and washings and dilute to volume in a 200-ml volumetric flask for the determination of total iron, titanium etc, if these are required.

DETERMINATION OF ALUMINIUM BY DIFFERENCE

The aluminium content of the rock material, as determined by the difference method, is obtained by subtracting from the total of mixed oxides expressed as a percentage of the sample taken, the combined total of other elements present, each determined separately. These elements include total iron, calculated as Fe_2O_3, titanium as TiO_2, phosphorus as P_2O_5, vanadium as V_2O_3, chromium as Cr_2O_3, zirconium as ZrO_2, that part of the manganese present in the residue, calculated as Mn_3O_4 and "silica traces" as SiO_2. Other elements in the ammonia precipitate are seldom present in amounts sufficient to be totalled in this way.

The determination of aluminium by difference has been shown to give rise to distorted values, and hence can no longer be recommended.

Determination of Calcium

Calcium is separated from the filtrates remaining after the collection of the "aluminium traces", by precipitation as calcium oxalate. Much of the strontium present in the rock material will also be precipitated. Although the classical scheme for silicate rock analysis includes provision for the separation and separate determination of strontium, this method is no longer considered adequate, and determination by atomic absorption spectroscopy in a separate portion of the rock material is recommended. The amount of strontium present in most silicate rocks does not introduce significant errors into the calcium determination by its precipitation as oxalate, and for most rocks, the determination of co-precipitated strontium is not necessary.

Any platinum salts that have separated out in the ammonia group filtrate should be removed by filtration prior to precipitating calcium.

Procedure. The combined filtrates and washings from the removal of the aluminium traces should have a volume of about 300 ml. Make this solution just alkaline to bromocresol purple indicator and heat to boiling. Add a solution of 5 g of ammonium oxalate in 100 ml of hot water, bring to the boil again, digest on a steam bath for 10 minutes and then allow to stand overnight.

Collect the precipitated calcium oxalate on a close-textured filter paper and wash five times with a cold solution containing 1 g of ammonium oxalate per litre. Reserve the filtrate and washings. Rinse the precipitate back into the beaker used for the precipitation and dissolve by warming with 5 ml of concentrated hydrochloric acid and 100 ml of water. Heat the solution to boiling and filter back through the paper, collecting the filtrate in a clean 400-ml beaker and washing the paper seven or eight times with warm water.

Add 2 g of ammonium oxalate dissolved in about 50 ml of warm water followed by concentrated ammonia solution until a slight precipitate forms that does not redissolve on stirring. Clear this precipitate with 2 drops of concentrated hydrochloric acid, heat the solution to boiling and precipitate calcium oxalate by adding 4 N ammonia until the solution is just alkaline to methyl red or bromocresol purple indicator. Transfer the beaker to a steam bath for 30 minutes then allow to cool and stand for 3-4 hours or overnight. Collect the precipitate on a close-textured filter paper and wash six times with the ammonium oxalate wash solution. Combine the filtrate and washings with those obtained from the first precipitation of calcium oxalate, and reserve for the determination of magnesium.

Transfer the filter paper and contents to a clean, ignited and weighed platinum crucible and dry in an electric oven set at 105°, or over a very low flame. Burn off the filter paper by heating over a Bunsen burner at dull red heat and then ignite over the full flame of a Meker burner. Cool, and weigh as calcium oxide, CaO. Calcium oxide residues are hygroscopic and if a muffle furnace is available, as an alternative method the crucible can be ignited at a temperature of 500°, and the residue weighed as calcium carbonate.

Determination of Magnesium

Magnesium is determined by precipitation as magnesium ammonium phosphate hexahydrate, and weighing as magnesium pyrophosphate:

$$2Mg(NH_4)PO_4.6H_2O = 13H_2O + Mg_2P_2O_7 + 2NH_3$$

Any manganese present in the solution will also be precipitated as an ammonium phosphate and be ignited and weighed as pyrophosphate, $Mn_2P_2O_7$. Other contaminants of the magnesium pyrophosphate include barium and traces of strontium and calcium not collected in the oxalate precipitate.

Although some analysts precipitate magnesium in the presence of the large amounts of ammonium salts that have been added to the rock solution in the course of the earlier separations, this cannot be recommended, as the ammonium salts tend to prevent the precipitation of small amounts of magnesium, and to give incomplete precipitation with larger amounts. In the procedure described, these are removed by evaporation with concentrated nitric acid.

The ignition of magnesium pyrophosphate is one of the most difficult operations in the classical scheme of silicate rock analysis. If the burning off is done at too high a temperature, some of the pyrophosphate may be reduced and the precipitate invariably becomes impregnated with carbon, which is then very difficult to burn off. The composition of the precipitate does not always correspond exactly to the composition $Mg(NH_4)PO_4.6H_2O$ - some $Mg(NH_4)_4(PO_4)_2$ may be included. On ignition this forms $Mg(PO_3)_2$, which can only be converted to pyrophosphate by ignition at temperatures in the range 1150° to 1200°. At these temperatures the pyrophosphate itself slowly loses P_2O_5. If the magnesium ammonium phosphate precipitate contains traces of calcium or other elements, the ignition may cause fusion of the residue, even at temperatures as low as 1000°.

Procedure. Combine the filtrates and washings from the calcium oxalate precipitations in a large beaker and evaporate on a steam bath to give a solution volume of about 200 ml. Rinse this into a 600-ml beaker, allow to cool and cover with a clock glass. Displace the clock glass slightly and add 100 ml of concentrated nitric acid. Replace the cover glass and return the beaker to the steam bath, heating until all reaction has ceased. Rinse and remove the cover glass and evaporate to dryness.

Dissolve the residue in about 100 ml of water, acidifying with a few drops of hydrochloric acid if necessary. Make the solution just alkaline to methyl red or bromocresol purple indicator, digest on a steam bath for 10 minutes and collect any small residue on an open-textured filter paper. Wash this residue with the ammonium nitrate wash solution, transfer to a platinum crucible, dry, ignite and weigh. The weight of residue should not exceed 2 mg, and is mostly alumina. Dilute the filtrate to about 250 ml with water and make just acid with hydrochloric acid. Add 6 g of diammonium hydrogen phosphate dissolved in about 50 ml of water, followed by 30 ml of concentrated aqueous ammonia, added with continuous stirring. Cover the beaker with a clock glass and set it aside overnight, preferably in a refrigerator. If the sample material contains only very small amounts of magnesium, allow to stand for 48 hours.

Collect the precipitate on a close-textured filter paper and wash six times with N ammonia solution (approximately 50 ml of concentrated ammonia per litre). Reserve the combined filtrate and washings. Rinse the precipitate back into the beaker used for the precipitation, add 1 ml of concentrated hydrochloric acid and 50 ml of water and warm to give a clear solution. Filter this solution back through the filter paper used for the first precipitate, and wash well with water. Collect the filtrate and washings in a 400-ml beaker and dilute to about 200 ml with water. Add 1 g of diammonium hydrogen phosphate dissolved in a little water, add then concentrated aqueous ammonia drop by drop until the precipitation of magnesium appears to be complete (or until the solution is alkaline, if only traces of magnesium are present), and then 10 ml of concentrated ammonia. Allow the solution to stand overnight.

Collect the precipitate on a close-textured filter paper and wash with dilute ammonia solution as before. The filtrate and washings can be combined with those obtained from the first precipitation of magnesium, and used for the determination of nickel, if this is required. Otherwise these filtrates and washings are discarded. Transfer the paper containing the magnesium ammonium phosphate precipitate to a clean, ignited and weighed platinum crucible and heat carefully over a Bunsen burner to dry the precipitate, char the paper and burn off the carbon at as low a temperature as possible. Heat over a full Bunsen flame until a completely white residue is obtained, then over a Meker burner or in an electric furnace set at a temperature of 1050° to obtain constant weight. Weigh as magnesium pyrophosphate, $Mg_2P_2O_7$, which contains 36.22% MgO.

DETERMINATION OF MANGANESE IN THE MAGNESIUM RESIDUE

The ignited magnesium pyrophosphate residue is fully soluble in dilute hydrochloric acid only with some difficulty. It is, however, soluble in concentrated sulphuric acid, to give a solution that can be readily used for the photometric determination of that portion of manganese not precipitated with ammonia and incorporated in the "mixed oxides".

Procedure. Moisten the magnesium pyrophosphate residue with 5 ml of water, add 2 ml of concentrated nitric acid and 5 ml of 20 N sulphuric acid. Transfer the crucible to a hot plate and evaporate to fumes of sulphuric acid. Allow to cool, dilute with water and rinse into a small beaker for the photometric determination of manganese by oxidation to permanganate with potassium periodate as described in Chapter 28.

References

1. JEFFERY P G and WILSON A D., Analyst (1959) 84, 663
2. HOFFMAN J I., J. Res. Nat. Bur. Stds. (1940) 25, 379
3. CHIRNSIDE R C., J. Soc. Glass Technol. (1959) 43, 5T
4. HARWOOD H F and THEOBALD L S., Analyst (1933) 58, 673
5. HOLT E V and HARWOOD H F., Mineral. Mag. (1928) 21, 318
6. HILLEBRAND W F and LUNDELL G E F., Applied Inorganic Analysis, 2nd ed., Wiley, New York, 1953, p.870
7. PECK L C and SMITH V C., U S Geol. Surv. Prof. Paper 424-D, p.401, 1961

CHAPTER 4

The Rapid Analysis of Silicate Rocks

The complete chemical analysis of silicate rocks using classical methods is very time consuming and requires the services of a skilled analyst. Such analyses are expensive. A geologist studying a particular exposure is likely to be restricted by cost to those analyses that are illustrative rather than to those that are informative. Steps towards cheaper and more rapid silicate analyses therefore were welcomed by both the field geologist and petrographer.

Classical rock analysis is based firmly upon the technique of gravimetric analysis. This itself is time consuming, but is rendered all the more so by the use of a particular sequence of lengthy precipitations, many of which have to be repeated to obtain even reasonably quantitative separation. Schemes for the rapid analysis of silicate and other rocks are based upon a replacement of these gravimetric procedures with other techniques that are simpler and more rapid in themselves. Even more important than this, these determinations are frequently made in the presence of other elements, avoiding lengthy and tedious separation stages.

What is surprising in some of the earlier schemes is that the authors, when compelled to use separations, have failed to take advantage of newer methods - ion exchange or solvent extraction, for example - but preferred to retain the imperfect separations used in the classical scheme, and sometimes even use them in a way that no classical rock analyst would contemplate or tolerate. This can be illustrated by a single example quoted by Chirnside.(1) What classical analyst would make a single precipitation of the elements of the ammonia group by adjusting the pH to 4.8-5.1 by dropwise addition of ammonia, and then hope to determine the whole of the calcium in the filtrate? (Shapiro and Brannock,(2) original scheme for rapid rock analysis.)

The classical scheme for the analysis of silicate rocks as we understand it today was not designed, it is the result of continuous evolution. Schemes for the rapid analysis of silicate rocks are also undergoing this process of evolution, and too much time should not be spent in considering and condemning the earlier attempts, at the expense of examining the latter schemes.

The points to look for in any new scheme for rock analysis must surely be:

A. Separation procedures should be avoided wherever possible; the determination of individual components should be made in the presence of the remaining components.

B. The method used for determining an individual component should be specific for that component and give results of acceptable accuracy and precision.

C. The number of sample portions taken for the analysis should be reduced to the minimum.

D. The scheme should be capable of application to a wide range of silicate rocks, and no commonly occurring silicate rock should give inaccurate, misleading or ambiguous results by the methods as described.

Thus the considerable amount of effort spent in devising new schemes for the analysis of silicate rocks based upon complex ion formation and ion-exchange separation,[3-5] although useful, may not be in the main stream of development in rapid, instrumental methods of rock analysis.

Schemes that have been introduced have a number of features in common. Firstly, although provision is made for the determination of a total of thirteen most abundant components of silicate rocks, "moisture", total water and ferrous iron are determined in separate portions of the rock material. The methods of analysis adopted for these determinations may well be the same as those used in the classical scheme.

Secondly, instrumental methods are employed wherever possible. These include flame photometric procedures for sodium and potassium and spectrophotometric procedures for silicon, aluminium, total iron, titanium, manganese and phosphorus. A number of different titrimetric procedures have been described for determining calcium and magnesium. Many of these are disappearing, being replaced largely by newer methods based upon atomic absorption spectroscopy.

Thirdly, although there are notable exceptions to this, most of the schemes for rapid analysis rely upon the decomposition of two sample portions for the determination of the ten components. Silicon and sometimes also aluminium are determined in one fraction, a small sample weight decomposed by fusion with sodium hydroxide, whilst alkali metals, iron, titanium, manganese, phosphorus, calcium and magnesium are determined in a somewhat larger sample portion decomposed by evaporation with hydrofluoric acid.

Finally, the techniques themsleves are suitable for "batch operation", that is six or eight rock samples can be processed at one time with very little extra time to that required for processing a single sample. To the beginner in rock analysis, or to the analyst who only occasionally is called upon to undertake this kind of work, the opportunity of including a previously analysed sample can be most valuable in preventing gross errors, such as those arising from faulty mathematics.

Although there are considerable varations in detail between one scheme for the rapid analysis of silicate rocks and another, in outline at least there is a striking uniformity. Thus the details shown in Fig. 2, although taken from a later version of the scheme by Shapiro and Brannock,[6] apply equally well to the scheme of Blanchet and Malaprade[7] and have more than a passing resemblance to those of Corey and Jackson,[8] Riley and Williams[10] and others.

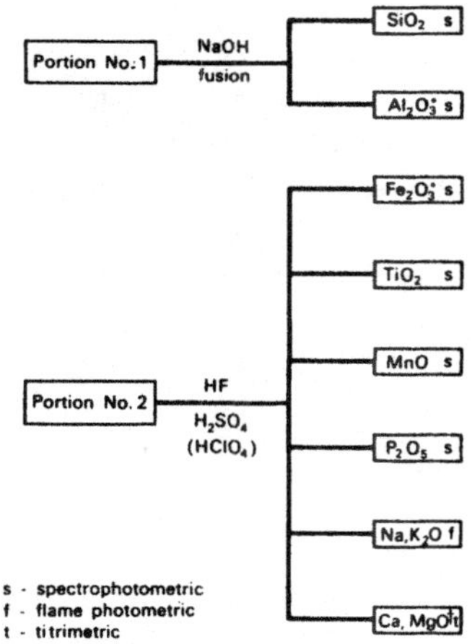

FIG 2. Scheme for the rapid analysis of silicate rocks.
*Also by titrimetry, +also by photometry.

<u>Portion No. 1</u>. This portion, used by Shapiro and Brannock[6] for the determination of both silica and alumina, and by Riley[9] for silica only, is prepared by fusing a small sample weight with sodium hydroxide in a nickel crucible. As silica is

always present in major amounts, and is frequently greater than all the remaining elements put together, great care is necessary if accurate results are to be obtained, both in the measuring of volumes and optical densities and in the preparation of the solution. This is especially necessary at the fusion stage where loss by "dusting" can occur. Even when this great care is taken, the accuracy is limited by the inherent inaccuracies of spectrophotometric methods (although some improvement can be obtained by making replicate analyses and by taking a series of readings on each sample).

Most authors have described molybdenum blue methods for silica, mostly resembling that described in detail, although methods based upon the formation of the yellow silicomolybdate have also been suggested.

There is more variety in the methods used for aluminium, which can be determined in either the first or the second portion of rock material taken for the analysis. These methods range from photometric determination with aluminon,[8] alizarin red-S,[6] or 8-hydroxyquinoline,[9] to titrimetric methods based upon complex formation with EDTA.[1 With many silicate rocks, aluminium is the most abundant constituent after silicon, and great care is therefore also required in making precise measurements. None of the reagents so far suggested is specific or even selective for aluminium. For this reason some authors separate aluminium from interfering elements,[8] whilst others add complexing reagents to limit or prevent this interference. The methods adopted in some schemes of rapid rock analysis for this determination have received considerable criticism, and a great deal more work is necessary before the determination of aluminium can be made as easily, precisely and accurately as many of the others.

Portion No. 2. This portion is used for determining total iron, titanium, manganese, phosphorus, calcium, magnesium and the alkali metals. It is prepared for the analysis by evaporation with hydrofluoric and either sulphuric or perchloric acid. Difficulties arise when, as frequently happens, a residue remains after this treatment. Oxide minerals such as chromite, rutile, or corundum, do not contain appreciable amounts of alkali metals, and no significant error will be introduced into their determination if such residues are discarded. This unattacked portion can, however, contain quite an appreciable amount of the total titanium content of the rock, together with significant amounts of other minor constituents. Iron, titanium and these other minor elements can be recovered following a fusion of the residue with sodium carbonate or potassium pyrosulphate.

Resistant silicate minerals are more difficult to prepare for analysis and the use of a PTFE- or platinum-lined bomb has been recommended for the acid-resistant fraction,[9] or the total rock sample.[11,12]

Photometric methods were commonly described for the determination of titanium, manganese and phosphorus, elements that are present in small or minor amounts, where the accuracy and precision of spectrophotometric methods is adequate. Hydrogen peroxide is the most frequently recommended reagent for titanium, but it is of low sensitivity. Tiron (catechol-3:5:disulphonic acid) and diantipyrylmethane are a great deal more sensitive, and are therefore better for acidic and intermediate rocks which contain only small amounts of titanium. Manganese is determined as permanganate after oxidation with either potassium periodate or ammonium persulphate. The two methods commonly used for phosphorus are based upon the formation of a yellow phosphovanadomolybdate and upon a molybdenum blue given by the reduction of phosphomolybdate. All these methods are described in detail later in this book.

Although photometric methods are frequently recommended for the determination of total iron, the precision obtainable is barely adequate for basic and other rocks rich in ferrous or ferric iron. For these a titrimetric method using potassium dichromate, potassium permanganate or ceric sulphate solution provides an acceptable alternative to the photometric methods suggested. For rocks containing only small amounts of iron, photometric methods using 2:2'-dipyridyl or 1:10-phenanthroline are preferred to thioglycollic acid,[13] hydrochloric acid,[2] tiron, salicyclic acid or other reagents suggested for this application.

The unsatisfactory nature of some of the determinations in schemes of rapid rock analysis has long been apparent - photometric methods for aluminium and titrimetric methods for calcium and magnesium being the most deserving of criticism. It is not surprising therefore that in the recently described schemes of rapid rock analysis, they have been replaced by methods based upon atomic absorption spectroscopy.

The decomposition procedures introduced for the earlier combined spectrophotometric/ titrimetric schemes, and gradually improved and refined by later workers, were found to be adaptable to this new technique, which now appears to have displaced entirely many of the earlier rapid techniques. As with other schemes, rapid rock analysis based upon atomic absorption must be supplemented with the more traditional methods for "moisture", total water and ferrous iron.

The determination of certain elements present to a minor or trace extent in silicate rocks is particularly sensitive by this technique. This has led to their inclusion in some of the proposed schemes. These include barium, vanadium, chromium, strontium and zinc.

Individual Schemes of Rapid Rock Analysis

SCHEMES OF SHAPIRO AND BRANNOCK(2,6,14)

Although not the first of their kind, the schemes of Shapiro and Brannock probably did more to whet the appetite of the geologist for this type of analysis than those of either their predecessors or their imitators. Criticism of certain of the methods has led to a great improvement in both technique and in the methods selected, although most of the procedures are now obsolete.

SCHEME OF RILEY(9)

This scheme differs from others proposed at or before that time in making good use of modern separation processes where they could conveniently contribute to the precision and accuracy of the determination. Thus for example before titrating calcium and magnesium with EDTA, the interfering elements are removed by extraction of their complexes with 8-hydroxyquinoline into chloroform. Similarly an anion exchange separation is used to remove iron, aluminium and titanium before the flame photometric determination of the alkali metals. Riley's scheme has been widely followed, possibly because the methods used were capable of somewhat greater accuracy and precision than some of the other methods suggested, but possibly also because a wide range of silicate rocks and minerals could be analysed without difficulty by following the detailed instructions given.

SCHEME OF LANGMYHR AND GRAFF(11)

This scheme is based upon the use of two sample portions for the total of ten constituents, but both portions are decomposed with hydrofluoric acid. The first portion, for silica only, is decomposed in a closed PTFE vessel at an elevated temperature and pressure with hydrofluoric acid alone. After adding aluminium chloride to complex the excess hydrofluoric acid, the silica is determined photometrically as the yellow silicomolybdate. The second portion is decomposed in an open PTFE vessel with a mixture of hydrofluoric and sulphuric acids. The sulphate solution obtained is used directly for the photometric determination of total iron, titanium, manganese and phosphorus, and to provide a reagent blank solution for the silicon determination.

In order to separate the alkali and alkaline earths from iron, aluminium and other elements of the ammonia group, Langmyhr and Graff employ a double precipitation with ammonia. The value of this separation stage is questionable since, as noted in the previous chapter, some of the aluminium passes into the filtrate from the ammonia precipitation, giving low values for the aluminium determination. Moreover, the distribution of manganese between the ammonia precipitate and the filtrate will certainly present difficulties in the subsequent determination of calcium, and possibly also that of aluminium.

SCHEME OF INGAMELLS[15,16]

A considerable advance in rapid rock analysis was that introduced by Ingamells, involving fusion of the rock material with anhydrous lithium metaborate ($LiBO_2$). The decomposition of a 0.1-0.2 g portion is complete in about 10 minutes, and the solution of the melt in nitric acid can be used for the photometric determination of silicon, aluminium, total iron, titanium, manganese and phosphorus, as well as nickel and chromium (which featured in a scheme for rapid rock analysis for the first time).

Sodium and potassium can be determined in the solution by flame photometry in the usual way, and the determination can be extended to include rubidium, and possibly also caesium after adding potassium.[17]

SCHEME OF SHAPIRO[18]

Shapiro has carried the scheme described by Ingamells a stage further and used the acid solution of the melt for the determination of calcium and magnesium (and also sodium and potassium) by atomic absorption spectroscopy. If an emission spectrograph is available, then strontium, barium, chromium, copper, zinc, nickel and zirconium can all be determined in this same solution using a rotating wheel electrode.[17]

As far as we have been able to determine, this scheme is the first to utilise atomic absorption methods for rapid rock analysis.

Although platinum crucibles were originally suggested for this fusion, the melt adheres to the metal and can only be removed from the crucible with difficulty. New graphite crucibles were however recommended by Shapiro,[18] as the melt does not then adhere to the graphite and can easily be removed for dissolution in dilute mineral acid. In order to prevent the polymerisation of silica in the acid extract, the silica concentration should not exceed about 150 ppm. This means that all other constituents of the rock material will also be present in the solution at high dilution and particularly sensitive methods of determination are required. For example, the phosphovanadomolybdate method for phosphorus used by Shapiro and Brannock[6] is no longer sufficiently sensitive, and is replaced by a molybdenum blue method. The following detailed instructions for the preparation of the rock solution have been adapted from the work of Shapiro[18] and Ingamells.[16]

Procedure. Accurately weigh approximately 0.1 g of the finely powdered silicate rock material into a new graphite ("vitreous carbon") crucible and add 0.6 g of anhydrous lithium metaborate. Allow to stand before the open door of a muffle furnace set at a temperature of 1000° for a few minutes, and then insert into the furnace on a silica tray. Fuse for 15 minutes, then remove the silica tray containing the crucible and allow to cool. The melt does not wet the crucible, and it can be readily detached from the graphite.

Transfer the melt from the crucible to a 1500-ml polythene or polypropylene beaker containing approximately 950 ml of water and 15 ml of concentrated hydrochloric acid. The melt will dissolve slowly over a period of 2 to 3 hours, but this process may be hastened by mechanical stirring. When dissolution is complete, transfer the solution to a 1-litre volumetric flask and dilute to volume with water. If the determination of silica is not to be made immediately, transfer the solution to a clean, dry, polythene bottle for storage.

This solution can be used for the determination of silica by a molybdenum blue method as described in Chapter 37, and other elements by atomic absorption spectroscopy or spectrophotometry as appropriate.

SCHEME OF BELT[19]

This, one of the earliest using atomic absorption spectroscopy, was developed before the widespread use of the high-temperature nitrous oxide-acetylene flame. A mixture of hydrofluoric, nitric and perchloric acids was used to effect a decomposition of the silicate matrix. The residue was taken up in hydrochloric acid, lanthanum added as releasing agent for calcium and magnesium in the presence of aluminium and phosphorus, and the determination of sodium, potassium, manganese, iron, magnesium and calcium made using an air-acetylene flame.

SCHEME OF BERNAS[20]

In this scheme a single portion of the rock material is used for the analysis, and the decomposition is accomplished with hydrofluoric acid in a sealed PTFE vessel held at a temperature of 110° for 30-40 minutes. Boric acid is added to complex excess fluoride ion and to dissolve insoluble fluorides. After dilution to a suitable volume, the solution is aspirated into a nitrous oxide-acetylene flame for the determination of silicon, aluminium, titanium, vanadium, calcium and magnesium, and into an air-acetylene flame for total iron, sodium and potassium. The concentrations of all these elements is determined by reference to calibration graphs or by narrow range bracketing.

Although it is known that most silicate rocks and minerals are completely decomposed by this hydrofluoric acid digestion, further work would appear to be desirable on its effectiveness in respect of the wide range of accessory minerals that occur in such rocks.

In view of the extent to which vanadium occurs in many silicate rocks (q.v.), the inclusion of this element in a general scheme by using a much larger sample portion is curious, especially in view of the absence of the much more abundant manganese from the scheme.

No attempt is made to prevent inter-element effects by the addition of releasing agents, nor to suppress ionisation of certain elements by alkali addition. Bernas regarded his fluoborate system as having the "ability to compensate for inter-element effects and thus to eliminate interference phenomena". Whilst this may be true for many commonly occurring silicates, it is doubtful if it holds for the full range of silicate and other rocks that may be encountered.

SCHEME OF LANGMYHR AND PAUS[21]

A variety of decomposition procedures were used by Langmyhr and Paus in their scheme for the analysis of silicate rocks, which includes the determination of silicon, aluminium, titanium, calcium and magnesium with a nitrous oxide-acetylene flame. The calibration is made by using a bracketing technique. Although the use of lanthanum as a releasing agent and alkali salts as ionisation suppressants are advocated in the paper, their use is not described in any detail.

SCHEMES OF ABBEY[22-26]

In a series of monographs published by the Geological Survey of Canada, Abbey describes both a hydrofluoric-, nitric-perchloric acid decomposition and a lithium borate fusion for the determination of selected groups of elements present in silicate rocks. These are summarised in Table 3.

TABLE 3. SCHEMES FOR DETERMINING SELECTED GROUPS OF ELEMENTS

Reference	Decomposition procedure	Elements determined
22	$HF/HNO_3/HClO_4$	Lithium, magnesium, zinc, iron
23	$HF/HNO_3/HClO_4$	Iron, magnesium, calcium, sodium, potassium
24	Lithium fluoborate	Silicon, aluminium, iron, magnesium, calcium, sodium, potassium
25	$HF/HNO_3/HClO_4$	Barium, strontium, plus iron, magnesium, calcium, sodium, potassium (and lithium, rubidium and caesium by flame photometry)
26	Lithium fluoborate	Silicon (by differential spectrophotometry) phosphorus (by spectrophotometry) together with aluminium, iron, manganese, titanium, chromium, nickel, calcium, magnesium, strontium, barium, sodium and potassium (by atomic absorption spectroscopy).

In these schemes, strontium is added as releasing agent in the determination of aluminium, manganese, chromium, nickel, approximate silicon, magnesium, calcium, sodium, potassium and total iron. Aluminium is added as a buffer for the titanium determination and sodium for barium and strontium.

Abbey's scheme for the determination of fourteen constituents of silicate rocks which is similar in many respects to that of Van Loon and Parissis[27] requires only the accompanying determination of ferrous iron, carbon dioxide, total carbon and sulphur, moisture and total water to give a complete analysis of most silicate rocks and minerals. Abbey himself refers to the determination of silica as being the 'weakest link' of those described in his paper, and suggests that more reliable determinations of silica may be made by a 'neo-classical' procedure involving fusion of the sample material with lithium metafluoborate followed by precipitation and dehydration of the silica along the lines of the classical procedure. The possibility exists of extending the lithium metaborate decomposition procedure[26] to the determination of cobalt, copper, lead, rubidium and zinc when present at reasonable levels in silicate rocks. The scheme is shown in outline in Figure 3. The procedure given below is based upon Abbey's work and the scheme shown in Figure 3.

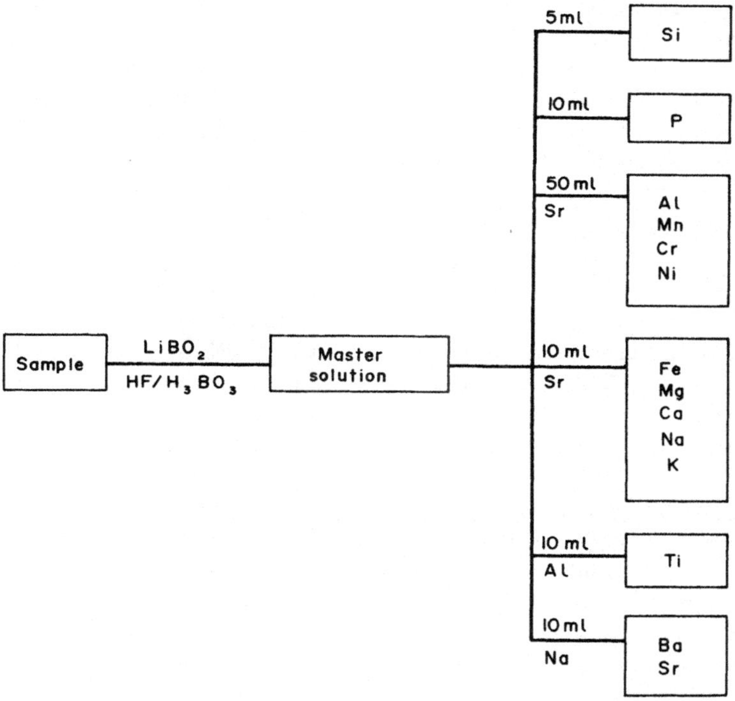

FIG 3. Scheme for the rapid analysis of silicate rocks (Abbey, Geol. Surv. Canada Paper 74-19, 1974). (All determinations by atomic absorption spectroscopy except silicon by differential spectrophotometry and phosphorus by spectrophotometry).

Procedure. Ignite a covered graphite crucible at a temperature of about 1000° for 15 to 20 minutes. Allow to cool and without disturbing the graphite dust within the crucible, add 1 g of lithium metaborate. Carefully brush 0.2 g of the sample material, accurately weighed, into the crucible and mix with the metaborate. Heat the crucible at 950° to 1000° in a preheated electric muffle furnace for 15 minutes. Rotate the crucible to ensure complete attack of any particles remaining on the walls of the crucible, and heat for a further 5 minutes. Remove the crucible from the furnace and pour the hot fusion into 40 ml of water contained in a plastic (trimethylpentene) jar. If, on cooling, the crucible contains any particles of the melt, brush these into jar.

Add a PTFE magnetic stirring bar, followed by 25 ml of hydrofluoric acid (diluted 6 + 19 with water) and immediately cap the jar. Stir until the fusion bead fragments have completely disintegrated then chill in a refrigerator. Open the jar and immediately add 100 ml of a solution containing 5 g boric acid per litre. Cap again and stir until a clear solution is obtained. Filter through a coarse textured paper to remove any suspended graphite, wash with water and dilute to volume in a 200-ml volumetric flask. Mix well and transfer to a dry polyethylene bottle. This now contains the "master solution".

Determination of aluminium, manganese, chromium, nickel and approximate silicon.

Transfer by pipette 20 ml of a solution containing 36 g of strontium nitrate per litre to a 100-ml volumetric flask, add by pipette 50 ml of the rock solution, dilute to volume with water and mix well. Use this solution for the determination of aluminium, manganese, chromium, nickel and approximate silicon using an atomic absorption spectrometer set at the appropriate wavelength and with the appropriate hollow cathode lamp as described in the following chapters.

Determination of magnesium, calcium, sodium, potassium and total iron. Transfer by pipette 10 ml of the above strontium solution to a 100-ml volumetric flask, add by pipette 10 ml of the sample solution, dilute to volume with water and mix well. Use this solution for the determination of magnesium, calcium, sodium, potassium and total iron using an atomic absorption spectrometer set at the appropriate wavelength and with the appropriate light source as described in the following chapters.

Determination of titanium. Transfer by pipette 1 ml of an aluminium solution (Note 1) to a clean, dry tube of about 15 ml capacity, add by pipette 10 ml of the sample solution and mix well. Use this solution for the determination of titanium by atomic absorption spectroscopy as described in Chapter 44.

Determination of barium and strontium. Transfer by pipette 0.5 ml of the sodium buffer solution (Note 2) to a clean, dry tube of about 15 ml capacity, add by pipette 10 ml of the sample solution and mix well. Use this solution for the determination of barium and strontium as described in Chapters 9 and 39.

Determination of phosphorus. Transfer by pipette 10 ml of the sample solution to a 50-ml volumetric flask. Add 5 ml of acid molybdate solution (Note 3) and 2 ml of a hydrazine solution containing 0.15 g per 100 ml. Dilute almost to volume and mix well. Heat for 20 to 30 minutes by immersion in a bath of boiling water. Allow to cool, dilute to volume and measure the optical density of the molybdenum blue produced, as described in Chapter 34.

Notes: 1. Dissolve 14 g of aluminium nitrate nonahydrate in 25 ml of water and evaporate with 100 ml of concentrated hydrochloric acid. Repeat the evaporation with hydrochloric acid to complete the decomposition allowing to dry overnight on a steam bath. Dissolve the dry residue in 25 ml of concentrated hydrochloric acid and a little water, repeat the evaporation once more then dissolve in 10 ml of hydrochloric acid plus 50 ml of water and dilute to 100 ml.

2. Dissolve 10 g of sodium chloride in water and dilute to 100 ml.

3. Dissolve 5 g of ammonium molybdate in about 200 ml of 5 M sulphuric acid and dilute to 250 ml with the same acid.

4. As shown in figure 3, the scheme described by Abbey includes the determination of silicon by differential spectrophotometry. The somewhat involved procedure, described in detail in the original paper, does not always give results of acceptable accuracy, and cannot therefore be recommended. For this reason, the procedure is not described here in detail and if the results obtained in the determination of "approximate silicon" as described above are not acceptable, then reference should be made to alternative methods, see Chapter 37.

The following published work is also of interest.

Barredo and Diez[28] use a flux composed of lithium carbonate and boric acid, and add EDTA to improve the stability of the rock solution and to reduce the inter-element effects in the use of atomic absorptiometric methods. Barium is added as a releasing agent.

Govindaraju[29] describes a scheme for the determination of 16 constituents of silicate rocks based mainly on ion-exchange dissolution and emission spectrometry.

Price and Whiteside[30,31] describe a general method for the analysis of siliceous materials. Samples are decomposed in a pressure vessel with hydrofluoric acid, boric acid is added to complex fluoride ion and dissolve precipitated fluorides and the determination completed by atomic absorption spectrophotometry using a nitrous oxide-acetylene flame. Releasing agents such as lanthanum and detailed matching of standards to samples were considered unnecessary.

Strelow, Liebenburg and Victor(32) decompose silicate materials with a mixture of hydrofluoric, hydrochloric and perchloric acids, and separate a total of 10 major and minor elements using cation exchange chromatography. Spectrophotometry, atomic absorption spectroscopy and titrimetry were used to complete the determinations.

Mazzucotelli, Frache, Dadone and Baffi(33) describe a procedure in which anion exchange followed by cation exchange chromatography is used to separate 15 major, minor and trace constituents of silicate materials. Atomic absorption spectroscopy (with spectrophotometry for titanium) was used to complete the determinations.

The use of a strontium metaborate flux is suggested by Jeanroy(34) and a mixed lithium and strontium metaborate flux by Aslin(35).

Flameless atomic absorption methods for the analysis of silicate rocks have been reviewed by Langmyhr(36). In this method, solid sample material is atomised directly into the absorption path of an atomic absorption spectrometer. The accuracy of analysis using this technique is claimed to be as good or better than that obtained by conventional methods. It has yet to be widely adopted.

It can be seen from the many papers that have been referred to that there is an abundant choice facing the analyst entering this field. To a large extent the choice of method will be dictated by limitations in the analysis and particularly the final determination step. Thus for silica, an acid decomposition using an open vessel clearly cannot be employed and a fusion such as that with lithium metaborate is the only alternative to a decomposition with hydrofluoric acid in a bomb. The need to remove fluorine and fluoride ion before the determination of aluminium and titanium is well understood. EDTA titration procedures cannot sensibly be used for the determination of calcium or magnesium when present as trace components in the presence of an excess of the other. The EDTA titration for calcium is particularly difficult in the presence of manganese etc.

It must be emphasised that it is comparatively easy to obtain results by rapid methods of analysis; it is more difficult to obtain results that are accurate as well as precise. As with all procedures in rock analysis, care must be taken with the manipulative aspects of each determination, particularly with pipetting and dilutions-to-volume. In using absorption and emission spectrophotometry and atomic absorption spectroscopy, the appropriate conditions (choice of wavelength, band width, choice of fuel gas, ionisation suppressor, releasing agent etc) must be rigidly controlled and not deviate without due cause and conscious decision to do so from those that have been published. Finally the need for repeated calibration is stressed and use of standard or reference sample material is strongly recommended.

References

1. CHIRNSIDE R C., J. Soc. Glass Technol. (1959) 43, 5T.
2. SHAPIRO L and BRANNOCK W W., U S Geol. Surv. Circ., 165, 1952.
3. OKI Y, OKI S and HIDEKATA S., Bull. Chem. Soc. Japan (1962) 35, 273
4. SHIBATA H., OKI Y and SAKAKIBARA Y., Chishitsugaku Zasshi (1960) 66, 195
5. MAYNES A D., Anal. Chim. Acta (1965) 32, 211
6. SHAPIRO L and BRANNOCK W W., U S Geol. Surv. Bull. 1144-A, 1962
7. BLANCHET M L and MALAPRADE L., Chim. Anal. (1967) 49, 11
8. COREY R B and JACKSON M L., Analyt. Chem. (1953) 25, 624
9. RILEY J P., Anal. Chim. Acta (1958) 19, 413
10. RILEY J P and WILLIAMS H P., Mikrochim. Acta (1959) (4), 516
11. LANGMYHR F J and GRAFF P R., A contribution to the analytical chemistry of silicate rocks: A scheme of analysis for eleven main constituents based on decomposition by hydrofluoric acid, Norges Geol. Undersokelse, 230, Oslo, 1965.
12. MAY I and ROWE J J., Anal. Chim Acta (1965) 33, 648
13. MERCY E P L., Geochim. Cosmochim. Acta (1956) 9, 161
14. SHAPIRO L and BRANNOCK W W., U S Geol. Surv. Bull. 1036-C, 1956
15. INGAMELLS C O., Talanta (1964) 11, 665
16. INGAMELLS C O., Analyt. Chem. (1966) 38, 1228
17. SUHR N H and INGAMELLS C O., Analyt. Chem. (1966) 38, 730
18. SHAPIRO L., U S Geol. Surv. Prof. Paper 575-B, p.187, 1967
19. BELT C B Jr., Analyt. Chem. (1967) 39, 676
20. BERNAS B., Analyt. Chem. (1968) 40, 1682
21. LANGMYHR F J and PAUS P E., Anal. Chim. Acta (1968) 43, 397
22. ABBEY S., Geol. Surv. Canada Paper 67-37 (1967)
23. ABBEY S., Geol. Surv. Canada Paper 68-20 (1968)
24. ABBEY S., Geol. Surv. Canada Paper 70-23 (1970)
25. ABBEY S., Geol. Surv. Canada Paper 71-50 (1972)
26. ABBEY S., Geol. Surv. Canada Paper 74-19 (1974)
27. VAN LOON J C and PARISSIS C M., Analyst (1969) 94, 1057
28. BARREDO F B and DIEZ L P., Talanta (1976) 23, 859
29. GOVINDARAJU K., Analusis (1973) 2, 367
30. PRICE W J and WHITESIDE P J., Analyst (1977) 102, 664

31. PRICE W J and WHITESIDE P J., Analusis (1977) 5, 275

32. STRELOW F W E, LIEBENBERG C J and VICTOR A H., Analyt. Chem. (1974) 49, 1409

33. MAZZUCOTELLI A, FRACHE R, DADONE A and BAFFI F., Talanta (1976) 23, 879

34. JEANROY E., Analusis (1973) 2, 10

35. ASLIN G E H., Geol. Surv. Canada (1974) paper 74-1, 49

36. LANGMYHR F J., Talanta (1977) 24, 277

CHAPTER 5

The Alkali Metals (Lithium, Sodium, Potassium, Rubidium, Caesium)

Sodium and potassium are always included in the list of elements determined in any 'complete' rock analysis. The range of alkali metal contents varies from 100 ppm or less in certain ultrabasic rocks, such as dunites and peridotites to as much as 10 per cent K_2O or 15 per cent Na_2O in rocks composed largely of felspar minerals.

Gravimetric methods have been used for the determination of all five of the alkali metals. These required a most careful separation of the total alkali metals from silica, aluminium, calcium and other elements present. The most frequently used procedure for this, described originally by Lawrence Smith,[1] involved decomposing the rock sample by ignition with ammonium chloride and calcium carbonate. The alkali metals were recovered by leaching with water, and were then separated from the small amount of calcium taken into solution. Sulphates were converted into chlorides and the introduced ammonium salts were removed by volatilisation. Some authors[2] considered that special precautions were necessary to ensure a complete recovery of lithium with the remaining alkali metals. The mixed chloride residue obtained after the expulsion of ammonium salts was carefully ignited and weighed before the separation of the individual alkali metals.

An alternative procedure for decomposing the silicate rock material and recovering the alkali metals as chlorides was based upon evaporation of the sample with hydrofluoric acid and precipitation of iron, aluminium, calcium and other elements with ammonia and ammonium carbonate, a method derived from work by Berzelius.[3]

The separation of lithium from the other alkali metal chlorides was based upon the solubility of lithium chloride in organic solvents such as isobutanol, pentanol, pyridine or ether-ethanol mixtures. The determination was often completed by converting the lithium chloride to sulphate prior to weighing. This gravimetric procedure for lithium is not sufficiently sensitive for application to the majority of silicate rocks, and lithium could only be reported in those samples rich in this element.

Potassium was usually determined following the precipitation with chloroplatinic acid, perchloric acid or sodium cobaltinitrite. The insoluble potassium salts were collected and weighed directly, although a number of indirect procedures were also in common use. Sodium was generally determined by difference, although some analysts preferred to precipitate as a triple acetate of uranium and zinc or other divalent metal. In all these precipitation procedures, corrections for the solubility of the potassium or sodium salts were necessary, and none of the weighing forms could be considered ideal.

Even when present in quantity, the separation of rubidium and caesium from potassium and from each other was seldom attempted, and on these few occasions the results obtained were not always very reliable.

All these gravimetric procedures for the five alkali metals have now been rendered obsolete by the development of instrumental methods, such as flame photometry and atomic absorption spectroscopy. The emission spectra of the alkali metals are simple, consisting of a prominent line or doublet known as the resonance line(s), corresponding to the transition between the lowest excited state and the ground state, together with weaker lines relating to other transitions. In many flames, lithium and sodium are only slightly ionised; this is in contrast to the remaining alkalis, where the degree of ionisation increases in the order K, Rb, Cs, until with caesium a large part of the metal in the flame is in the ionised state. This accounts for the poor sensitivity to caesium, and the general decrease of sensitivity from sodium to caesium. Lithium is about as sensitive as rubidium.

One way of increasing sensitivity to a particular element in a given solution is to introduce a second, more easily ionised element such as another alkali metal into the flame. This serves to decrease the extent of ionisation of the required element and thus increase the number of atoms available for the transition which gives rise to the resonance line(s). At high alkali concentrations some loss of emission (and consequent calibration curvature) occurs by self absorption. This effect has been noted particularly with lithium and sodium.[4]

Early designs of flame photometer used low temperature flames such as air-propane, or air-coal gas, together with a filter system for wavelength selection. These instruments could not be used for the determination of the rarer alkali metals (lithium, rubidium and caesium) as the characteristic emission could not be isolated by the filters available. Considerable improvements were obtained with the introduction of prism monochromators, enabling all five metals to be determined although, as described below, interferences between the metals gave rise to difficulties in determining lithium and rubidium. These could be minimised by plotting emission spectra thus enabling background corrections to be made. These difficulties disappear with the increased wavelength resolution that is obtainable using a grating monochromator.

In general terms, the chemical interferences in the determination of the alkali metals are best minimised by using either a very cool (air-propane) or a very hot (acetylene-nitrous oxide) flame. Much work remains to be done on the determination of alkalis in silicate rocks, particularly following that of Hildon and Allen[5] and Luecke.[6]

A number of authors have described the determination of sodium and potassium in silicate rocks by atomic absorption spectroscopy, as well as the determination of lithium[7] and rubidium.[8] As with flame photometric methods it is necessary to add potassium to the solutions before determining rubidium. Vosters and Deutsch[8] also recommended adding lanthanum as a flame buffer, a practice commonly used to limit inter-element effects. Rock solutions can be prepared in the same way as those for the flame photometric determination of the alkali metals.

One of the most successful ways of separating mixtures of the alkali metals is by ion-exchange chromatography. Resins such as Dowex 50[9] or Amberlite 120[10] can be used for this separation, which requires the use of a cation-exchange resin as all five elements form well-characterised positive ions in solution. The separation of sodium from potassium is well established,[11] and the successful separation of rubidium and caesium has also been reported.[12] However, none of these procedures is ideal for the routine separation of all five alkali metals, and are now little used.

Determination of Sodium and Potassium in Silicate Rocks by Flame Spectrophotometry

The method described in detail below is based upon work by one of the authors[13] and by Eardley and Reed.[14] The rock material is decomposed by evaporation with hydrofluoric, nitric and sulphuric acids. Sulphuric acid is added to assist in the removal of fluoride ion which will otherwise depress the flame emission of potassium. In most cases the small amount of residue that remains after this treatment can be ignored, although for really accurate work this residue should be collected and analysed separately. Caesium sulphate solution is added as an ionisation buffer.

Method

Reagents: Sodium sulphate, anhydrous.

Potassium sulphate.

Caesium sulphate solution, dissolve 0.205 g of pure caesium sulphate, Cs_2SO_4, in 500 ml of water.

Procedure. Accurately weigh approximately 0.1 g of the finely powdered silicate rock material into a platinum dish (Note 1), moisten with a little water and add 10 ml of concentrated nitric acid, 5 ml of concentrated sulphuric acid and 10 ml of hydrofluoric acid. Place the dish on a cool hotplate and evaporate to fumes of sulphuric acid. Cool, rinse down with a little water and add a further 5 ml of hydrofluoric acid and evaporate until the contents of the dish are fuming well. Again allow to cool, rinse down with water, evaporate, this time until all fumes of sulphuric acid have ceased and a dry residue is obtained.

Add 1 ml of concentrated nitric acid to the cool residue, followed by 25 to 30 ml of water. Warm the dish to detach the residue and rinse the contents into a 400-ml beaker. Digest until all soluble material has passed into solution, and collect any small residue on a small close-textured filter paper and wash with a little water. Discard the residue (Note 2) and dilute the filtrate and washings to volume with water in a 250-ml volumetric flask.

Transfer 25 ml of this solution (Note 3) by pipette to a 50-ml volumetric flask, add 5 ml of the caesium sulphate solution, dilute to volume with water and mix well. Prepare also a reagent blank solution in the same way as this sample solution, using the same quantities of reagents but omitting the powdered rock material.

Set the flame spectrophotmeter base line whilst spraying water, and the sensitivity at 10 ppm full-scale deflection spraying the 10 ppm standard solution, with the wavelength control set at 589 nm for sodium and 766.5 nm for potassium. The adjustment of sensitivity will affect the base line position and it may be necessary to reset the controls alternatively two or three times. Once this adjustment is complete, spray also the reagent blank solution and the sample solution, noting the flame emission recorded at both wavelengths. The flame emission of a series of standards should also be measured. This should span the composition range 0 to 10 ppm. The calibration is not a straight line.

Where the flame spectrophotometer is fitted with a wavelength drive motor and a pen recorder, the flame response for either element can be obtained by scanning the spectrum. The reagent blank solution and the standard solutions should similarly be examined. For sodium scan from about 560 nm to 620 nm and for potassium from 730 nm to 820 nm, (Note 4).

Calibration. Dry small quantities of both sodium and potassium sulphates (Note 5) in an electric oven at a temperature of 120° and allow to cool in a desiccator.

Weigh 0.229 g of the anhydrous sodium sulphate and 0.185 g of the potassium sulphate into the same beaker, dissolve in water and dilute to volume in a litre volumetric flask. This solution contains 100 μg Na_2O and 100 μg K_2O per ml. Prepare a 20 ppm standard by diluting 50 ml of this stock solution, measured with a pipette, and 1 ml of concentrated nitric acid to 250 ml with water and mixing well. Transfer 25 ml of this solution by pipette to a 50-ml volumetric flask, add 5 ml of the aluminium sulphate solution and 5 ml of the caesium sulphate solution and dilute to volume with water. This gives a 10 ppm standard for setting the sensitivity control of the flame spectrophotometer.

Notes: 1. The weight of sample material taken for the analysis and the volume of the final solution may be varied in relation to the alkali content of the sample, but the final solution for measurement should be of the correct acid concentration and contain the recommended concentration of caesium sulphate.

2. This small residue contains much of the zircon and some of the tourmaline present in the rock, together with varying proportions of other minerals that are difficult to decompose with hydrofluoric acid. These minerals seldom contain more than trace amounts of alkali metals and the error involved in neglecting these small amounts is often within the limits of the experimental error of the method. If the amount of residue indicates that an appreciable part of the sample has not been attacked, an alternative procedure, such as fusion with lithium borate, should be used for the sample decomposition. Obviously lithium cannot then be determined in this sample portion.

3. This volume of solution is suitable for determining up to 5 per cent of sodium and potassium oxides in the rock sample.

4. There are a number of ultrabasic rocks that contain only small amounts of sodium and potassium. The importance of these rocks is out of all proportion to the extent to which they occur. If reliable results are to be obtained for the alkali metals at these levels, particular attention must be paid to the details of the determination. It has been pointed out(15) that at the high instrument sensitivities required for determining these small amounts of alkali metals, a large percentage of the total light emission can be due to background radiation. The simplest way of obtaining a background correction is to record the flame emission over the ranges involved using an instrument fitted with a motorised wavelength drive and a pen recorder. More dilute standards will also be required for these materials.

5: Laboratory reagent grade sodium and potassium sulphates were found to be sufficiently free from cross contamination to be used as standards in this way. For more rigorous standards Specpure carbonates may be used.

Determination of Potassium in Micas

The need for very accurate potassium determinations has been underlined by the simple way in which geological ages can be obtained from an accurate knowledge of the $A^{40}:K^{40}$ ratios. The small amounts of argon present can readily be recovered and determined isotopically with a mass spectrometer. Potassium can also readily be determined by flame spectroscopy, but the accuracy and precision obtained by the usual procedures such as the one given in detail above, although adequate for most petrographic purposes, do not match that obtainable for argon.

A flame photometric procedure for this accurate determination has been given by Abbey and Maxwell;[16] the sample material is decomposed with hydrofluoric acid and sulphuric acid, and the excess hydrofluoric acid removed by fuming. After the addition of a known excess of magnesium sulphate, the excess sulphuric acid is in turn removed by evaporation to dryness. The sulphates are ignited, the cooled residue is leached with warm water, and the resulting suspension filtered, giving a neutral sulphate solution containing the alkali metals, magnesium and little else. The approximate relative proportions of the alkali metals are then determined and a neutral sulphate standard solution prepared approximating in composition to that of the sample solution. The final potassium determination is made with a flame photometer using this solution with added potassium as standard.

The need to eliminate small systematic errors in the potassium determination was also underlined by Rice[17], who compared dissolution methods using atomic absorption spectroscopy for the final determination. A cation-exchange separation was used based on the work of Strelow et al.[18]

Determination of Lithium

The simple method described above for sodium and potassium can be extended to include also the lithium present in the rock material. A somewhat greater instrument sensitivity-setting is required, but this is well within the range of most modern instruments. The solution prepared as described above for sodium and potassium, can be used directly for the lithium measurement, but for rocks containing only a few ppm of lithium, a more concentrated solution is desirable.

Of all the alkali metals, lithium has the least effect on the flame response of the others, and is least affected by them.[19] For most purposes this small effect upon the lithium response can be ignored, but for accurate determinations, the amounts of sodium and potassium present in the rock solution can be measured and similar amounts added to the standard solutions used for the lithium calibration.

The most sensitive lithium line is the resonance line at 670.8 nm. Filter-type instruments cannot be used for measurement at this wavelength as emission from potassium will also be recorded. Similarly emission from sodium will provide a sloping background against which the lithium emission can be recorded if a prism monochromator is used for wavelength selection. These problems do not arise when an instrument incorporating a grating monochromator is used. A procedure for determining lithium, together with rubidium and caesium, as described by Abbey[(20)] is given below.

The use of atomic absorption spectroscopy for the determination of lithium in silicate rocks has been described by a number of authors including O'Gorman and Suhr,[(21)] Stone and Chesher,[(22)] Zelyukova et al[(23)] and Abbey.[(24)] The rock solution for the determination can be prepared by either acid digestion (HF, H_2SO_4, etc) or by borate fusion. The procedure given below, adapted from that given by O'Gorman and Suhr,[(21)] uses a fusion with sodium borate.

Procedure. Accurately weigh approximately 0.1 g of the finely powdered silicate rock material into a graphite fusion crucible, add 0.5 g of anhydrous sodium borate ($Na_2B_4O_7$), and fuse for 10 minutes in an electric muffle furnace at a temperature of 1000°. Pour the molten bead directly into a beaker containing exactly 50 ml of a 3 per cent nitric acid solution. Add a magnetic stirring bar and stir until dissolution is complete. Use this solution without further treatment for direct nebulisation into the flame of the atomic absorption spectrophotometer in the usual way.

To prepare standard solutions use portions of a silicate rock known to be essentially lithium-free (eg USGS Peridotite PCC-1) fused, as above, with sodium borate and add aliquots of a lithium standard solution prepared from pure dried lithium carbonate. Two ranges of standards are recommended equivalent to 0-200 ppm and 0-2000 ppm Li in the silicate rock.

Matrix problems that arise with the use of a graphite furnace for the determination of lithium by atomic absorption spectroscopy have been recorded by Katz and Tartel.[(25)]

Determination of Rubidium and Caesium

In theory there is no reason why the methods described above for the determination of sodium and potassium should not be used for rubidium and caesium. However since these elements are seldom present in more than trace amounts, and since their flame response is very dependent on other elements present in the solution, a modified technique is preferred.

Horstman[26] has described a method based upon the precipitation of oxide elements (iron, aluminium etc) with calcium carbonate, and of calcium by precipitation as sulphate in aqueous alcoholic solution. The sodium and potassium contents of the rock solution (and lithium where this is present in quantity) are determined, and similar amounts added to the rubidium and caesium standards used to calibrate the flame spectrophotometer. This is given in detail below.

Method

Procedure. Accurately weigh a sample of 0.5-2 g of the finely powdered silicate rock into a large platinum dish, moisten with water and add sufficient sulphuric and hydrofluoric acids to give complete decomposition of the sample. Add 2-3 drops of concentrated nitric acid and evaporate on a hot plate to fumes of sulphuric acid. Allow to cool, add 5 ml of water and again evaporate, this time until all excess sulphuric acid is removed, leaving only a moist residue. Allow the dish to cool and dissolve this residue, by warming if necessary, in 25 ml of water. Transfer the

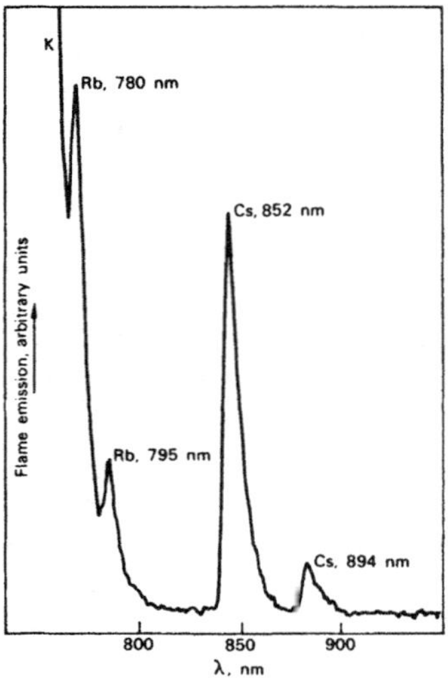

FIG. 4 Flame emission of rubidium and caesium from pollucite.

solution with any unattacked mineral grains (and sulphate precipitate if any) to a 150-ml beaker and dilute with water to a total volume of 70-80 ml. Add a few drops of bromothymol blue indicator solution and neutralise by adding solid calcium carbonate. Allow the precipitate to stand overnight.

Heat the solution and precipitate to boiling and filter on to a medium-textured filter paper. Wash the residue well with hot water until the volume of combined filtrate and washings reaches about 150 ml. Discard the residue and evaporate the solution to a volume of 50 ml, cool and add 50 ml of ethanol to precipitate calcium sulphate. Allow the precipitate to stand overnight and then collect on a close-textured filter paper. Wash the precipitate with a mixture of equal volumes of water and ethanol until the volume of the combined filtrate and washings again reaches 150 ml. Discard the sulphate residue and evaporate the solution to dryness on a steam bath. Dissolve the dry residue in a little water and dilute to volume in a 50-ml volumetric flask.

Using a flame spectrophotometer set according to the manufacturer's instructions, determine the sodium and potassium content of the solution and also the approximate values of lithium, rubidium and caesium. For rocks containing only traces of the rarer alkali metals prepare a series of calibration solutions for each of these metals spanning the approximate values and containing the same amounts of sodium and potassium as the rock solution. For those rocks and minerals containing appreciable amounts of rubidium or caesium add each of these elements to the standard solutions used to calibrate for the other.

The flame photometer trace shown in Fig. 4 was obtained with an instrument fitted with a prism monochromator and a solution prepared from a sample of pollucite, decomposed as described above. It illustrates the difficulty of determining small amounts of rubidium in the presence of potassium with this type of instrument. Clearly the improved wavelength resolution obtainable with a grating instrument can be used with advantage for this determination.

The determination of alkali metals by direct injection of the powdered silicate rock material into a flame has been described by Lebedev.(27) More recently this has been used with atomic absorption spectrophotometry by Govindaraju et al.(28) for rubidium and by Langmyhr and Thomassen(29) for rubidium and caesium. This is a method that should be capable of further development but its application to a wider range of rocks and minerals needs further study.

The method given in detail below is that due to Abbey.[20] The rock sample is decomposed by a conventional acid digestion procedure and there are no chemical separations. Following the addition of potassium to suppress ionisation, lithium, rubidium and caesium are determined by direct emission or absorption (lithium and rubidium only) in an air-acetylene flame. When required a standard addition technique can be used to overcome matrix effects. Aliquots of the same sample solution can be used for the determination of strontium and barium by atomic absorption spectroscopy with a nitrous oxide-acetylene flame.

Procedure. Accurately weigh approximately 0.5 g of the finely powdered rock material in duplicate into 100-ml platinum dishes and add to each 5 ml of concentrated nitric acid, 2 ml of concentrated perchloric acid and 5 ml of hydrofluoric acid. Transfer to a hot plate and evaporate to copious fumes of perchloric acid. Allow to cool, rinse down the walls with a little water and evaporate on the hot plate once more, this time to complete dryness. Allow to cool, add 2 ml of concentrated hydrochloric acid, again rinse down the walls with a little water and evaporate to complete dryness (Note 1).

Using a safety pipette, add 2 ml of concentrated hydrochloric acid, rinse down with a little water, warm to dissolve the chlorides and transfer with water to a 50-ml volumetric flask. To one of the duplicates add alkali solutions containing 25 μg lithium, 250 μg rubidium and 2.5 μg caesium (Note 2). Add 1 ml of potassium buffer (76.3 g potassium chloride per litre) to both spiked and unspiked samples and dilute each to volume with water (Note 3).

Determine the flame emission at 670.8 nm for lithium, 780.0 nm for rubidium and 852.1 nm for caesium, preferably by recording over a wavelength range containing the complete emission peak. An air-acetylene flame should be used. Lithium and rubidium can be determined in the same solution by atomic absorption spectroscopy, also using an air-acetylene flame.

Notes: 1. If barium is to be determined in the same sample solution - as described by Abbey - the residue should be collected, fused with a little anhydrous sodium carbonate and extracted with water. Discard the aqueous extract and acidify the residue with a very little diluted hydrochloric acid. This can then be added to the main rock solution before its final evaporation to dryness.

2. In the final 50-ml volume, these quantities correspond to an additional 50 ppm lithium, 500 ppm rubidium and 5.0 ppm caesium.

3. If barium and strontium are to be determined in this solution, pipette

10 ml into a clean 50-ml volumetric flask, add 2 ml of potassium buffer solution, 10 ml of a lanthanum buffer solution (50 g La_2O_3 per litre), dilute to volume with water, mix well and determine the barium and strontium by atomic absorption using a nitrous oxide-acetylene flame.

References

1. SMITH J L, Amer. J. Sci. 2nd Ser. (1871) 50, 269
2. HERING H, Anal. Chim. Acta (1952) 6, 340
3. BERZELIUS J J, Lehrbuch der Chemie. 3rd ed. (1841) 10, 46
4. DEAN J A, Flame Photometry, McGraw-Hill, New York, 1960
5. HILDON M A and ALLEN W J F, Analyst (1971) 96, 480
6. LUECKE C, Chem. Erde (1979) 38, 1
7. OHRDORF R, Geochim. Cosmochim. Acta (1968) 32, 191
8. VOSTERS M and DEUTSCH S, Earth Planet Sci. Lett. (1967) 2, 449
9. BEUKENKAMP J and RIEMAN W, Analyt. Chem. (1950) 22, 582
10. RIECHEN L E, Analyt. Chem. (1958) 30, 1948
11. ELLINGTON F and STANLEY L, Analyst (1955) 80, 313
12. COHN W E and KOHN H W, J. Amer. Chem. Soc. (1948) 70, 1986
13. JEFFERY P G, unpublished work
14. EARDLEY R P and REED R A, Analyst (1971) 96, 699
15. EASTON A J and LOVERING J F, Anal. Chim. Acta (1964) 30, 543
16. ABBEY S and MAXWELL J A, Chem. in Canada (1960) 12, 37
17. RICE T D, Anal. Chim. Acta (1977) 91, 221
18. STRELOW F W E, TOERIEN Von S and WEINERT C H S W, Anal. Chim. Acta (1970) 50, 399
19. GROVE E L, SCOTT C W and JONES F, Talanta (1965) 12, 327
20. ABBEY S, Geol. Surv. Canada Paper No 71-50 (1972)
21. O'GORMAN J V and SUHR N H, Analyst (1971) 96, 335
22. STONE M and CHESHER S E, Analyst (1969) 94, 1063
23. ZELYUKOVA Yu V, NIKONOVA M P and POLUEKTOV N S, Zhur. Anal. Khim. (1966) 21, 1407
24. ABBEY S, Geol. Surv. Canada Paper 67-37 (1967)
25. KATZ A and TARTEL N, Talanta (1977) 24, 132
26. HORSTMAN E L, Analyt. Chem. (1956) 28, 1417
27. LEBEDEV V I, Zhur. Anal. Khim. (1969) 21, 337
28. GOVINDARAJU K, MEVELLE G and CHOUARD C, Chem. Geol. (1971) 8, 131
29. LANGMYHR F J and THOMASSEN Y, Z. Anal. Chem. (1973) 264, 122

CHAPTER 6

Aluminium

Of all the elements present in the igneous rocks forming the crust of the earth, only oxygen and silicon are more abundant than aluminium. Dunites and peridotites, the first rocks of the magmatic sequence to crystallise contain very little aluminium. Most of the remaining magmatic rocks contain 22 to 12 per cent Al_2O_3, decreasing as differentiation proceeds. Andalusite, sillimanite and kyanite, industrially important aluminosilicates found in metamorphic rocks, may contain up to 60 per cent Al_2O_3.

In the classical procedure for determining aluminium in silicate rocks, iron, aluminium and other elements of the ammonia group were precipitated together and weighed as "mixed oxides". The elements other than aluminium present in this residue were then determined separately, leaving the aluminium content to be obtained by difference. This procedure involved the accurate determination of iron, titanium, vanadium, chromium, phosphate and that part of the manganese (and nickel if present in more than trace amount) precipitated with the elements of the ammonia group.

This determination of aluminium by difference is unsatisfactory in that a small amount of aluminium usually escapes precipitation by ammonia and can be recovered from the filtrate, and that any error in determining the remaining elements of the ammonia group will be accumulative and be reflected in a corresponding error in the aluminium content. Traces of silica (remaining in the solution after the dehydration and removal of the main silica fraction) will be carried down in the ammonia precipitate and, if not recovered and determined, will be counted as alumina.

In dilute acetic acid solution aluminium forms an insoluble yellow complex with 8-hydroxy-quinoline that can be used for the gravimetric determination of aluminium. The complex can be dried to constant weight at a temperature of 130-140°, when it has the composition $Al(C_9H_6ON)_3$ with 11.10 per cent Al_2O_3.

The chief advantage of 8-hydroxyquinoline as a reagent for aluminium is that it gives a direct, positive determination. Unfortunately a large number of elements form similar complexes with the reagent, and a separation procedure must therefore first be applied.

8-Hydroxyquinoline can also form the basis of an indirect titrimetric determination of aluminium using bromination of the reagent. The precipitated aluminium complex is dissolved in dilute mineral acid and titrated with a standard solution of potassium bromate. Methyl red, which is destroyed by an excess of the oxidising reagent, is used to indicate the approximate end point. This leaves a slight excess of bromate which can then be measured by adding potassium iodide and continuing the titration with standard sodium thiosulphate solution until the blue colour given with starch just disappears. As with the gravimetric method based upon 8-hydroxyquinoline, a prior separation from interfering elements is necessary, and high values will result from the small amount of co-precipitation of the reagent.

Aluminium forms a very stable complex with EDTA, but the reaction is extremely slow at room temperatures; complex formation is achieved however within a few minutes at the boiling point. For this reason and also because suitable indicators are lacking, most of the titrimetric EDTA methods for aluminium are based upon adding excess reagent, boiling the solution to complete complex formation, and then titrating the excess reagent with another metal ion for which an indicator is available. A variety of metals have been suggested for this, zinc with xylenol orange as indicator,(1) zinc with dithizone,(2) thorium with alizarin red S,(3) lead with PAR,(4) copper with pyrocatechol violet(5) and others.

Some improvement can be obtained by replacing EDTA with CyDTA (cyclohexanediamine-tetraacetic acid) which reacts much more readily with aluminium;(6) however, as suitable indicators are lacking, a back-titration of the excess complexone is still the preferred method.

It is usually necessary to separate aluminium from other elements that react with EDTA and CyDTA, including iron and titanium, as well as vanadium, manganese, nickel and chromium, all of which can reach minor-element proportion in some silicate rocks. Evans(7) has proposed a procedure involving two CyDTA back-titrations that do not require a prior separation. The first titration gives the sum of iron, aluminium and titanium. In the second, iron alone is titrated - aluminium and titanium being masked with fluoride ion. The titanium content is then determined photometrically in a separate aliquot of the rock solution, enabling the aluminium content to be calculated by difference. Any nickel present in the sample will be reported as iron, and any chromium or zirconium reported as aluminium.

Although numerous reagents have been suggested for photometric determination, none is specific or even selective for aluminium. Procedures based upon the formation of coloured lakes with certain organic compounds such as aluminon(8) (aurine tricarboxylic acid), eriochrome cyanine R,(9) or alizarin red S,(10, 11) although advocated for several years, are subject to interference from iron and other elements when applied to silicate rocks, and are not as precise as one would wish.

The complex formed between aluminium and 8-hydroxyquinoline, used for both gravimetric and titrimetric determination of aluminium, can also be used for a photometric determination. The yellow solution has a maximum optical density at a wavelength of about 400 nm, and the Beer-Lambert Law is valid up to at least 200 μg Al_2O_3 per 25 ml of chloroform. Some fading occurs, particularly when solutions are exposed to direct sunlight, and Riley[12] has recommended storage in a dark cupboard. The stability depends to some extent upon the quality of the chloroform used.[13]

Riley[12] has described the application of this photometric method to silicate rocks using 2:2'-dipyridyl to complex iron present in the solution, and applying a correction for titanium. Manganese does not interfere. Low results can be obtained if fluoride ions are not completely removed, although small traces of fluoride can be complexed by adding beryllium.

Aluminium forms a blue-coloured complex with pyrocatechol violet that has been used by Wilson and Sergeant[14] for the determination of aluminium in silicate rocks and minerals. The recommended pH is 6.1-6.2, obtained with an ammonium acetate-acetic acid buffer. The colour development occurs over a period of about 1 hour, after which the colour is virtually constant. The Beer-Lambert Law is valid up to 80 μg Al_2O_3 per 100 ml of solution, and maximum values of optical density are obtained at a wavelength of 580 nm. Many elements interfere with the determination, but as a routine method where extreme accuracy is not important, the addition of a solution of 1:10-phenanthroline will serve to complex iron, leaving titanium as the only major interference for which a correction is necessary. In a more rigorous procedure, given in detail below, iron, titanium, vanadium and zirconium can be removed by a cupferron extraction.

Although a large number of other reagents have been suggested for the photometric determination of aluminium, only a very few appear to have been applied to the analysis of silicate rocks. Aluminium is one of the more difficult elements to determine by atomic absorption spectroscopy. Flame conditions are particularly critical, as is flame stoicheiometry. Appreciable ionisation occurs in the high temperature flames (nitrous oxide-acetylene is commonly employed[15,16]), and a large excess of an ionisation buffer such as potassion chloride has been recommended.[17] The degree of ionisation is also affected by the nature and quantity of the anions present. Interelement effects can also be severe. The most sensitive lines are at 309.27 and 309.28 nm.

The determination of aluminium in silicate rocks by atomic absorption spectroscopy is used as part of a scheme for the determination of a number of elements, and is described in greater detail in the chapter concerned with rapid methods of analysis.

Gravimetric Determination with 8-Hydroxyquinoline

In the procedure given in detail below, iron, titanium and other elements are removed by an extraction of their cupferrates into chloroform, followed by a direct precipitation of aluminium in the aqueous phase with 8-hydroxyquinoline. The precipitate is collected, dried to constant weight in an electric oven, and weighed as $Al(C_9H_6ON)_3$.

N-Nitrosophenylhydroxylamine (used as the ammonium salt, with the trivial name of "cupferron") forms insoluble complexes with ferric iron, titanium and vanadium. These can be removed from aqueous acid solution by extraction with chloroform.[18] Aluminium remains in the aqueous phase and can, after extraction of the excess reagent and removal of residual chloroform, be determined gravimetrically, titrimetrically or photometrically with 8-hydroxyquinoline. Solutions of cupferron in chloroform are unstable and must be kept ice-cold or, as preferred, prepared from the solid reagent in situ. Even the solid reagent deteriorates on keeping, and only fresh, good-quality cupferron should be used.

Method

Reagents:

Cupferron

8-Hydroxyquinoline solution, dissolve 2.5 g of the reagent in 100 ml of 2 N acetic acid.

Bromocresol purple indicator solution, dissolve 0.1 g in 100 ml of ethanol.

Procedure. Accurately weigh approximately 0.5 g of the finely powdered silicate rock material into a platinum dish, moisten with water and add 1 ml of concentrated nitric acid, 10 ml of 20 N sulphuric acid and 10 ml of concentrated hydrofluoric acid. Transfer the dish to a hot plate and evaporate to fumes of sulphuric acid. Allow to cool, rinse down the sides of the dish with a little water, add 5 ml of water and 5 ml of concentrated hydrofluoric acid and again evaporate to fumes of sulphuric acid. Allow to cool, dilute with 5 ml of water and again evaporate to fumes. Repeat the evaporation once more to ensure expulsion of residual fluorine and to remove most of the remaining sulphuric acid.

Transfer the moist residue to a 250-ml beaker using about 100 ml of 2 N hydrochloric acid and heat on a hot plate until all soluble material has dissolved. Collect any insoluble material on a small close texture paper, wash with a little water and transfer to a small platinum crucible. Dry and ignite the residue and fuse with 0.5 g of a flux containing anhydrous sodium carbonate and borax glass (Note 1). Allow to cool, dissolve the melt in 2 N hydrochloric acid and add to the main rock solution.

Transfer the solution to a 250-ml volumetric flask and dilute to volume with 2 N hydrochloric acid. Pipette 100 ml of this solution into a 250-ml beaker, cover with a clock glass and allow to stand in a refrigerator until a temperature of less than about 8° is reached, and then transfer to a 500-ml separating funnel with about 100 ml of ice cold water (Note 2). Extract iron, titanium etc, by adding about 0.5 g of solid cupferron and 20 ml of chloroform. Stopper the funnel, invert and release the pressure by opening the tap momentarily. Close the tap and shake the funnel for about 1 minute. The organic layer becomes coloured with the iron cupferrate, and can be run off and discarded. Repeat the extraction with 0.25 g portions of cupferron and 20 ml aliquots of chloroform until the extracts are no longer coloured, and then with pure chloroform two or three times to remove all excess cupferron from the aqueous phase. Discard the organic extracts and run the aqueous solution into a 400-ml beaker. Rinse the funnel with a little N hydrochloric acid and add to the main solution.

Heat to boiling to remove all traces of chloroform remaining in the solution and add 8 g of ammonium acetate, a few drops of bromocresol purple indicator solution and aqueous ammonia drop by drop until the yellow colour imparted by the indicator is just replaced by purple. Heat the solution to about 60° and precipitate the aluminium by slowly adding 20 ml of the 8-hydroxyquinoline solution (Note 3). Bring the solution just to boiling and allow to stand for 30 minutes (Notes 4 & 5).

Collect the precipitate on a weighed, sintered glass crucible of medium porosity, wash with cold water, dry in an electric oven set at a temperature of $130-140^{\circ}$ and weigh as $Al(C_9H_6ON)_3$. This precipitate contains 11.10 per cent Al_2O_3.

Notes: 1. Any corundum present in the rock material is particularly resistant to decomposition unless borax or boric acid is added. For most rocks 0.5 g of a flux containing 10 per cent borax glass in anhydrous sodium carbonate will be sufficient.

2. Failure to keep the temperature down will result in the formation of a gummy organic product on the addition of cupferron.

3. Sufficient reagent should be added to provide 15-20 per cent excess over that required as calculated from the composition of the aluminium precipitate.

1 ml of a 2.5 per cent solution of 8-hydroxyquinoline will precipitate 2.9 mg of Al_2O_3.

4. The precipitated complex of aluminium should be pure yellow in colour, but may be tinged green by the presence of a trace of iron remaining after the cupferron extraction. The residue can be removed from the sintered glass crucible with 6 N hydrochloric acid, and the crucible cleaned with concentrated nitric acid before the next determination.

5. Some analysts prefer to remove the bulk of mineral acid from the solution after the extraction of the cupferrates by evaporation with concentrated nitric acid with a small amount of sulphuric acid present. This serves also to destroy any organic material remaining in the solution.

Titrimetric Determination with CyDTA

This procedure is similar to that described by Mercy and Saunders,[19] and is based upon the removal of interfering elements by a chloroform extraction of their cupferrates.

Method

Reagents:

Cyclohexanediaminetetraacetic acid solution (CyDTA), add approximately 8 g of reagent to 100 ml of water and dissolve by adding sodium hydroxide solution drop by drop. Dilute to 1 litre with water and store in a polyethylene bottle.

Hexamethylenetetramine solution, saturated solution in water.

Xylenol orange indicator solution, dissolve 0.25 g in 50 ml of water.

Lead nitrate solution, dissolve 6.5-6.7 g of lead nitrate in 1 litre of water.

Aluminium chloride standard solution, accurately weigh about 0.25 g of pure clean aluminium foil or wire into a small beaker and dissolve in about 40 ml of 3 N hydrochloric acid. Transfer to a 1 litre volumetric flask and dilute to volume with water. This solution contains about 500 µg Al_2O_3 per ml, the exact concentration can be calculated from the weight of the aluminium taken.

Standardisation. Transfer 25 ml of the standard aluminium solution to a 500-ml conical flask, add 5 ml of N hydrochloric acid, 150 ml of water and 25 ml of CyDTA solution. Add sufficient hexamethylenetetramine buffer solution to bring the pH of the solution to a value in the range 5.0-5.5, add a few drops of indicator solution and titrate the excess CyDTA solution with the lead nitrate solution to an end-point given by the permanent appearance of a violet colour.

Also transfer 15 ml of the CyDTA solution to a separate 500-ml conical flask, add 5 ml of N hydrochloric acid, 150 ml of water, sufficient hexamethylenetetramine buffer solution to bring the pH to 5.0-5.5 and a few drops of indicator solution. Again titrate with lead nitrate solution to a violet end-point. Calculate the equivalence of the lead solution to the CyDTA solution, and hence the equivalence of the CyDTA solution to that of the standard aluminium.

Procedure. Prepare a solution of the rock material by evaporation with hydrofluoric nitric and sulphuric acids as described above (method based upon precipitation with 8-hydroxyquinoline,) but using only 0.1 g of the finely ground sample material. Decompose any residue by fusion, add the acid extract to the main rock solution to give a solution volume of about 30 ml and extract iron, titanium and vanadium as above with a chloroform solution of cupferron. Remove all organic material by extraction with chloroform containing a little acetone.

Transfer the aqueous solution to a 400-ml beaker, rinse the funnel with a little water and add the washings to the solution in the beaker. Add 5 ml of concentrated perchloric acid and 5 ml of concentrated nitric acid and evaporate to fumes of perchloric acid. As soon as fumes appear add a few drops of concentrated hydrochloric acid and continue the evaporation (Note 1). After about 5 minutes fuming add a further few drops of concentrated hydrochloric acid, and repeat after fuming for a further 5 minutes. Continue the evaporation to incipient crystallisation, then allow to cool, add 150 ml of water, warm to give a clear solution and allow to cool.

Add 25 ml of the CyDTA solution (Note 2), add buffer solution to bring the pH to 5.0-5.5, and a few drops of indicator solution (Note 3). Titrate the excess CyDTA with the lead nitrate solution to a violet end-point and hence calculate the aluminium content of the solution (Note 4).

Notes: 1. This evaporation with concentrated hydrochloric acid serves to volatilise any chromium as chromyl chloride, and can be omitted for rocks low in chromium.

2. This provides 10-15 ml of CyDTA solution in excess for rocks containing about 10 per cent Al_2O_3, and should be increased for those rocks containing more alumina.

3. According to Mercy and Saunders,[19] vanadium is not completely removed in the cupferron extraction. Interference from this small amount can be prevented by adding 3 drops of concentrated (100-vol) hydrogen peroxide to the solution before adding the indicator prior to titrating with lead nitrate solution.

4. If manganese or nickel are present in more than trace amounts, these should be determined separately and corrections applied to the aluminium value as determined. 1 mg MnO is equivalent to 0.72 mg Al_2O_3 and 1 mg NiO to 0.68 mg Al_2O_3.

Spectrophotometric Determination with 8-Hydroxyquinoline

This procedure is based upon that given by Riley.[12] Apart from the removal of silica, no prior separations are made. Beryllium is used to complex any remaining trace of fluoride ion. Analysts using this method should be aware of the hazards of this material and the safety precautions that should be taken.

Method

Reagents: 8-Hydroxyquinoline solution, dissolve 1.25 g of the reagent in 250 ml of a pure grade of chloroform and store in a refrigerator. This solution deteriorates slowly giving a brown coloration, when it should be discarded.

Complexing reagent solution, dissolve 1 g of hydroxylamine hydrochloride, 3.6 g of sodium acetate trihydrate and 0.4 g of beryllium sulphate tetrahydrate in 50 ml of water. Add 0.04 g of 2:2'-dipyridyl dissolved in 20 ml of 0.2 N hydrochloric acid and dilute to 100 ml with water.

Standard aluminium stock solution, dissolve 0.106 g of pure aluminium wire or foil in dilute hydrochloric acid, avoiding an excess, transfer to a 1 litre volumetric flask and dilute to volume with water. This solution contains 200 μg Al_2O_3 per ml.

Standard aluminium working solution, pipette 50 ml of the stock solution into a 500-ml volumetric flask and dilute to volume with water. This solution contains 20 μg Al_2O_3 per ml.

Procedure. Decompose a 0.1 g portion of the finely powdered silicate rock material by evaporation with 5 ml of hydrofluoric, 1 ml of nitric and 5 ml of 20 N sulphuric acids as described earlier in this chapter (gravimetric determination with 8-hydroxyquinoline). Decompose any residue remaining after this evaporation by fusion with sodium carbonate as described, and dilute the combined extracts to 500 ml with water.

Pipette 5 ml of this rock solution into a 75- or 100-ml separating funnel, add 5 ml of water and 10 ml of the complexing reagent solution (measured with an "automatic pipette" or similar device and NOT by mouth-operated pipette) and allow to stand for a few minutes. Then add 20 ml of the 8-hydroxyquinoline reagent solution. Stopper tightly, invert the funnel and release the pressure by momentarily opening the tap.

Close the tap and shake for 5 to 8 minutes, releasing the pressure at intervals. Allow the layers to separate and run the organic layer through a small wad of filter paper inserted in the stem of the funnel into a 25-ml volumetric flask (Note 1). Rinse the separating funnel three times with a total of about 4 ml of chloroform and add these washings to the solution in the 25-ml flask. Dilute to volume with chlorofor

Measure the optical density in a 1-cm cell, preferably stoppered, using chloroform as the reference solution, with the spectrophotometer set at a wavelength of 410 nm. The complex is light sensitive, and measurement should therefore be completed as soon as possible after dilution to volume. Measure also the optical density of a reagent blank extract, prepared in the same way as the sample extract, but omitting the rock material, and calculate the alumina content from the calibration graph (Note 2).

Calibration. Transfer aliquots of 0-10 ml of the standard aluminium solution containing 0-200 μg Al_2O_3 to separate 75- or 100-ml separating funnels and dilute each solution to 10 ml with water if necessary. Add complexing reagent and 8-hydroxyquinoli solution and extract the aluminium complex as described above. Measure the optical density of each extract relative to chloroform in 1-cm cells at a wavelength of 410 nm, as for the sample extract and plot the relation of optical density to alumina concentr- ation.

Notes: 1. The aluminium complex is pure yellow in colour, and gives a pure yellow solution in chloroform. If the pH of the aqueous solution is too low, not all the iron will be reduced by the hydroxylamine and a green coloured extract will be obtained. This can be avoided by increasing the complexing reagent concentration to bring the pH to a value between 4.9 and 5.0.

2. A small correction is necessary for any titanium present in the sample material. This can be determined by adding 10 μg TiO_2 (equivalent to 1 per cent TiO_2 in the sample), to 100 μg Al_2O_3 and when extracting with 8-hydroxyquinoline as for the calibration solutions, measuring the increased optical density due to titanium alone.

Spectrophotometric Determination with Pyrocatechol Violet

This procedure, due to Wilson and Sergeant,[14] is also based upon the use of cupferron for the removal of interfering elements.

Method

Reagents:

Pyrocatechol violet solution, dissolve 0.075 g of reagent in 50 ml of water.

Hydroxylamine hydrochloride solution, dissolve 10 g of reagent in 100 ml of water.

Buffer solution, dissolve 50 g of ammonium acetate in 450 ml of water, bring the pH to 6.2 by adding acetic acid (use a pH meter) and dilute to 500 ml with water.

Cupferron

Chloroform.

Standard aluminium working solution, dilute the stock solution obtained as described above, with water to give a working solution containing 4 µg Al_2O_3 per ml.

Procedure. Decompose 0.1 g of the finely powdered silicate rock material by evaporation with hydrofluoric, nitric and sulphuric acids as described and extract iron, titanium, vanadium and zirconium with cupferron. Transfer the aqueous solution to a 250-ml volumetric flask and dilute to volume with water. Pipette an aliquot of this solution containing not more than 40 µg of alumina into a 100-ml beaker and dilute to 20 ml with water. Add 2 ml of hydroxylamine hydrochloride solution, 2 ml of pyrocatechol violet solution and 5 ml of the buffer solution. Mix well and bring the pH to a value of 6.1-6.2 by careful addition of ammonia (use a pH meter), taking care not to allow the solution to become distinctly alkaline at any time.

Rinse the solution into a 100-ml volumetric flask with a little water, add 50 ml of the buffer solution and dilute to volume with water. Set aside for 2 hours and then measure the optical density in 1-cm cells with the spectrophotometer set at a wavelength of 580 nm. Determine also the optical density of a reagent blank solution similarly prepared but omitting the sample material.

Calibration. Transfer aliquots of 0-20 ml of the standard aluminium solution containing 0 to 80 µg Al_2O_3 to separate 100-ml beakers and dilute each solution to 20 ml as necessary. Add hydroxylamine hydrochloride, pyrocatechol violet and buffer solutions and continue as described above for the sample solution. Plot the relation of optical density to aluminium concentration.

References

1. CLULEY H J, Glass Technol. (1961) 2, 71
2. VOINOVITCH I A and LEFRANC-KOUBA A, Chim. Anal. (1960) 42, 543
3. TER HAAR K and BAZEN J, Anal. Chim. Acta (1954) 10, 23
4. LANGMYHR F J and KRISTIANSEN H, Anal. Chim. Acta (1959) 20, 524
5. KRIZ M, Sklar a Keramik (1956) 6, 140
6. PRIBIL R and VESELY V, Talanta (1962) 9, 23
7. EVANS W H, Analyst (1967) 92, 685
8. COREY R B and JACKSON M L, Analyt. Chem. (1953) 25, 624
9. GLEMSER O, RAUELF E and GRISEN K, Zeit. Anal. Chem. (1954) 141, 82
10. BEHR A, BLANCHET M L and MALAPRADE C, Chim. Anal. (1960) 42, 501
11. SHAPIRO L and BRANNOCK W W, US Geol. Surv. Bull. 1144-A, (1962)
12. RILEY J P, Anal. Chim. Acta (1958) 19, 413
13. LINNELL R H and RAAB F H, Analyt. Chem. (1961) 33, 154
14. WILSON A D and SERGEANT G A, Analyst (1963) 83, 109
15. CAPACHO-DELGADO L and MANNING D C, Analyst (1967) 92, 553
16. RAMAKRISHNA T V, WEST P W and ROBINSON J W, Anal. Chim. Acta (1967) 39, 81
17. MARKS J Y and WELCHER G G, Analyt. Chem. (1970) 42, 1033
18. MILNER G W C and WOODHEAD J L, Anal. Chim. Acta (1955) 12, 127
19. MERCY E L P and SAUNDERS M J, Earth Planet. Sci. Lett (1966) 1, 169

CHAPTER 7

Antimony

Neither gravimetric nor titrimetric methods are of sufficient sensitivity for the determination of antimony at the level at which it occurs in silicate rocks, and recourse must be made to more sensitive techniques such as neutron activation,[1] spectrophotometry or atomic absorption spectroscopy. Onishi and Sandell[2] used rhodamine B as a photometric reagent for antimony. Special care was required to ensure that a complete isolation of antimony was obtained, separating it particularly from iron, gallium, thallium, tungsten and gold, all of which give sensitive colour reactions with rhodamine B. Copper sulphate was added to a solution of the rock material in sulphuric acid and the antimony and copper precipitated with hydrogen sulphide. This separation is far from complete and additional separation stages were then required to remove the final traces of iron, gallium, thallium and gold. This method is limited more by poor reproducibility at the ppm level than by sensitivity. The authors claim a limit of determination of 0.1-0.2 ppm Sb - the concentration that they observed in many igneous rocks.

Rhodamine B was also used by Ward and Lakin[3] to determine antimony in soils and rocks. Their procedure incorporated an extraction into isopropyl ether, followed by the formation of the coloured complex with rhodamine B in the etherial solution. Iron (III), arsenic, gold, tin and thallium are also extracted into isopropyl ether and could give rise to serious errors in antimony determinations, particularly at the ppm level. The procedure was devised primarily for soil samples containing relatively large amounts of antimony and is not directly applicable to rock material containing much less than 0.5 ppm Sb.

The reaction between chloroantimonate ions and brilliant green has also been used to determine antimony.[4] Soil and rock samples can be decomposed either by fusion with potassium hydrogen sulphate or by heating with ammonium chloride. The complex with brilliant green is formed in hydrochloric acid solution containing sodium hexametaphosphate, and is extracted into toluene for photometric measurement. The lower limit of determination appears to be about 0.4 ppm Sb.

The procedure given below is based upon that described by Schnepfe.[5] Antimony in the sample solution is reduced and volatilised as stibine, SbH_3, which is absorbed in mercuric chloride solution, oxidised with ceric sulphate and extracted as rhodamine B chloroantimonate. Arsenic distils as arsine, AsH_3, but the successful recovery of 1 μg Sb in the presence of 1000 μg As suggests that arsenic is unlikely to interfere in the determination. If it is present in high concentration in the rock sample, additional oxidant will be required.

Spectrophotometric Determination of Antimony in Silicate Rocks

A variety of procedures were used by Schnepfe[5] in silicate sample decomposition including fusion with a mixture of potassium and sodium carbonates, potassium hydroxide, potassium hydroxide followed by sodium peroxide and potassium pyrosulphate. Only fusion with potassium pyrosulphate was successful in recovering antimony added to silicate rocks, and this procedure is used below. It should be noted however that much of the silicate matrix remains undecomposed in this fusion, and that any antimony present within these silicate structures is unlikely to be recovered.

Method

Apparatus. For the reduction of antimony, a hydride generator is required (also known as an arsine generator). A simple form of this is shown in Fig. 5.

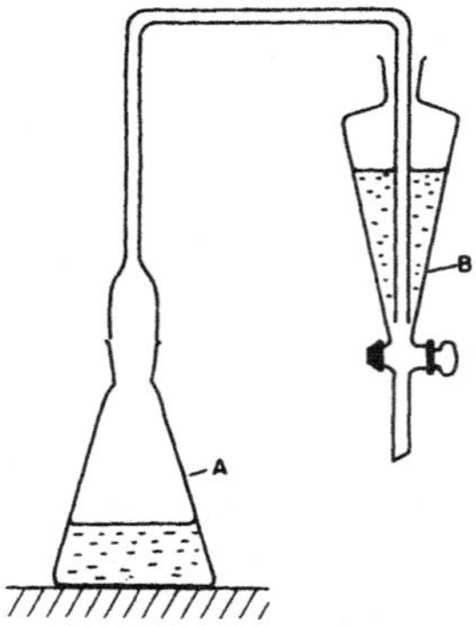

FIG. 5 A simple arsine generator

Reagents: Tartaric acid solution, dissolve 10 g in water and dilute to 100 ml.

Zinc shot, low in antimony.

Cadmium sulphate solution, dissolve 5 g in water and dilute to 100 ml.

Mercuric chloride solution, dissolve 5 g of mercuric chloride in 1 litre of 6 M hydrochloric acid.

Ceric sulphate solution, dissolve 10.6 g of tetrasulphateoceric acid, $H_4Ce(SO_4)_4$ in 80 ml of water, add 5.6 ml of 20 N sulphuric acid and dilute to 100 ml.

Rhodamine B solution, dissolve 0.2 g in 100 ml of water.

Rhodamine B-hydroxylamine hydrochloride solution, dilute 50 ml of the rhodamine B solution with 50 ml of water, add 1 g of hydroxylamine hydrochloride. Prepare freshly each day.

Standard antimony, stock solution, 100 µg Sb per ml. Dissolve 0.274 g of potassium antimonyl tartrate hemihydrate in 1 litre of N sulphuric acid. Prepare working standards by dilution of this stock solution with water, as required.

Procedure. Weigh approximately 1.5 g of potassium pyrosulphate into a small fused silica crucible and fuse gently to ensure removal of all moisture and give a quiescent melt. Allow to cool and accurately weigh approximately 0.2 g of the finely powdered silicate rock material into the crucible. Cover the crucible with a silica lid and fuse gently for 10 to 15 minutes. Allow to cool. Add 14 ml of water and 1 ml of tartaric acid solution and dissolve all soluble constituents of the melt by heating on a steam bath, stirring occasionally to assist dissolution.

Transfer the solution and residue to the hydride generator, Fig. 5 , rinsing the crucible and lid with 4 ml of water followed by 40 ml of 20 N sulphuric acid. Add the washings to the generator followed by 1 ml of the cadmium sulphate solution. Place a small filter funnel in the neck of the flask and bring the liquid gently to the boil. Do not prolong the boiling, but cool under a tap to bring to a temperature of 20° to 25°.

Add 15 g of zinc shot to the generator and immediately complete the assembly of the apparatus as shown in the figure, with the delivery tube in the 60-ml separating funnel which contains 20 ml of the mercuric chloride solution. Allow to react for 2 hours, then remove the generator and delivery tube, rinsing any solution adhering to the tube back into the separating funnel. Discard the contents of the flask.

Using a safety pipette, add exactly 10 ml of benzene to the separating funnel followed by 0.5 ml of the cerium solution. Stopper the funnel and shake for about a minute. Add 10 ml of the rhodamine B-hydroxylamine hydrochloride solution, again stopper and shake for 1 minute. Allow the phases to separate, drain off and discard the lower aqueous phase. Wash the benzene layer by shaking twice with 10 ml of 4 M hydrochloric acid and discard the washings. Drain the benzene layer into a small glass beaker and transfer it to a 15-ml centrifuge tube. Centrifuge to give a clear solution and then measure the optical density of the solution relative to benzene at a wavelength of 550 nm.

Prepare working curves for 0 to 1.0 μg Sb and 0 to 10 μg Sb using aliquots of a working standard solution. Transfer each aliquot to the hydride generator, add 1.5 g of potassium pyrosulphate, 1 ml of tartaric acid solution, 18 ml of water, 1 ml of cadmium solution and 40 ml of 20 N sulphuric acid. Then proceed as described above.

Determination of Antimony by Atomic Absorption Spectroscopy

The antimony values encountered in most silicate rocks are below those that can readily be determined by atomic absorption spectroscopy using an aqueous solution of the rock material. Terashima,[6] however describes a method in which antimony present in silicates, converted to stibine, SbH_3, introduced directly into an argon-hydrogen flame for absorption measurement at a wavelength of 217.6 nm. Satisfactory results are claimed for the range 0.06 to 4.0 ppm Sb.

Procedures based on the evolution of stibine using a flame-heated silica tube[7] and absorption on a glycomethacrylate gel with bound thiol groups[8] have yet to be adapted for application to rock material.

The methods described by Welsch and Chao[9] and McHugh and Welsch[10] were devised for the rapid determination of antimony in samples obtained in the course of geochemical exploration where consistency is perhaps more important than precision and accuracy. These procedures are based upon a decomposition of the antimony-containing minerals by heating with solid ammonium iodide solution in diluted hydrochloric acid and extraction of the antimony into a trioctylphosphine-methylisobutyl ketone solvent. The extract can be aspirated directly into an air-acetylene flame and the antimony absorption measured at 217.6 nm. For practical purposes the limit of determination is 0.25 ppm, representing 1 ppm Sb when a 0.5 g sample is extracted into 2 ml of TOPO-MIBK solvent.

References

1. LOMBARD S M, MARLOW K W and TANNER J T., Anal. Chim. Acta (1971) 55, 13
2. ONISHI H and SANDELL E B., Anal. Chim. Acta (1954) 11, 444
3. WARD F N and LAKIN H W., Analyt. Chem. (1954) 26, 1168
4. STANTON R E and McDONALD A J., Analyst (1962) 87, 299
5. SCHNEPFE M M., Talanta (1973) 20, 175
6. TERASHIMA S., Japan Analyst (1974) 23, 1331
7. HON P K, LAU O W, CHEUNG W C and WONG M C., Anal. Chim. Acta (1980) 115, 355
8. SLOVAK Z and DOCEKAL B., Anal. Chim. Acta (1980) 117, 293
9. WELSCH E P and CHAO T T., Anal. Chim. Acta (1975) 76, 65
10. McHUGH J B and WELSCH E P., Bull. U S Geol. Surv. No 1408, (1975)

CHAPTER 8

Arsenic

Distillation as arsine, AsH_3, has long been used for the recovery of arsenic from a wide variety of materials. It was earlier noted that the reagents used for the determination may contain similar amounts of arsenic to the rock samples but this is probably no longer true. Another distillation procedure that has been widely used is that of arsenious chloride from aqueous hydrochloric acid solution at a temperature not exceeding 108°. At this temperature neither antimony nor tin will distil. Any germanium present in the sample material will accompany the arsenic, but is not likely to interfere. Small amounts of selenium may also be recovered in the distillate.(1)

The extraction of arsenious iodide with carbon tetrachloride from solutions 9-12 M in hydrochloric acid(2) does not appear to have been applied to silicate or carbonate rocks, but could provide an alternative separation procedure to distillation or a co-crystallisation technique as described by Portmann and Riley.(3)

In the initial stages of rock decomposition, loss of arsenic as the volatile arsenious fluoride may occur if hydrofluoric acid is used, although Sanzolone <u>et al</u>(4) reported that in an oxidising medium no significant loss occurred. Terashima(5) added potassium permanganate to avoid loss of arsenic at this stage. An alkali fusion can also be used and Rader and Grimaldi(1) describe both fusion and acid digestion techniques to determine arsenic in marine shales.

Spectrophotometric Determination of Arsenic

A variety of procedures are in common use for the photometric determination of arsenic, many of them based upon the formation of an arsenomolybdate complex which is then reduced to a molybdenum blue. Silica and phosphorus, which form similar yellow-coloured complexes that can be reduced to molybdenum blues, will interfere unless removed in the earlier stages of the analysis. For this reason it is recommended(3) that volumetric flasks used should be filled with concentrated sulphuric acid and allowed to stand overnight.

Stannous chloride has been used(6,7) to reduce the arsenomolybdate, but the blue colour produced is relatively unstable. Hydrazine(8) and ascorbic acid(9) have been reported to give stable blues, but the reduction is slow and requires several hours for complete colour development. Portmann and Riley(3) have suggested using a mixed colour-reagent containing ammonium molybdate, ascorbic acid and an antimony salt, to produce an arsenic-molybdenum blue by a reaction analogous to that described for the determination of phosphorus.[10]

It is necessary to effect a compromise between a satisfactory sensitivity to arsenic (coupled with a low reagent blank) obtained at high acid concentrations, and rapid colour development which is promoted by low acidities. A major factor in the formation of molybdenum blue is the relative proportion of acid and molybdenum present, the amount of reducing agent being only of minor importance provided that it is above a critical level.[3]

The absorption spectrum of a molybdenum blue solution resembles that obtained by reduction of silicomolybdate or phosphomolybdate solutions, with maximum absorption occurring at a wavelength of about 835 nm. With their antimonial reagent, Portmann and Riley[3] recommended a somewhat higher wavelength of 866 nm.

An alternative spectrophotometric method for arsenic is that based upon the red colour given with the silver salt of diethyldithiocarbamate in pyridine solution.[11,12] To form the complex it is necessary to reduce the arsenic to arsine, for which arsenic-free zinc and hydrochloric acid should be used. Antimony and germanium form similar complexes[13] of somewhat lower sensitivity and can interfere, although the small amounts of antimony and germanium present in silicate rocks do not give rise to significant errors. Absorption spectra of the three complexes are shown in Fig 6.

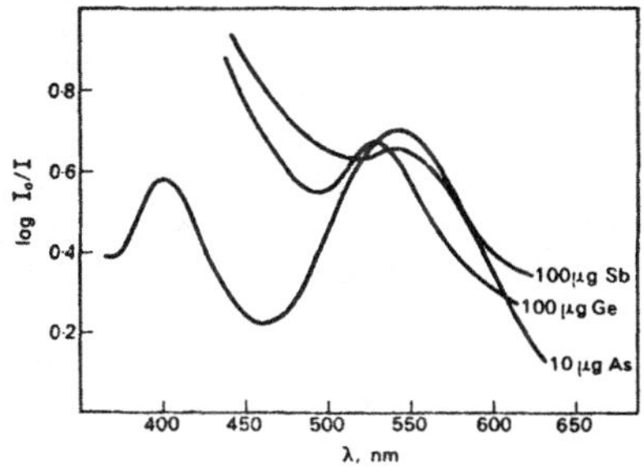

FIG. 6 Absorption spectra of arsenic, antimony and germanium complexes of diethyldithiocarbamate.

A MOLYBDENUM BLUE METHOD USING ACID DIGESTION

Rader and Grimaldi[1] have reported that a simple acid digestion technique can be used to recover arsenic from certain marine shales. The rock material is not appreciably attacked, although all sulphides present are completely decomposed. This procedure, avoiding the use of an alkaline fusion, cannot therefore be recommended where the arsenic is present in very small quantities, nor when present in the silicate matrix.

Method

Apparatus. See Fig. 7, consists of a round-bottomed flask with thermometer as shown, tap funnel to contain concentrated hydrochloric acid, condenser and receiving flask.

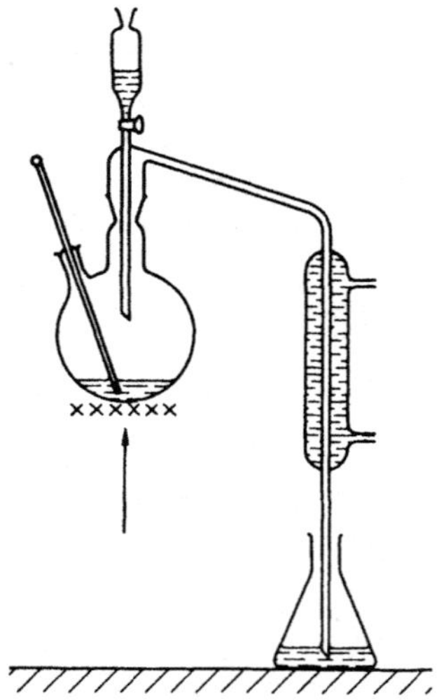

FIG. 7 Apparatus for the distillation of arsenic

Reagents: Hydrazine sulphate.

Potassium bromide.

Ammonium molybdate reagent solution, prepare two separate solutions, A and B as follows. For A, dissolve 10 g of crushed ammonium molybdate in 250 ml of 20 N sulphuric acid and dilute to 1 litre with water. For B, dissolve 0.75 g of hydrazine sulphate in 500 ml of water. Prepare the colour reagent as required by diluting 50 ml of solution A to 450 ml with water, adding 15 ml of solution B and diluting to 500 ml with water.

Standard arsenic stock solution, dissolve 0.132 g of pure arsenious oxide in 5 ml of approximately 5 N sodium hydroxide solution, make just acid with dilute hydrochloric acid and dilute to 1 litre with water. This solution contains 0.1 mg As per ml.

Standard arsenic working solution, dilute 5 ml of the stock solution to 500 ml with water. This solution contains 1 μg As per ml.

Procedure. Accurately weigh approximately 1 g of the finely powdered rock material into a 100-ml beaker, add 10 ml of concentrated nitric acid, gently swirl the contents to moisten the whole of the sample and digest on a steam bath for 15 minutes.

Allow to cool, add 10 ml of 20 N sulphuric acid and 7 ml of concentrated perchloric acid and evaporate on a hot plate to fumes of sulphuric acid. Allow to cool. If any organic matter remains after this treatment, repeat the evaporation with further additions of nitric and perchloric acids. Allow to cool and repeat the evaporation to fumes of sulphuric acid to complete the expulsion of nitric and perchloric acids. Again allow to cool, add 15 ml of water, stir to break up any residue, and using 20 ml of concentrated hydrochloric acid transfer the solution and residue to a distillation flask, Fig. 7. Add 0.5 g of potassium bromide and 1 g of hydrazine sulphate and assemble the apparatus. Transfer 10ml of water to the receiver, and concentrated hydrochloric acid to the tap funnel.

Distil the sample solution, keeping the tip of the delivery tube below the surface of the water in the receiver and, by adding hydrochloric acid ensure that the temperature in the distilling flask does not rise above 108°. Continue until approximately 25 ml of distillate has been collected (35 ml in the receiver); this usually takes 30 to 35 minutes. Remove the receiver, add 10 ml of concentrated nitric acid and evaporate to dryness on a hot plate. Place the flask in an electric oven at a temperature of 130°, and leave for 30 minutes to complete the expulsion of all traces of nitric acid. Add exactly 25 ml of the colour reagent to the flask, cover and heat on a steam bath for 20 minutes to develop the colour. Remove the flask, allow to cool, and measure the optical density in 1-cm cells using the spectrophotometer set at a wavelength of 840 nm

(Note 1). Measure also the optical density of a reagent blank solution prepared in the same way as the sample solution, but omitting the sample material.

Calibration. Transfer aliquots of 5-50 ml of the standard arsenic solution containing 5 to 50 μg As, to separate 100-ml conical flasks. Add 10 ml of concentrated nitric acid to each and evaporate to dryness on a hot plate. Proceed as described above for colour development and then measure the optical density of each solution. Plot the relation of optical density to arsenic concentration.

Note: 1. The direct measurement of optical density can be made if the sample solution contains less than 50 μg As. If the sample material contains more arsenic than this, transfer the solution to a 250-ml volumetric flask and dilute to volume with 0.5 N sulphuric acid. Additional colour reagent will not be required unless the arsenic content of the solution exceeds 400 μg.

A MOLYBDENUM BLUE METHOD USING A FUSION

The flux used consists of a mixture of potassium carbonate and magnesium oxide. Sodium carbonate should not be used, as this will result in the precipitation of sodium chloride during the distillation. No provision is made for removing silica, which tends to precipitate towards the end of the distillation.

Method

Reagents: Hydrobromic acid, concentrated.
Fusion mixture, mix one part by weight of magnesium oxide with three parts by weight of potassium carbonate.

Procedure. Accurately weigh 0.25-0.5 g of the finely powdered rock material into a 30-ml platinum crucible and add 2-3.5 g of the fusion mixture. Mix the charge carefully and cover with an additional 0.5 g of the fusion mixture. Cover the crucible and heat in an electric furnace set at 650° for a period of 30 minutes. Gradually raise the temperature of the furnace to 900° and then heat at this temperature for a further period of 30 minutes or until all organic matter has been destroyed. Allow to cool. Place the crucible in a beaker and add 20 ml of water (but no alcohol, even if manganate has been formed in the fusion). Cover the beaker and carefully add 60 ml of concentrated hydrochloric acid, gently agitating the solution. Allow the melt to disintegrate in the cold. Remove the crucible and rinse it with a further 10 ml of concentrated hydrochloric acid.

Transfer the solution to the distilling flask, rinse the beaker with a little hydrochloric acid and add this to the flask, giving a total volume of about 100 ml. Add 2 ml of concentrated hydrobromic acid and 0.5 g of hydrazine sulphate. Distil the solution, collecting about 50 ml of distillate in a beaker containing about 50 ml of water and kept cold by immersion in an ice bath. Add 25 ml of concentrated nitric acid to the distillate and evaporate to dryness on a steam bath. Transfer the beaker to an electric oven to complete the expulsion of nitric acid and then complete the determination of arsenic in the distillate as described in the previous method.

AN ARSINE-MOLYBDENUM BLUE METHOD

In this method the silicate rock material is decomposed by an alkaline fusion and the arsenic present is reduced to arsine with zinc and hydrochloric acid. The procedure described below is a modification of that given by Onishi and Sandell[14] involving an oxidation of the arsine and subsequent reduction of arsenomolybdate to molybdenum blue for photometric measurement.

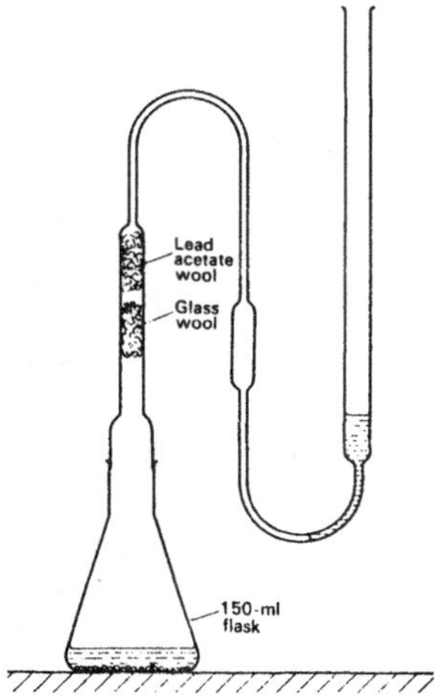

FIG. 8 . Apparatus for the evolution and collection of arsine.

Method

Apparatus. This is shown in Fig. 8 . The plug of glass wool prevents drops of liquid from being carried over. The second plug of lead acetate wool is used to remove from the gas stream hydrogen sulphide which otherwise interferes with the determination.

Lead acetate wool is prepared by soaking cotton wool in an aqueous solution of lead acetate containing 200 g of the reagent per litre, and drying in a current of warm air.

Reagents: Sodium hydroxide wash solution, dissolve 1 g of reagent in 100 ml of water.

Hydrochloric acid, "arsenic free".

Zinc metal, 20-30 mesh, "arsenic free".

Potassium iodide solution, dissolve 15 g of reagent in 100 ml of water.

Stannous chloride solution, dissolve 40 g of stannous chloride dihydrate in 100 ml of concentrated hydrochloric acid.

Iodine solution, dissolve 0.25 g of iodine and 0.4 g of potassium iodide in water and dilute to 100 ml.

Sodium bicarbonate solution, dissolve 4.2 g of the reagent in 100 ml of water.

Sodium metabisulphite solution, prepare as required by dissolving 0.5 g of reagent in 10 ml of water.

Ammonium molybdate reagent, when required for use, mix 10 ml each of two solutions A and B. For A, dissolve 1.0 g of crushed ammonium molybdate in 10 ml of water and add 90 ml of 6 N sulphuric acid. For B, dissolve 0.15 g of hydrazine sulphate in 100 ml of water.

Procedure. Fuse approximately 2.5 g of sodium hydroxide in a 10-ml nickel or zirconium crucible, and allow to cool. Accurately weigh approximately 0.5 g of the finely powdered rock material on to the solidified sodium hydroxide and add approximately 0.25 g of sodium peroxide (Note 1). Cover the crucible and gently fuse the contents for 20 to 30 minutes or until complete decomposition has been obtained. Care should be taken, particularly in the early stages of the fusion as a vigorous reaction may occur if the sample contains much sulphide. Allow to cool.

Fill the crucible with approximately 8 ml of water and warm on a hot plate or steam bath until the melt is completely disintegrated. Filter using a small medium-textured paper and wash the residue well with about ten 1-2 ml portions of the sodium hydroxide wash solution, collecting the filtrate and washings in the flask used for the arsenic reduction. Discard the residue. The solution should not exceed 25 ml in volume. Cool the solution and add 11 ml of concentrated hydrochloric acid drop by drop while swirling. Silica may separate from the solution, but this can be ignored. Now add

2 ml of the potassium iodide solution and 1 ml of the stannous chloride solution. Allow the solution to stand at room temperature for 15 to 30 minutes.

Pipette 1.0 ml of the iodine solution and 0.2 ml of the sodium bicarbonate solution into the absorption arm of the apparatus. Now add 2 g of zinc metal to the solution in the flask and immediately complete the assembly of the apparatus as in Fig. 8. Allow the gases evolved at room temperature to bubble gently through the iodine solution for about 30 minutes. At the end of this period the solution in the absorber should still be faintly yellow in colour, indicating an excess of iodine. Disconnect the flask and pour the contents of the absorber into a 10-ml volumetric flask. Rinse the absorber with no more than 1.0 ml of water and add this to the volumetric flask. Add a single drop of sodium metabisulphite solution to the flask to remove the excess iodine and then add 5.0 ml of the ammonium molybdate reagent solution. Heat the flask on a steam bath for 15 minutes, cool and dilute to volume with water.

Measure the optical density of the solution relative to water in 4-cm cells, using the spectrophotometer set at a wavelength of 840 nm. Determine also the optical density of a reagent blank solution prepared in the same way as the sample solution but omitting the rock material. The arsenic content of the sample can then be obtained by deduction of the blank and reference to the calibration graph (Note 2).

Calibration. Transfer aliquots of 1-3 ml of the standard arsenic solution containing 1 to 3 μg arsenic to separate 10-ml volumetric flasks and add 1 ml of the iodine solution and 0.2 ml of the sodium bicarbonate solution to each. Swirl gently to mix and then add 1 drop of sodium metabisulphite solution and 5 ml of the ammonium molybdate reagent solution to each flask. Heat on a water bath for 15 minutes, cool, dilute to volume with water and measure the optical density of each solution in 4-cm cells as described above. Plot the relation of these values to arsenic concentration.

Notes: 1. The addition of sodium peroxide is not essential, but assists the decomposition, particularly in the presence of sulphide minerals.

2. It is advisable to check the recovery of arsenic by adding a suitable aliquot of the standard arsenic solution to a rock material of known arsenic content, and determining the total arsenic content by the procedure given. Onishi and Sandell[14] suggest that the recovery of arsenic from normal silicate rocks is usually better than 95 per cent.

METHOD USING SILVER DIETHYLDITHIOCARBAMATE

For this method the rock material can be decomposed by fusion with sodium hydroxide as described in the previous section. The arsenic present in solution is then reduced to arsine with metallic zinc and hydrochloric acid in the apparatus shown, prior to reaction with the silver reagent solution. A somewhat similar method but based upon an initial attack of the material with a nitric-perchloric acid, or a hydrofluoric-hydrochloric-nitric acid mixture has been described by Marshall.[15]

Method

Reagents: Silver diethyldithiocarbamate solution, dissolve 0.5 g of the reagent in 100 ml of pyridine.

Hydrochloric acid, zinc, potassium iodide and stannous chloride as in the previous section.

Procedure. Prepare a solution of the silicate rock by fusion with sodium hydroxide as described in the previous section, and reduce any arsenic present to arsine with zinc and hydrochloric acid in the apparatus shown in Fig 8, passing the gases through 3 ml of the silver diethyldithiocarbamate solution placed in the absorption arm. Allow the reduction to proceed for 45 to 60 minutes at room temperature and then remove the reaction flask. Gently tilt the absorber to and fro to dissolve any complex adhering to the walls of the tube, transfer the pyridine solution to a 1-cm cell and measure the optical density with the spectrophotometer set at a wavelength of 540 nm. Before re-using the apparatus, rinse it out thoroughly with acetone, allow to dry and repack with fresh lead acetate wool.

Calibration. Transfer aliquots of 1-10 ml of the standard arsenic solution containing 1-10 µg As to the reaction flask, add hydrochloric acid, potassium iodide solution and arsenic-free metallic zinc, as for the sample solution. Assemble the apparatus with 3 ml of the silver diethyldithiocarbamate solution in the absorption arm, and collect the arsine as described above. Measure the optical density of each solution and plot the relation of optical density to arsenic concentration.

Determination of Arsenic by Atomic Absorption Spectroscopy

Distillation as arsine, AsH_3 can also be used as a means of isolating or concentrating arsenic prior to its determination by atomic absorption spectroscopy.[5,16] Problems occur in ensuring that all the arsenic is recovered in a reasonable time and that the gases evolved are presented to the flame of the instrument in a reproducible manner. Using an argon hydrogen flame, a 10-cm flame length and measurement at 193.7 nm, Terashima[16] reported a sensitivity of 0.04 ppm As, based on a 1 g sample weight.

Sanzolone et al[4] absorbed the generated arsine in a dilute silver nitrate solution and used this for arsenic determination by flameless atomic absorption spectroscopy ("carbon rod atomiser"). This allowed the determination of 0.1 to 5 μg As in the solution, giving a lower limit of 1 ppm As in a 0.1 g sample weight.

An alternative collecting medium for arsine - a glycolmethacrylate gel containing bound thiol groups ("Spheron Thiol") as suggested by Slovak and Docekal[17] has yet to be applied to the determination of arsenic in rock material.

References

1. RADER L F and GRIMALDI F S., U S Geol. Surv. Prof. Paper 391-A, (1961)
2. MILAEV S M and VOROSHINA K P., Zavod. Lab. (1963) 29, 410
3. PORTMANN J E and RILEY J P., Anal. Chim. Acta (1964) 31, 509
4. SANZOLONE R F., CHAO T T and WELSCH E P., Anal. Chim. Acta (1979) 108, 357
5. TERASHIMA S., Anal. Chim. Acta (1976) 86, 43
6. TUROG E and MEYER A H., Ind. Eng. Chem, Anal. Ed. (1929) 1, 136
7. HERON A E and ROGERS D., Analyst (1946) 71, 414
8. MORRIS H J and CALVERY H O., Ind. Eng. Chem, Anal. Ed. (1937) 9, 447
9. JEAN M., Anal. Chim. Acta (1956) 14, 172
10. MURPHY J and RILEY J P., Anal. Chim. Acta (1962) 27, 31
11. VASAK V and SEDIVEC V., Chem. Listy (1952) 46, 341
12. LIEDERMAN D, BOWEN J E and MILNER O I., Analyt. Chem. (1959) 31, 2052
13. FOWLER E W., Analyst (1963) 88, 380
14. ONISHI H and SANDELL E B., Mikrochim. Acta (1953), 34
15. MARSHALL N J., J. Geochem. Expl. (1978) 10, 307-313
16. TERASHIMA S., Japan Analyst (1974) 23, 1331
17. SLOVAK Z and DOCEKAL B., Anal. Chim. Acta (1980) 117, 293

CHAPTER 9

Barium

Gravimetric methods have long been used for the determination of barium in silicate rocks. They are applicable to those rocks containing 100 ppm or more, although the precision is poor at the lower level. Flame emission of calcium as the oxide bands interferes with the flame emission measurement of barium at the resonance line of 553.6 nm. However, by using a nitrous oxide-acetylene flame, adding potassium ion and carefully setting the flame conditions, it has been found possible to minimise the interferences in flame methods.

Gravimetric Determination of Barium in Silicate Rock

The residue obtained on decomposing silicate rocks by evaporation with concentrated hydrofluoric and sulphuric acids contains the barium present in the rock material as the insoluble sulphate. Accessory minerals that are incompletely attacked remain with the barium and must be separated before the residue can be weighed. After dehydration with concentrated sulphuric acid, ferric and aluminium sulphates will appear in the residue, but these will pass into the solution on prolonged digestion in dilute sulphuric acid. This method can be combined with the determination of manganese, total iron, titanium and phosphorus. Carbonate rocks and minerals are readily decomposed with hydrochloric acid, but subsequent precipitation of the barium with dilute sulphuric acid gives a residue that may be heavily contaminated with calcium, strontium and accessory minerals.

As an alternative procedure, silicate rocks can be decomposed by fusion with sodium carbonate, (eg as described by Bennett and Pickup[1], where the determination of barium can be combined with those of chromium, vanadium, sulphur, chlorine and zirconium) or by sinter with sodium peroxide. This has the advantage that the accessory minerals are more likely to be decomposed, and the disadvantage that the collected barium sulphate is likely to be more contaminated with other elements. After leaching and filtering, the insoluble residue can be cautiously acidified with hydrochloric acid and sulphuric acid added to precipitate the barium as sulphate, in the presence of added calcium if there is little or none in the silicate rock material.

One procedure commonly employed for the purification of the barium residue makes use of the solubility of barium sulphate in concentrated sulphuric acid, from which it may be recovered by dilution with water. An alternative procedure[2] makes use of the solubility of barium sulphate in ammoniacal EDTA solution. The calcium and strontium EDTA complexes are stable over a wide pH range, in contrast to the barium complex, which is stable only in alkaline solution. Barium sulphate is therefore precipitated on acidification.[3] The precipitated barium sulphate is collected, dried, ignited and weighed. Barium sulphate precipitates are not easy to collect. The particle size is very small and a close-textured filter paper is essential. Even so, some small part of the precipitate may pass through the paper and give rise to low and erratic results.

Method

Reagents: Sodium carbonate wash solution, dissolve 10 g of anhydrous reagent in 500 ml of water.

Acid EDTA solution, suspend 10 g of ethylenediaminetetraacetic acid (ie the free acid, not the sodium salt) in 50 ml of water, add 10 ml N sulphuric acid, neutralise and dissolve the EDTA by adding ammonia, make just acid to methyl red indicator and dilute to 100 ml with water.

Alkaline EDTA solution, suspend 0.5 g of ethylenediaminetetraacetic acid (free acid) in 50 ml of water, add 10 ml of N sulphuric acid 10 ml of concentrated ammonia solution and dilute to 100 ml with water.

Procedure. Accurately weigh approximately 2 g of the finely powdered rock material into a 3-inch platinum dish and add 1 ml of concentrated nitric acid (Note 1) 15 ml of 20 N sulphuric acid and 25 ml of concentrated hydrofluoric acid. Transfer the dish to a hot plate and evaporate to fumes of sulphuric acid. Allow to cool, rinse down the sides of the dish with a little water, add an additional 10 ml of water and again evaporate to fumes of sulphuric acid. Allow to cool and rinse the contents of the dish into a 250-ml beaker (preferably one free from scratches), containing about 100 ml of water. Transfer the beaker to a hot plate or steam bath and digest until all soluble material has passed into solution. Ferric and aluminium sulphates may take a little while to do this. Allow the solution to cool and stand, preferably overnight. Barium sulphate will appear as a fine white precipitate, best seen by gently stirring the solution once or twice with a glass rod.

Collect the small insoluble residue on a small close-textured filter paper, after first decanting the bulk of the solution. Wash the residue well with small quantities of cold water. The filtrate and the washings may be combined and allowed to stand overnight to ensure complete collection of the barium sulphate precipitate. It may then be used for the determination of manganese, titanium, total iron, phosphorus and other elements if required. Transfer the filter and residue to a small platinum crucible, dry and ignite over a burner set to give only a faint dull red colour to the bottom of the crucible. Allow to cool, add a small quantity of anhydrous sodium carbonate (Note 2) and fuse over a Bunsen burner at full heat for 30 minutes. Allow to cool. Extract the melt with hot water and rinse the solution and residue into a small beaker. Collect the residue on a small close-textured filter paper and wash it with a little hot sodium carbonate wash solution. Discard the filtrate and washings. Using a fine jet of water rinse the residue back into the beaker and dissolve in a slight excess of dilute hydrochloric acid. Rinse the platinum crucible used for the fusion with a little dilute acid and add this to the solution in the beaker. Filter this solution through the paper used for the previous filtration, and wash well with small quantities of water. Collect the filtrate and washings, which should have a volume of 30 to 35 ml, in a 50-ml beaker, add a few drops of 20 N sulphuric acid, stir and set aside to stand overnight.

Collect the precipitated barium sulphate on a small close-textured filter paper, washing first with 0.1 N sulphuric acid (Note 3) and then with a little water. Transfer the filter to a weighed platinum crucible, dry and char the paper, then burn off and finally ignite over the full flame of the burner in the open crucible (Note 4). Weigh as barium sulphate $BaSO_4$ (Note 5).

Notes: 1. Care must be taken to avoid loss of sample by spitting if the rock material contains carbonate.

2. The amount of sodium carbonate required will depend upon the amount of residue remaining. For most rocks 0.25-0.5 g will suffice, but if the weight of the residue exceeds more than about 0.1 g, between 6 and 10 times as much flux as the weight of the residue should be taken.

3. At room temperature barium sulphate is soluble in water to the extent of 2.3 mg per litre. The solubility is much less in dilute sulphuric acid solution.

4. An oxidising atmosphere is required to burn off all the carbon of the filter at as low a temperature as possible. Failure to do this may result in some reduction of the barium sulphate giving low results.

5. It should not be necessary to undertake purification of the ignited barium sulphate. If the residue is not pure white in colour, the following purification step may be used.

Add 2 ml of 20 N sulphuric acid to the ignited residue contained in the small platinum crucible. Transfer to a hot plate and heat until all the barium sulphate has passed into solution. (Any black specks are probably particles of carbon from the ignited filter paper and may be cleared by adding a little concentrated nitric acid to the COLD sulphuric acid solution, and evaporating to fumes of sulphuric acid). Allow to cool, rinse the solution into a small beaker, dilute to 25 ml with water and allow to stand overnight. Collect the precipitate, ignite and weigh as described above.

Determination of Barium by Atomic Absorption Spectroscopy

A systematic study of the barium spectrum was made by Dean *et al*[4] who found that the presence of alkali metals and other alkaline earth elements tended to enhance the barium emission, whilst aluminium, iron and titanium tended to depress the emission. The ease and convenience of atomic absorption spectroscopic techniques has resulted in a number of methods in which barium is determined in the presence of other elements.[5,6,7] However interference with the atomic absorption spectroscopic determination of barium was reported by Cioni *et al*[8,9] from the same group of elements as noted above, particularly alkali metals, alkaline earths, aluminium and iron. These workers investigated in detail earlier suggestions that the releasing agents strontium or lanthanum could be used to enable barium to be determined by this technique. A "standard additions" method was also investigated, but the authors concluded that neither technique gave accurate results. They concluded that a prior separation from interfering elements was essential.

A procedure for this, based upon the work of Frache and Mazzucotelli[10] is given below. An alternative procedure using flameless atomic absorption spectroscopy has been described.[8]

Carbon furnace atomic emission spectrometry has been used by Hutton *et al*[11] for the determination of barium in carbonate rocks. The interference from calcium is said to be low, and is removed by background correction.

Method

Reagents:

Hydrochloric acid, 3M in ethanol 20 per cent v/v.

Potassium chloride solution, approximately 10 per cent, dissolve 19 g potassium chloride in water and dilute to 1 litre.

Standard barium stock solution, dissolve 1.44 g of dried barium carbonate in water containing 5 ml of concentrated hydrochloric acid and dilute to volume in a 1 litre volumetric flask. This solution contains 1000 μg barium per ml. Prepare a solution containing 100 μg per ml by dilution with water as required.

Ion Exchange Column: Set up an ion exchange column in a borosilicate glass tube to accommodate a resin bed approximately 18 cm in length and 2 cm in diameter. Slurry Dowex 50 x 8 cation exchange resin, 200-400 mesh, with 300 ml of 3 M hydrochloric acid and transfer to the column. Elute the hydrochloric acid and wash the resin bed with water until the eluate no longer gives an acid reaction.

Procedure. Weigh approximately 1 g of the finely powdered rock material into a platinum or PTFE dish, moisten with water, add 10 ml of concentrated hydrofluoric acid and 10 ml of 70 per cent perchloric acid, stir with a platinum or PTFE rod and allow to stand overnight. Evaporate to incipient dryness. Add 10 ml of 70 per cent perchloric acid and again evaporate to incipient dryness. Add 5 ml of 70 per cent perchloric acid and once again evaporate to dryness. Add 20 ml of water and 5 ml of concentrated hydrochloric acid and warm to ensure all soluble material passes into solution (Notes 1 and 2).

Dilute the solution to approximately 100 ml with water and allow to flow through the cation exchange column at a rate of about 2.5 ml per minute. Elute with 300 ml of 3 M hydrochloric acid containing 20 per cent (v/v) of ethanol. (The collected eluate contains the iron, aluminium, magnesium, calcium, potassium and sodium present in the rock material and can be used for the determination of these elements. Strontium present will be partially retained on the column).

Elute the barium from the column with 300 ml of 3 M hydrochloric acid, and evaporate the eluate to a volume of about 50 ml. Transfer to a 100-ml measuring flask, add 10 ml of the potassium chloride solution and dilute to volume with water.

Prepare also a reagent blank solution using this procedure but omitting the sample material.

Set an atomic absorption spectrometer at a wavelength of 553.5 nm and, using a barium hollow cathode lamp, measure the absorption of the solution aspirated into a nitrous oxide-acetylene flame. Measure also the absorption of the reagent blank and a set of standard barium solutions prepared from the diluted standard stock solution, containing 10 ml of the potassium chloride solution and 100 to 1000 μg Ba per 100 ml.

Notes: 1. Any residue should be recovered, ignited in a small platinum crucible, and fused with a little flux consisting of equal weights of sodium carbonate and borax. The melt should be disintegrated in water, the insoluble material collected, washed free of sodium salts and dissolved in the minimum quantity of hydrochloric acid. This solution can then be added to the main rock solution before dilution to 100 ml, prior to the ion-exchange separation.

2. Sample materials composed of minerals largely resistant to decomposition with hydrofluoric and perchloric acids as described, should be decomposed by fusion with 2.5 g of the sodium carbonate - borax flux, and the procedure followed as described in Note 1. The reagent blank should be similarly prepared.

References

1. BENNETT W H and PICKUP R, Colon. Geol. Min. Res. (1952) 3, 171
2. BUSEV A I and KISELEVA L V, Vestnim Moskov. Univ. Ser. Mat, Mekd, Astron., Fiz. i Khim (1957) 12, 227
3. WILSON A D, unpublished work.
4. DEAN J A, BURGER J C, RAINS T C and ZITTEL H E, Analyt. Chem. (1961) 33, 1722
5. WARREN J and CARTER D, Can. J. Spectrosc. (1975) 20, 1
6. ABBEY S, Geol. Surv. Pap. Can. (1970) No 72-73
7. ABBEY S, LEE N J and BOUVIER L J, Geol. Surv. Pap. Can. (1974) No 74-19
8. CIONI R, MAZZUCOTELLI A and OTTONELLO G, Analyst (1976) 101, 956
9. IBID. Anal. Chim. Acta (1976) 82, 415
10. FRACHE R and MAZZUCOTELLI A, Talanta (1976) 23, 389
11. HUTTON R C, OTTAWAY J M, RAINS T C and EPSTEIN M S, Analyst (1977) 102, 429

CHAPTER 10

Beryllium

Beryllium can be precipitated as the ammonium phosphate $Be(NH_4)PO_4$ at a pH of 5.2 [1,2] and ignited to pyrophosphate as a weighing form by a procedure similar to that used for magnesium. Acetoanilide has been used[3] to precipitate beryllium in the form of a 2:1 complex containing 2.49 percent of metal. This complex can be dried at 100° and weighed directly. Interference from iron and alumina can be prevented by precipitating the complex at pH 8 in the presence of EDTA. 2-methyl-8-hydroxy-quinoline[4] has also been used to precipitate beryllium. EDTA, sodium potassium tartrate, potassium cyanide and ascorbic acid are added to mask interferences. The complex may be weighed directly, or dissolved in acid and the organic portion coulometrically brominated.

Many photometric procedures have been devised for beryllium. Reagents in common use include p-nitrobenzeneazo-orcinol,[5] acetyl-acetone,[6] beryllon II,[7] beryllon III[8] chrome azurol S[9] and fast sulphon black F.[10] Morin has been used extensively in fluorimetric methods[11-14]. The procedure given in detail below is based upon beryllon II, and is applicable to beryl ores and silicates rich in beryllium.

The determination of beryllium by atomic absorption spectroscopy is relatively free from interferences. Aluminium has been noted as a depressant, as also have silicon and magnesium. A nitrous oxide-acetylene flame is commonly used, and the absorption measurement made at a wavelength of 234.9 nm. An increase of sensitivity is obtained by using a graphite furnace. This forms the basis of the procedure given in detail below, due to Campbell and Simon.[15] A somewhat similar procedure has been described by Korkisch and Sorio,[16] but with an additional ion-exchange separation using a strongly acidic ion-exchange resin to remove aluminium. An earlier procedure by Sighinolofi[17] also used a graphite furnace for the determination of beryllium.

Spectrophotometric Determination of Beryllium with Beryllon II

Beryllon II, the trivial name for the tetra-sodium salt of 8-hydroxy-naphthalene-3:6-disulphonic acid-1-azo-2'-naphthalene-1':8'-dihydroxy-3':6'-disulphonic acid, is a deep violet colour in aqueous alkaline solution whilst the beryllium complex is blue. This difference in colour was used by Karanovich[7] as the basis for his determination of beryllium. Only chromium, molybdenum, platinum, thorium and the rare earths are stated to interfere,[2] and then only at ratios of 500:1. These elements are not normally present in this amount in beryllium ores, concentrates or minerals, and chemical separation is therefore not necessary.

In order to increase the selectivity of the reagent, Karanovich[7] added EDTA to complex other metals, and ascorbic acid to reduce iron to the ferrous state. Even so interference from iron does occur when it is present in large amounts, and if more than a trace is present it may be necessary to replace the EDTA with N:N-di-2-hydroxyethyl glycine ("Nervanaid F") which gives a clear colourless solution with ferric iron in strongly alkaline solution. An alternative procedure for sample materials containing much iron is to separate the beryllium by a prior precipitation as phosphate using titanium as carrier.

The colour of beryllium-beryllon II complex fades slowly, and the optical density measurements should not be unduly prolonged. Ideally the sample and standard solutions should be prepared together and measured at the same time - usually after standing for 15 to 30 minutes. The precision is very poor when less than about 8 μg of beryllium is present. Reagent solutions have an appreciable absorption at the wavelength used for the beryllium complex, and the reagent solutions must therefore be measured with some care.

Method

Reagents:

Ethylenediaminetetraacetic acid solution, dissolve 5 g of the disodium salt in 100 ml of water.

Ascorbic acid solution, dissolve 0.75 g in 100 ml of water. Prepare freshly every few days.

Beryllon II reagent solution, dissolve 0.1 g of the solid reagent in 100 ml of water.

Standard beryllium stock solution, dissolve 1.97 g of beryllium sulphate tetrahydrate in 2 N sulphuric acid and dilute to 100 ml. This solution contains 1 mg Be per ml.

Standard beryllium working solution, dilute a suitable aliquot of the stock solution to give a solution containing 4 μg Be per ml in 0.5 N sulphuric acid.

Procedure. Accurately weigh 0.1-0.2 g of the finely powdered low-grade beryllium ore or silicate rock into a platinum crucible, add 2-3 g of potassium fluoride and fuse over a burner to give a clear melt. Allow to cool and add 4-6 ml of concentrated sulphuric acid. Transfer the crucible to a hot plate under an infrared heating lamp, heat slowly until all hydrofluoric acid is removed, then more strongly to evaporate sulphuric acid, and finally over a burner to give a clear melt at dull red heat. Allow to cool and then lay the crucible on its side in a 250-ml beaker containing 50-60 ml of water and heat on a hot plate until complete solution is obtained (Note 1). Rinse and remove the platinum crucible. Cool the solution, transfer quantitatively to a 100-ml volumetric flask, make up to the mark with water and mix well.

Pipette a suitable aliquot of 10 ml or less (Note 2) of this solution into a 100-ml volumetric flask, and a second aliquot equal in volume to the first into a 50-ml conical flask. Add also to each flask any additional beryllium required as described in Note 2. Neutralize by titrating the contents of the conical flask with N sodium hydroxide solution using phenol phthalein as indicator and record the titre. Now add to the volumetric flask 2 ml of the ascorbic acid solution, replace the stopper and swirl to mix well. Add 5 ml of the disodium EDTA solution and again swirl to mix. Now add N sodium hydroxide solution equivalent in volume to the titre, together with 5 ml in excess. Replace the stopper and again mix well.

Dilute the contents of the flask to a volume just short of 45 ml (previously mark the flask at this level) with water, add 5 ml of the beryllon II reagent solution, dilute to volume with water and mix well. Allow this solution to stand for 15 minutes and then measure the optical density in 1-cm cells against the reagent blank solution (Note 3) using a spectrophotometer set at a wavelength of 630 nm. Plot a curve relating the optical density of the standard solution to the beryllium concentration and hence calculate the beryllium content of the sample material.

<u>Notes</u>: 1. Any unattacked mineral grains can be filtered off and discarded.

2. This aliquot should contain 8-16 μg Be. It has been noted that the precision of the determination is very poor if less than 8 μg is taken. If the 10-ml aliquot contains less than 8 μg Be, add 2 ml of the standard beryllium solution to the volumetric flask. This standard solution is 0.5 N in sulphuric acid, and it will therefore be necessary to determine the volume of N sodium hydroxide solution required to neutralize the 2-ml aliquot.

3. Prepare also a reagent blank solution by transferring 10 ml of water to a separate 50-ml volumetric flask, adding ascorbic acid and other reagents as described above. The standard solutions are obtained by transferring 2, 3 and 4 ml of the standard beryllium solution to separate 50-ml volumetric flasks, and following the procedure given above. In each case the volume of sodium hydroxide solution required to neutralize the added sulphuric acid must be included.

Determination of Beryllium by Atomic Absorption Spectroscopy

The procedure described below is directly applicable to most silicate rocks. Generally a sample weight of 100 mg is sufficient but 250 mg portions have been used for basic rocks containing only 0.02 ppm Be. A PTFE-lined pressure vessel is used to assist in the decomposition of any beryl that may be present.

Method

Reagents:

EDTA, 0.3 M solution, dissolve 111.7 g of the disodium salt of ethylenediaminetetraacetic acid in water with sufficient ammonia to bring the pH to 8. Dilute to 1 litre.

Bromophenol blue indicator solution, dissolve 0.125 g of the solid reagent together with 0.1 g of sodium hydroxide in 250 ml of water.

Acetylacetone solution, add 10 ml of acetylacetone to 90 ml of xylene.

Standard beryllium stock solution, dissolve 1.97 g of beryllium sulphate tetrahydrate in water, add 2 ml of concentrated hydrochloric acid and dilute to 100 ml with water. Prepare working solutions by dilution as required. This solution contains 1 mg Be/ml.

Procedure. Accurately weigh 0.1 to 0.25 g of the finely powdered rock material into a PTFE-lined pressure decomposition vessel, add 2 ml of concentrated nitric acid and 4 ml of concentrated hydrofluoric acid. Close the vessel and heat at a temperature of 225° for 16 hours. Allow to cool, transfer the contents of the vessel to a platinum dish and evaporate to dryness. Cool, add 4 ml of 20 N sulphuric acid and evaporate to fumes of sulphuric acid. Allow to cool once more, add 4 ml of 3 M hydrochloric acid and 7 ml of the EDTA solution, warming if necessary to give a clear solution. Transfer to a 50-ml separating funnel.

Add bromophenol blue indicator solution followed by 12 M ammonia solution dropwise until the indicator colour remains a permanent blue, (Note 1) with 2 drops in excess. Add 10 ml of the acetylacetone solution and extract beryllium by shaking for six minutes. Discard the aqueous layer and strip the beryllium from the organic layer by shaking with 4 ml of 3 M hydrochloric acid for 10 minutes. Transfer the aqueous layer to a 5-ml volumetric flask and dilute to volume with 3 M hydrochloric acid.

Transfer 10 µl of this solution to a graphite furnace attached to an atomic absorption spectrometer. Dry the sample at a temperature of 110° for 20 seconds, char at 800° for 20 seconds and atomise at 2700° for 6 seconds (Note 2). A beryllium hollow cathode lamp should be used, argon gas for the drying and charring stages, and the absorption measured with the argon gas interrupted at a wavelength of 234.9 nm. Calibrate the procedure using aliquots of the standard beryllium working solution that have been put through the solvent extraction procedure.

Notes: 1. A pH of greater than 8 is recommended for this extraction.

2. The exact instrumental conditions used should depend upon the manufacturer's instructions. However, it should be noted that loss of beryllium has been recorded by charring at temperatures in excess of 800°.

References

1. HURE J, KREMER M and LEBERQUIER F, Anal. Chim. Acta (1952) 7, 57
2. The Determination of Beryllium, National Chemical Laboratory, HMSO, 1963
3. DAS J and BANERJEE S, Zeit. Anal. Chem. (1961) 184, 110
4. BACON J R and FERGUSON R B, Analyt. Chem. (1972) 44, 2149
5. POLLOCK J B, Analyst (1956) 81, 45
6. ADAM J A, BOOTH E and STRICKLAND J D, Anal. Chim. Acta (1952) 6, 462
7. KARANOVICH G G, Zhur. Anal. Khim. (1956) 11, 417
8. PAKALNS P and FLYNN W W, Analyst (1965) 90, 300
9. PAKALNS P, Anal. Chim. Acta (1964) 31, 576
10. CABRERA A M and WEST T S, Analyt. Chem. (1963) 35, 311
11. SANDELL E B, Anal. Chim. Acta (1949) 3, 89
12. MAY I and GRIMALDI F S, Analyt. Chem. (1961) 33, 1251
13. SILL C W, WILLIS C P and FLYGARE J K, Analyt. Chem. (1961) 33, 1671
14. SILL C W and WILLIS C P, Geochim. Cosmochim. Acta (1962) 26, 1209
15. CAMPBELL E Y and SIMON F O, Talanta (1978) 25, 251
16. KORKISCH J and SORIO A, Anal. Chim. Acta (1976) 82, 311
17. SIGHINOLOFI G P, Atom. Absorpt. Newsl. (1972) 11, 96

CHAPTER 11

Bismuth

A variety of "chemical" procedures have been described for determining bismuth in silicate and other rocks, but these are applicable primarily to rock materials that are enriched in bismuth. These include photometric methods based upon dithizone after extraction of bismuth iodide into isoamyl acetate,[1] and on extraction as bismuth diethyldithio-carbamate[2], an atomic absorption method involving extraction of soluble bismuth into nitric acid and nebulisation into an air-acetylene flame[3] and a polarographic method applied after an extraction with diethylammonium diethyldithiocarbamate.[4] None of these methods is of sufficient sensitivity for determining bismuth in normal silicate rocks. Methods that can be used, based upon flameless atomic absorption spectroscopy, have been described by Heinrichs[5] and Kane[6]. The procedure described below has been adapted from the latter paper.

Method

Reagents: Hydroxyammonium chloride solution, dissolve 1 g of the solid reagent in 100 ml of water.

Potassium iodide solution, dissolve 1.0 g of potassium iodide and 1.0g of hydroxyammonium chloride in 1 litre of water.

EDTA stock solution, dissolve 3.72 g of the disodium salt of ethylenediamine tetraacetic acid in 100 ml of water.

EDTA working solution, dilute 1 ml of the stock solution to one litre with water, to give a 10^{-5}M solution as required.

Standard bismuth solution, dissolve 0.1 g of metallic bismuth in 5 ml of diluted (1+9) nitric acid by digesting at a temperature of 100°C. Cool, transfer to a 100-ml volumetric flask. This solution contains 1 mg Bi per ml. Prepare by dilution with water as required, solutions containing 100, 50 and 10 µg per ml.

Procedure. Accurately weigh 0.1 g of the finely ground rock or ash material (Note 1) into a PTFE beaker. Add 3 ml of concentrated perchloric acid and 10 ml of concentrated hydrofluoric acid. Prepare the rock solution and also a reagent blank and a series of standards covering the range 2 to 50 ng Bi, using the same quantities of the acids by the following procedure. Digest and evaporate to dryness at a temperature of 115-120°. Dissolve the residue by gentle warming with 2 ml of concentrated perchloric acid and 5 ml of water. Cool and transfer to a 60-ml separating funnel.

Add 1 ml of the hydroxyammonium chloride solution to reduce ferric iron, then 5 ml of methylisobutyl ketone. Add 5 ml of the potassium iodide solution and shake well to mix the phases. Allow to separate and discard the aqueous phase. Rinse the organic layer with 2.5 ml of water and 2.5 ml of the potassium iodide solution, again discarding the aqueous layer.

Strip the bismuth from the organic layer by shaking twice with separate 1-ml volumes of the 10^{-5}M EDTA solution. Combine the aqueous extracts. Transfer 20 µl to the graphite furnace of the spectrometer and determine the absorption of bismuth in accordance with the instrument makers instructions (Note 2).

Notes 1: If the rock material is of high organic carbon content, ash a weighed amount in a muffle furnace by raising the temperature from 100° to 500° over a period of 3 to 4 hours. Ignite at 500° for at least 12 hours. The ash content of the material should then be calculated from the loss in weight.

2: Peak height measurements may be used to construct the calibration graph from which the concentration of bismuth in the original rock material can be calculated.

References

1. MOTTOLA H A and SANDELL E G., Anal. Chim. Acta (1961) 25, 520
2. STANTON R E., Proc. Aust. Inst. Min. Metal (1971) No 240, 113
3. WARD F N and NAKAGAWA H M., U S Geol. Surv. Prof. Paper 575-D (1967) 239
4. RUSSEL H., Zeit. Anal. Chem. (1962) 189, 256
5. HEINRICHS H, Z. Anal. Chem. (1979) Feb
6. KANE J S, Anal. Chim. Acta (1979) 106, 325

CHAPTER 12

Boron

In acidic rocks boron frequently occurs in the mineral tourmaline, and high-boron granites usually contain abundant quantities of this mineral. Tourmaline contains about 3 per cent boron, 10-11 per cent B_2O_3. A procedure is given below for the determination of boron in tourmaline, which can be adapted for other minerals high in boron, such as datolite (6% B, 20-22% B_2O_3).

Much of the earlier data on the occurrence of boron is unreliable owing to the lack of accurate and suitable methods of analysis. However, renewed interest has led to a re-examination of older methods of separation involving distillation as methyl borate, and the devising of new methods based upon pyrohydrolysis, solvent extraction and ion-exchange. A number of colour-forming reagents have been described for the photometric determination which, except for those minerals high in boron, have displaced the earlier titrimetric procedures.

Distillation as methyl borate is regarded as the classical procedure for the separation of boron, but conflicting reports exist in the literature regarding the adequacy of this method. Difficulties can arise from the presence of boron in almost all laboratory glassware, and the distillation apparatus itself should be made from fused silica. Another difficulty that has been noted is the need for maintaining anhydrous or near anhydrous conditions during the esterification and evolution of the borate.

Ion-exchange separation procedures are generally more rapid and simpler than distillation methods. Silicate rocks and minerals are brought into solution by an alkaline fusion, and the aqueous extract acidified with hydrochloric acid. The passage of this solution through a column of cation exchange resin removes iron and other interfering ions to give a clear solution that can be used directly for the determination of boron by either titrimetry or spectrophotometry. In a version of this procedure Fleet(1) dispenses with the use of a column and adds the resin directly to the acid extract of the fused rock melt. This procedure is described in detail below for application to silicate rocks.

Spectrophotometric Determination of Boron in Silicate Rocks

Of the many reagents described for the photometric determination of boron only curcumin, dianthrimide (1,1'-iminodianthraquinone) and carmine (carminic acid) have been extensively used.

Curcumin, the active colour principle of the vegetable product turmeric, has long been used for the detection and determination of small amounts of boron. Considerable difficulties were initially experienced in obtaining repeatable quantitative results, but conditions necessary for a reliable procedure have now been established.[2] Alonso and Sanchez[3] described its application to the determination of boron in geological materials after a preliminary separation using Dowex 50W-X8 cation exchange resin.

The use of dianthrimide for the photometric determination of boron has been carefully examined by Danielsson.[4] This reagent is more sensitive than carmine, but less so than curcumin. It gives a linear calibration in concentrated sulphuric acid solution. The reagent itself has an absorption band with a maximum value below 400 nm, clearly separated from that of the boron complex which has a maximum value at 620 nm. The rate of reaction of boron with dianthrimide is strongly dependent upon temperature. Optical density values are also temperature dependent, although this effect is related to the acid concentration.

Carmine, the name given to a naturally occurring dyestuff from cochineal, is a calcium-aluminium compound of carminic acid, which is a derivative of anthraquinone. Both carmine and carminic acid react with boron in concentrated sulphuric acid solution to give blue coloured complexes, but as carminic acid is deliquiescent, carmine is preferred.[5] In the absence of boron the colour of the dye at pH 6.2 is bright red, but in the presence of boron this changes to blue. The wavelength of maximum absorption changes from 520 nm for the reagent to 585 nm for the boron complex.[6] The Beer-Lambert Law is obeyed over the concentration range 0-10 ppm of boron. The colour development characteristics may vary with the brand of reagent used.[1] Hatcher and Wilcox[6] have reported that the coloured complex with boron can be measured after 45 minutes and then shows no appreciable change at the end of 4 hours. Fleet,[1] however, noted that the absorption reached a maximum after 40 minutes and thereafter decreased. The procedure, described below by Fleet uses a cation exchange resin to separate interfering ions from boron.

An alternative collection procedure using pyrohydrolysis has been described by Farzaneh, Troll and Neubauer.[7] The sample material was mixed with pure calcium fluoride and the BF_3 liberated was collected in a sodium hydroxide-sodium carbonate solution prior to photometric determination with carminic acid.

Method

Reagents:

Mannitol solution, dissolve 1 g of reagent in 100 ml of water.

Carmine solution, dissolve 50 mg of the reagent in 100 ml of concentrated sulphuric acid.

Hydrochloric acid, 0.6 N.

Sodium hydroxide solution, 0.1 N.

Standard boron stock solution, dissolve 0.5716 g of recrystallised boric acid in water and dilute to 1 litre. This solution contains 100 μg boron per ml.

Standard boron working solutions, dilute aliquots of the stock solution with water to give three new solutions containing 5, 10 and 20 μg boron per ml respectively.

Cation exchange resin, wash 50 ml of a strongly acid resin such as Amberlite IR 120(H) or Dowex 50W-X8 with 6 N hydrochloric acid and then water until the eluate is free from acid.

Procedure. Accurately weigh approximately 0.2 g of the finely powdered rock sample (or a smaller amount if the sample material contains more than 200 ppm boron) into a 10-ml platinum crucible and add 1.25 g of potassium carbonate. Mix, and fuse over a Bunsen burner for 1 hour. Allow the crucible to cool, loosen the melt by warming with a small amount of water, and transfer the solution and residue to a 50-ml polypropylene beaker. Cover the beaker and add 2 ml of mannitol solution, followed by 20 ml of the cation exchange resin and 2 ml of 0.6 N hydrochloric acid. Break up any lumps of residue, mix with the ion exchange resin and add enough water to give a slurry. Allow to stand overnight.

Collect the ion-exchange resin and any precipitated silica on a small medium-textured filter paper, wash well with water and discard it. Collect the filtrate and washings in a 400-ml polypropylene beaker, add 23 ml of 0.1 N sodium hydroxide solution and carefully evaporate to dryness on a steam bath. Allow to cool and add by pipette 5 ml of 0.6 N hydrochloric acid. When the residue has dissolved, pour this solution into a centrifuge tube and centrifuge.

Pipette 2 ml of the clear solution into a 50-ml polypropylene beaker, add 2 drops of concentrated hydrochloric acid and with great care add 10 ml of concentrated sulphuric acid. Allow the solution to cool, then add 10 ml of the carmine reagent solution. Swirl gently to mix the contents of the beaker and allow to stand for 40 minutes. Measure the optical density of the solution in 1-cm cells with the spectrophotometer set at a wavelength of 585 nm.

For the reference solution, transfer 2 ml of water to a 50-ml polypropylene beaker and add concentrated hydrochloric acid, sulphuric acid and carmine reagent as described for the sample solution. A reagent blank should also be prepared from 1.25 g of potassium carbonate fused without rock material in a separate platinum crucible, and carried through the procedure as described. A series of three standards can be used for calibration by transferring 2 ml aliquots of the three working solutions, containing 10, 20 and 40 µg boron respectively, to separate beakers and proceeding as described above.

Titrimetric Determination of Boron in Tourmaline

One of the earliest procedures to be devised for the determination of boron in silicate rocks and minerals was that of titrating liberated boric acid with standard alkali in the presence of a polyhydric alcohol - mannitol being now commonly employed. The boric acid-mannitol complex acts as a strong monobasic acid. When combined with an ion-exchange separation, this procedure can be simply and easily applied to the analysis of tourmaline and other silicate minerals containing boron as a major component. The procedure described here has been adapted from that given by Kramer.(8)

Method

Reagents: Sodium hydroxide solutions, approximately 5 M, and also 0.05 M standardised by titration against standard hydrochloric acid in the usual way.

Mannitol

Ion-exchange resin, strongly acid, cationic resin such as Amberlite IR 120(H) or Dowex 50W-X8, in the form of a bed 2 cm in diameter and 25 cm in length. The column should be of borosilicate glass (attack of the glass is negligible) or polypropylene tubing. To prepare the column for use or to regenerate for further use wash the bed with 100 ml of 6 N hydrochloric acid followed by water until the eluate is free from acid.

Procedure. Accurately weigh approximately 0.5 g of the finely powdered tourmaline or other boron mineral into a small platinum crucible, mix with 3 g of anhydrous sodium carbonate and fuse over a Bunsen burner for 30 minutes. Transfer the crucible to a Meker burner and continue the heating for a further 30 minutes. Allow the melt to cool, spreading it around the sides of the crucible in the usual way.

Place the crucible on its side in a 100-ml polythene or polypropylene beaker containing 20 ml of water and add concentrated hydrochloric acid down the sides of the beaker until there is an excess of about 1 ml above the amount required to neutralise the alkali carbonate used for the fusion. Warm the solution and allow to stand until the melt has completely decomposed and all soluble material has passed into solution. Rinse and remove the platinum crucible and lid. At this stage no unattacked mineral grains should be present, and the only residue a few flakes of silica precipitated from the solution. Filter the solution through a 9-cm open-textured paper supported in a polythene funnel into a polythene beaker, and wash the residue with hot water to give a solution volume of about 50 ml. Discard the residue.

Add 5 M sodium hydroxide solution drop by drop until the formation of a precipitate that only just fails to dissolve on warming. Clear this precipitate with a few drops of hydrochloric acid, and transfer the solution to the ion-exchange column, previously washed with water. Allow the eluate to collect at a rate of between 30 and 40 ml per minute in a 400-ml polypropylene beaker, and wash the resin with about 200 ml of water. Add 0.5 ml of concentrated hydrochloric acid and boil the solution for 1 minute but not longer (Note 1), to expel any carbon dioxide, and cool the solution to room temperature.

Using a magnetic stirrer and a pH meter, add sodium hydroxide solution, first concentrated then diluted, drop by drop, until the pH of the solution reaches 7. Now add 20 g of solid mannitol and titrate the boric acid with 0.05 M standard sodium hydroxide solution until the pH again reaches 7. Subtract a reagent blank value from the titre before calculating the results. The reaction that occurs may be expressed by the equation

$$H_3BO_3 + NaOH = NaH_2BO_3 + H_2O$$

so that 1 ml of 0.05 M sodium hydroxide solution is equivalent to 1.741 mg B_2O_3 (Note 2).

Notes: 1. Prolonged boiling of hydrochloric acid solutions will result in substantial loss of boron.[9]

2. Where boron is included in the summation of the rock or mineral, it is customary to express the results as per cent boric oxide, B_2O_3. Where boron occurs only as a trace constituent, parts per million boron is used.

References

1. FLEET M E., Analyt. Chem. (1967) 39, 253
2. HAYES M R and METCALFE J., Analyst (1962) 87, 956
3. ALONSO S J and SANCHEZ G A., An. Quim. (1972) 68, 335
4. DANIELSSON L., Talanta (1959) 3, 138
5. DANIELSSON L., Organic Reagents for Metals, Hopkin & Williams Ltd, Chadwell Heath, Essex. Ed. JOHNSON W C., Vol 2, p.32, 1964
6. HATCHER J T and WILCOX L V., Analyt. Chem. (1950) 22, 567
7. FARZANEH A, TROLL G and NEUBAUER W., Z. Anal. Chem. (1979) 296, 383
8. KRAMER H., Analyt. Chem. (1955) 27, 144
9. FELDMAN C., Analyt. Chem. (1961) 33, 1916

CHAPTER 13

Cadmium

Simple spectrographic, spectrophotometric and atomic absorption spectrometric methods are not sufficiently sensitive to determine the levels of cadmium encountered in normal silicate rocks. The procedure described below, using flameless atomic absorption spectroscopy, is based on the work of Sighinolfi and Santos,[1] Langmyhr *et al*[2] and Gong and Suhr.[3] A prior extraction of cadmium from hydriodic acid solution is used, following work by McDonald and Moore.[4] Flameless atomic absorption has been used also by Heinrichs,[5] and Bea Barredo *et al*.[6]

Method

Reagents: Xylene, reagent grade.

Ethylene-diamine solution, add 8 ml to 92 ml of water and shake to dissolve.

Amberlite LA-2, a secondary amine available from Rohm and Haas and laboratory suppliers. Add 25 ml to 225 ml of Xylene and convert the amine to the iodide salt as follows. Shake for 5 minutes with an equal volume of aqueous sodium carbonate (10g/litre). Separate the organic layer and extract three times with equal volumes of 6 M potassium iodide solution. Centrifuge to remove all aqueous phase and use the xylene solution.

Equipment: An atomic absorption spectrometer fitted with a graphite furnace. The operating conditions should be as recommended by the manufacturers.

Procedure. Accurately weigh approximately 0.2 g of the finely powdered rock material into a platinum or PTFE dish and evaporate to dryness with 10 ml of concentrated (40%) hydrofluoric acid and 0.5 ml of concentrated (70%) perchloric acid. Moisten the residue with water, add 0.5 ml of concentrated perchloric acid and again evaporate to dryness. Convert the residue to the iodide salts by adding 2 ml of concentrated hydriodic acid. Dissolve in water (Note 1) and dilute to a volume of 25 ml (Note 2).

Transfer the solution to a 60-ml separating funnel, add 2 ml of the Amberlite LA-2 solution in xylene and shake for 3 to 4 minutes to extract cadmium. Allow the layers to separate and discard the lower, aqueous layer. To the organic layer add 2 ml of the ethylenediamine solution, shake for 3 to 4 minutes and again allow the layers to separate. Collect the aqueous layer. Repeat the extraction with a further 2 ml of the ethylenediamine solution and combine the extracts. If necessary, centrifuge the combined extracts to remove all trace of the organic phase.

Carefully acidify the extract by adding concentrated sulphuric acid to give a final acidity of approximately 5% by volume. Pipette a suitably sized aliquot of this solution into the graphite furnace. Using nitrogen gas at a flowrate of about 2 litres per minute, evaporate and char at a temperature of 200°C for 1 minute. Atomise at a temperature of 2500°C for 10 seconds and measure the cadmium absorption at a wavelength of 228.8 nm.

Measure also the absorption of a reagent blank at this wavelength, similarly prepared and atomised. For calibration a standard solution of cadmium should be prepared by serial dilution of a stock standard solution containing 1 g cadmium in 1 litre dilute sulphuric acid. The calibration should cover the range 10 to 1000 ppb.

The final acidity of the standards should be the same as the samples, ie 5% v/v with respect to sulphuric acid.

Notes: 1. If the sample material is not completely decomposed by this procedure, the residue should be recovered, ignited in platinum, fused with a little sodium carbonate, dissolved in water, just acidified with hydriodic acid and added to the main rock solution before dilution.

2. For most rocks, the whole of this solution should be taken for the extraction of cadmium. For rocks rich in this element, the solution should be diluted to volume in a 25-ml volumetric flask and an aliquot taken for the analysis.

References

1. SIGHINOLFI G P and SANTOS A M., Mikrochim. Acta (1976) 477
2. LANGMYHR F J, STUBERGH J R, THOMASSEN Y, HANSSEN J E and DOLEZAL J., Anal. Chim. Acta (1974) 71, 35
3. GONG H and SUHR N H., Anal. Chim. Acta (1976) 81, 297
4. McDONALD C W and MOORE F L., Analyt. Chem. (1973) 45, 983
5. HEINRICHS H., Z. Anal. Chem. (1979) 294, 345
6. BEA BARREDO F, POLO POLO C and POLO DIEZ L., Anal. Chim. Acta (1977) 94, 283

CHAPTER 14

Calcium

Peridotites and dunites, the earliest rocks to crystallise, contain only small amounts of calcium which, with the precipitation of olivine and enstatite, tends to be concentrated in the remaining melt. The succeeding stages of magmatic emplacement involve the crystallisation of much of the monoclinic pyroxenes and the calcium-rich felspars, giving rocks containing a great deal more calcium and culminating in the gravity separation of anorthosites which may contain up to 20 per cent CaO. The residual magma is depleted in calcium and the succeeding rocks contain successively less. Late stage granites contain only small amounts.

Gravimetric Determination of Calcium

After the removal of iron, aluminium and other elements of the ammonia group, any calcium present in the filtrate can be precipitated as oxalate, accompanied by much of the small amount of strontium found in most silicate rocks. In the classical procedure for determining calcium, the first oxalate precipitate is redissolved in dilute hydrochloric acid and then re-precipitated from a smaller volume of solution. This gives a precipitate that is almost completely free from magnesium and manganese[(1)] that can be ignited to oxide in a platinum crucible.

After ignition, calcium oxide tends to increase in weight by absorption of water and carbon dioxide. This does not usually give rise to any serious error if cooled in a desiccator and weighed without undue delay, but can be avoided by igniting the precipitate at a temperature of only 500°, converting the oxalate to carbonate, in which form it is weighed.

Although it is customary to remove iron, aluminium, titanium and phosphorus before precipitating calcium this is not absolutely necessary as calcium oxalate can be precipitated quantitatively from a weakly acid solution containing citric or other organic acid preventing precipitation of the ammonia group elements. Manganese interferes and difficulties also occur if the rock material is rich in magnesium or titanium.

In the classical scheme of rock analysis the ignited calcium oxalate precipitates were used for the determination of strontium, and a correction was then applied to obtain the "true" calcium content. However, none of the chemical methods gives a perfect separation and values for calcium are likely to be almost as much in error after the correction as before. The most frequently used method for this separation was based upon the solubility of calcium nitrate in concentrated nitric acid. Strontium nitrate is relatively insoluble and can be collected and weighed on a sintered glass or silica crucible and subtracted from the total weight of oxides. This gravimetric procedure for the determination of strontium (and hence correction of the gravimetric calcium value), has now been displaced by flame photometric and atomic absorption methods that do not require any extensive separation stages.

As an alternative to the gravimetric measurement the calcium oxalate can be dissolved in dilute sulphuric acid and the liberated oxalic acid titrated with standard potassium permanganate solution (see below). The small amount of strontium present in silicate rocks and collected largely in the oxalate precipitate will also be counted as calcium.

Method

Reagents: Ammonium oxalate wash solution, dissolve 1 g of reagent in 500 ml of water and make just alkaline to methyl red indicator.
Potassium permanganate 0.1 N solution, standardise by titration with sodium oxalate or arsenious oxide.

Procedure. Precipitate the calcium as oxalate as described in Chapter 3. Collect the precipitate on a close-textured filter paper and wash with the appropriate wash solution as described. Dissolve the calcium oxalate from the filter with a small amount of hot dilute hydrochloric acid. Add 100 ml of 3 N sulphuric acid, heat to a temperature of 60-70° and titrate with standard 0.1 N potassium permanganate solution.

1 ml of 0.1 N potassium permanganate is equivalent to 2.80 mg calcium oxide, giving a titration of about 36 ml for a 1 g portion of silicate rock containing 10 per cent CaO. Thus for anorthosites and similar rocks containing greater amounts of calcium, 0.5 g sample portions should be taken for the determination. For carbonate rocks a 1 g sample weight should be used, and the rock solution diluted to volume in a 250-ml volumetric flask. A 50-ml aliquot can then be taken for the precipitation of calcium and subsequent titration as described below.

Titrimetric Determination of Calcium with EDTA

Ethylenediaminetetraacetic acid forms complexes with most metals and cannot be used for the titrimetric determination of calcium unless special precautions are taken to avoid interference from trivalent and other divalent elements. In the analysis of silicate rocks this interference is largely from iron, aluminium, manganese and magnesium. Iron and aluminium can be precipitated with ammonia, but traces of aluminium and some part of the manganese can then always be recovered in the filtrate. Small amounts of both calcium and magnesium are usually co-precipitated with the ammonia precipitate, but these can be recovered in a subsequent filtrate following re-precipitation with ammonia. Alternatively iron and aluminium can be removed from the rock solution by extraction of the complexes with 8-hydroxyquinoline into chloroform as described by Cluley(2) for the analysis of glass. The interference from any remaining traces of iron and aluminium can be considerably reduced by adding triethanolamine. In order to titrate calcium in the presence of magnesium a pH of about 12 is used; at this pH magnesium is precipitated as hydroxide and does not seriously interfere. A detailed procedure for this determination is given in the chapter dealing with magnesium where it is combined with the titrimetric determination of calcium plus magnesium in order to obtain values for both elements.

Although suitable for routine analysis this procedure, in common with many others that have been suggested, is subject to certain errors. The end-point of the calcium determination is particularly difficult to determine, especially in the presence of manganese or iron which affect the indicator, even with the addition of triethanolamine as complexing agent. Only small amounts of ammonium salts can be tolerated, as these prevent the complete precipitation of magnesium, which is then titrated with the calcium. Certain indicators cannot be used in the presence of magnesium, although these are undoubtedly the best for pure calcium solutions. In the absence of magnesium, as for example in certain limestones and marbles, acid alizarin black SN (mordant black 25, C.I. 21725),(3,4) metalphthalein (phthalein complexone) screened with naphthol green B,(5) and methyl thymol blue(6) all give sharp, easily identified end-points. A procedure for this determination is given below.

The difficulties that arise with indicators in the presence of manganese are particularly acute with high concentrations of this element. When so present it can be oxidised and precipitated as MnO_2 with potassium bromate.(7)

An alternative approach to the determination of both calcium and magnesium in silicate rocks is that based upon ion-exchange separation from all other interfering elements and from eath other. Once this separation has been made there is no difficulty in determining calcium, as a somewhat lower pH of about 10-10.5 can be used with erichrome black T as indicator for both calcium and magnesium. This procedure, devised by

Abdullah and Riley,[8] takes several days for the complete separation, but most of this time can be used for other determinations.

Determination of Calcium in Carbonate Rocks (Low in Magnesium)

In this procedure any calcium present in the acid-insoluble fraction is discarded, and only the soluble calcium titrated. Up to about 4 per cent MgO can be tolerated. Acid-soluble iron, aluminium and other metals are not likely to be present in more than trace amounts, and these traces can be complexed by the addition of potassium cyanide and triethanolamine.

Method

Reagents:

EDTA 0.02 M standard solution, dissolve 7.4 g of the disodium salt of EDTA in 1 litre of water and standardise by titration using standard calcium solution.

Triethanolamine-potassium cyanide (care - POISON) solution, dissolve 6.4 g of potassium cyanide in 60 ml of water and mix with 40 ml of triethanolamine.

Hydroxylamine hydrochloride solution, dissolve 10 g of the reagent in 100 ml of water.

Sodium hydroxide solution, dissolve 30 g of the reagent in water and dilute to 100 ml.

Acid alizarin black SN indicator, grind together 0.2 g of the reagent with 10 g of sodium chloride.

Standard calcium solution, dissolve 0.500 g of pure, dry calcium carbonate in the minimum amount of dilute hydrochloric acid, transfer to a 500-ml volumetric flask and dilute to volume with water.

Procedure. Accurately weigh approximately 0.5 g of the finely powdered limestone rock into a 400-ml beaker of the "tall" or "conical" pattern, and moisten with water. Cover the beaker with a clock glass and add dilute perchloric acid down the side of the beaker, until all solid material has dissolved, avoiding an excess. Boil the solution to expel carbon dioxide, allow to cool and dilute with water to volume in a 500-ml volumetric flask. If any residue remains, collect this on a filter paper, wash with water and transfer the combined filtrate and washings to the 500-ml volumetric flask before dilution to volume with water. Discard the residue.

Pipette 50 ml of this limestone solution into a 250-ml conical flask, add 5 ml of hydroxylamine hydrochloride solution followed by 5 ml of the triethanolamine-cyanide solution (NB: use a measuring cylinder!), 10 ml of the sodium hydroxide solution and enough of the indicator to give a reasonably strong red-to-purple colour to the solution. Titrate with standard EDTA solution until the indicator is pure blue with no trace of a pink colour.

Photometric Determination of Calcium

Very few reagents are known that give colour reactions which are specific or even selective for the calcium ion. One of the most interesting of these few is calcichrome, believed to be cyclo-tris-7-(1-aso-8-naphthalene-3:6-disulphonic acid) used as an indicator for the titration of calcium with EDTA.[9] This reagent has been used for the photometric determination of calcium,[10] but does not appear to have been applied to this determination in silicate or carbonate rocks, possibly because of interference from magnesium. Murexide (ammonium purpurate) and glyoxal bis(2-hydroxy-anil) have also been suggested as photometric reagents for calcium, but also do not appear to have been used for rock analysis. Leonard[11] has, however, used glyoxal bis(2-hydroxyanil) for determining calcium in magnesium carbonate and his method can probably be adapted for use with magnesites.

Determination of Calcium by Flame Photometry

The spectrum obtained when calcium salts are aspirated into a suitable flame is relatively simple, consisting of a resonance line at 422.7 nm and band systems with maxima at wavelengths of 544, 606 and 622 nm. There is further emission in the near infrared, and a doublet at 393/97 nm due to calcium ions present in high temperature flames. Sodium interferes with the determination of calcium by contributing to the background emission at the wavelength of the resonance line, but this can be overcome by using a recording instrument and tracing the emission from about 410 nm to 440 nm.

At high concentration the alkali elements also interfere by reducing the calcium emission, but this type of effect is a more serious problem in the presence of iron, aluminium, sulphate and phosphate. These elements form compounds with calcium, particularly in low-temperature flames. This interference can be completely prevented by separating the calcium by precipitation as oxalate, as in the classical procedure. As a double precipitation of the ammonia group elements is necessary, this method is long and tedious. An alternative rapid procedure is to add an excess of each of the interfering elements to both the rock solution and the calcium standard solutions. The amounts added are such as to give limiting calcium suppression. This technique has been reported by Kramer[12] for the determination of calcium in silicate rocks and minerals.

Determination of Calcium by Atomic Absorption Spectroscopy

As with flame emission photometry, interference with the determination of calcium arises from the presence of aluminium, iron and other elements that form compounds with calcium in the flame. This interference is very much less than that recorded with emission photometry, and can be reduced still further by using a nitrous oxide-acetylene flame. Under these conditions, the only serious interference with the determination in silicate rocks is from aluminium. This can be overcome by adding potassium to the solution. The determination of calcium by this technique is usually combined with that of magnesium and one rock solution can be prepared for both determinations, see chapter dealing with magnesium.

For a review of the mutual interferences between the atomic absorption determination of calcium and some other elements, see Harrison and Ottaway.[13]

References

1. JEFFERY P G and WILSON A D, Analyst (1959) 84, 663
2. CLULEY H J, Analyst (1954) 79, 567
3. BELCHER R, CLOSE R A and WEST T S, Chemist Analyst (1958) 47, 2
4. BELCHER R, CLOSE R A and WEST T S, Talanta (1958) 1, 238
5. TUCKER B M, J. Austr. Inst. Agr. Sci. (1955) 21, 100
6. KORBL J and PRIBIL R, Chem. and Ind. (1957) p.233
7. BORISSOVA-PANGAROVA R and MITROPOLITSKA E, Dokl. Bolg. Akad. Nauk, (1977) 30, 395
8. ABDULLAH M I and RILEY J P, Anal. Chim. Acta (1965) 33, 391
9. CLOSE R A and WEST T S, Talanta (1960) 5, 221
10. HERRERO-LANCINA M and WEST T S, Analyt. Chem. (1963) 35, 2131
11. LEONARD M A, J. Pharm. Pharmacol. (1962) 14 (suppl.) 63T
12. KRAMER H, Anal. Chim. Acta (1957) 17, 521
13. HARRISON A and OTTAWAY J M, Proc. Soc. Analyt. Chem. (1972) 9, 205

CHAPTER 15

Carbon

Occurrences of graphite, diamond and the various forms of fossil fuel - anthracite, coal, lignite, bitumen and petroleum - have long been of interest economically, but have seldom been the concern of the rock analyst. Of more extensive occurrence and also of considerable industrial importance are the deposits of limestone, other carbonate rocks and the carbonate ore minerals. These materials do not usually present major problems in analysis, as the methods used for the determination of most elements in silicates can usually be adapted to carbonate rocks. Igneous carbonates, "carbonatites" with abnormal concentrations of certain rare elements, may present unusual problems and require special analytical procedures.

The carbon dioxide content of some common rocks and minerals is shown in Fig. 9. The maximum theoretical content of over 50 per cent is observed in some samples of magnesite, whilst rocks containing from 35 to 50 per cent exist in many extensive natural formations composed of chalk, limestone, dolomite and dolomitic limestone. From 5 to 35 per cent carbon dioxide has been noted in carbonatite and metamorphic rocks, whilst smaller amounts are by no means uncommon in many silicate rocks. Jeffery and Kipping[1] succeeded in determining parts per million of carbon dioxide in a number of rocks from which it had not previously been reported, and it appears likely that most silicate rocks contain small amounts of this constituent.

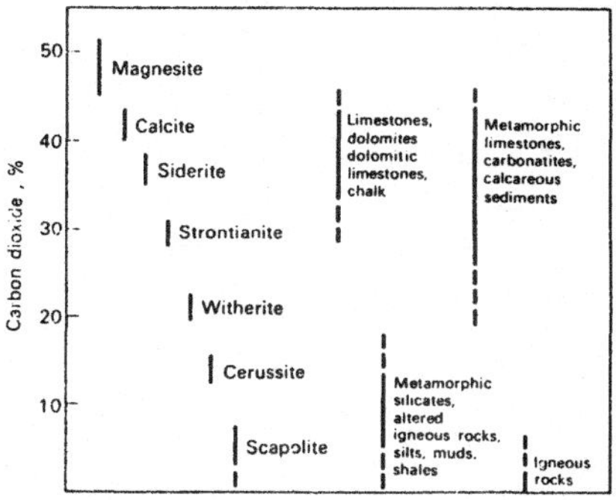

FIG. 9. The carbon dioxide content of rocks and minerals

The presence of carbonate minerals in silicate rocks is indicated by the effervescence that is obtained by warming with dilute acid. About 1 g of the powdered sample, freed from air bubbles by boiling with a little water, is acidified with a few millilitres of hydrochloric acid, allowed to stand for a few minutes and after examination, is gently warmed. As little as 0.1 per cent carbon dioxide can be detected as bubbles of gas. Calcite is decomposed in the cold, and the liberated carbon dioxide is easily seen before warming. Other carbonate minerals, particularly siderite, ankerite, and dolomite, will evolve carbon dioxide only on heating. Care must be taken not to confuse hydrogen with carbon dioxide. (Hydrogen may be formed by reduction of the hydrochloric acid if the sample contains any tramp iron introduced in the preparation of the material for analysis.) Gaseous hydrogen sulphide may also be liberated from certain sulphite minerals, but this is not likely to be confused with carbon dioxide because of its smell.

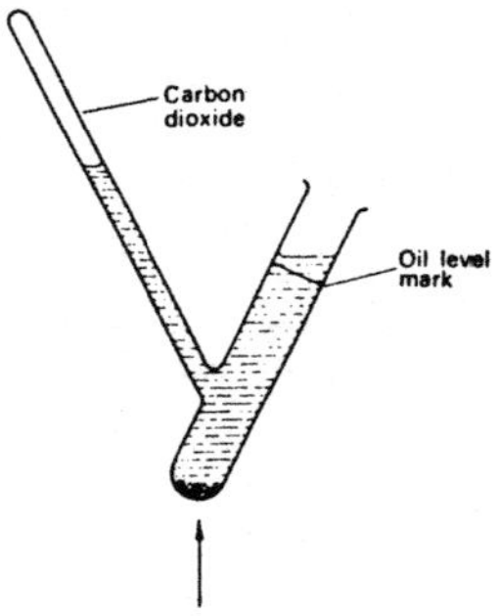

FIG. 10 Shapiro and Brannock's apparatus for the determination of carbon dioxide in silicate rocks.

Determination of Carbon Dioxide

The simplest procedures for determining carbon dioxide are based upon a measurement of the volume of gas liberated on heating the rock material with mineral acid. These procedures are not the most accurate, and gravimetric procedures are generally to be preferred. A gas chromatographic procedure can be used for those rocks containing only traces of carbon dioxide.

METHOD OF SHAPIRO AND BRANNOCK[2]

Following Fahey[3], Shapiro and Brannock have devised a special apparatus (Fig.10) in which the carbon dioxide is liberated and measured. It consists of a boiling tube with a sealed-in side arm. The size and shape represents a compromise between accuracy (requiring a long thin side arm as the measuring tube) and loss of carbon dioxide by solution in the oil (which increases with the length of travel of the gas bubbles). This apparatus should not be used for carbonate rocks, and an upper limit of about 6 per cent carbon dioxide has been suggested. Some samples, particularly those that have been crushed or ground using iron machinery, may contain small amounts of tramp iron. This gives gaseous hydrogen on heating with mineral acid, and an error of as much as 0.2 per cent can be obtained. This error can be avoided by adding mercuric chloride to the sample, converting any tramp iron to ferrous iron without liberation of hydrogen.

Method

Reagents: Mercuric chloride, dry powder.
Mercuric chloride solution, saturated aqueous.
Paraffin oil, a heavy fairly viscous oil. That sold in the United Kingdom as "liquid paraffin" is suitable.
Carbon dioxide standard, a silicate rock that has been analysed by a gravimetric method, containing about 2 per cent carbon dioxide is recommended for the calibration of the side arm of the apparatus.

Procedure. Weigh a suitable amount of the finely powdered rock material into the dry tube, taking care that no material adheres to the sides. Add mercuric chloride solution and tap the tube gently to remove all air bubbles. Fill the tube to the mark with oil, and rotate the tube to displace all the air from the side arm and finally bring the tube to a position with the side arm pointing vertically upwards.

Add 2 ml of 6 N hydrochloric acid to the tube and immerse the lower part of the apparatus in an oil bath maintained at a temperature in the range 110-120°. Keep the tube at this temperature for at least 5 minutes and until all the liberated bubbles of gas have collected in the side arm. Remove the tube from the water bath and cool the side arm by running cold water over it for a short while. Using a millimeter scale, measure the length of the side arm occupied by carbon dioxide gas.

Repeat this operation using known weights of the carbon dioxide standard in order to calibrate the side arm.

METHOD USING SCHROTTER'S APPARATUS

This apparatus, shown in Fig. 11 is designed to determine the loss in weight that occurs when carbon dioxide is removed from carbonate minerals. The method is not capable of great accuracy, although with experience, precise and reasonably accurate results can be obtained for carbonate rocks and minerals. It should not be used for samples containing less than about 10 per cent carbon dioxide. The greatest error arises from the failure to completely dry the gases that are evolved on warming.

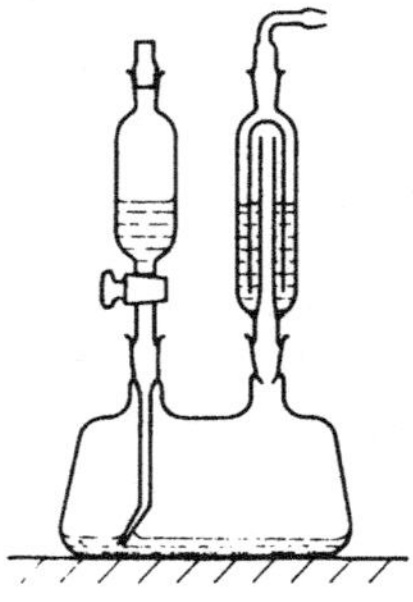

FIG. 11 Schrotter's apparatus for the determination of carbon dioxide

Method

Apparatus: This consists of a flat-bottomed, 100-ml flask fitted with a tap funnel containing phosphoric acid and a gas bubbler for concentrated sulphuric acid. The use of PTFE sleeves between the ground glass joints is advantageous.

Reagents: Phosphoric acid, diluted 1 + 2.

Procedure. Clean and dry the complete apparatus as shown in Fig. 11 and weigh the empty flask. Transfer a 1-5 g portion of the ground sample material to the flask, and reweigh to give the weight of the sample used. Add about 5 ml of water to the flask, and reassemble the apparatus with 5 ml of phosphoric acid in the tap funnel and 2 ml of concentrated sulphuric acid in the bubbler. Close both the bubbler and the tap funnel with the correct stoppers, and weigh the assembled apparatus.

Remove the stoppers from the bubbler and the tap funnel, and open the tap to allow the phosphoric acid to run slowly into the flask, taking care that the carbon dioxide is evolved slowly, and bubbles gently through the sulphuric acid, interrupting the supply of phosphoric acid, and cooling the flask if the reaction becomes too vigorous. Finally close the tap of the funnel just before the phosphoric acid is completely transferred to the flask.

When the evolution of gas has ceased, gently swirl the flask, place it on a hot plate, and heat the contents to a temperature of about 90°, but do not allow to boil. Allow the apparatus to cool, open the tap and, using a water pump, draw a current of dried air through the apparatus for a period of 2 minutes to displace carbon dioxide remaining in the flask. Replace the stoppers and reweigh the assembled apparatus. Report the percentage loss in weight as the carbon dioxide content of the sample material.

GRAVIMETRIC METHOD USING ABSORPTION TUBES AND ACID DECOMPOSITION

Where greater accuracy is required the following gravimetric method may be used. Carbon dioxide evolved on decomposition with phosphoric acid is dried by passage through an absorption tube packed with anhydrous magnesium perchlorate, and the carbon dioxide absorbed in a weighed tube containing soda-asbestos. A copper phosphate layer or packing is used to retain any hydrogen sulphide that may be formed by action of the acid on certain sulphide minerals that may be present in the sample material. The apparatus required is shown in Fig. 12.

Borgstrom(4) has recommended that for certain scapolites where the carbon dioxide content is not readily liberated with hydrochloric acid (or, for that matter, with phosphoric acid), a mixture of hydrochloric and hydrofluoric acids should be used. Clearly this is not possible in the glass apparatus commonly used for this determination. Low reagent blanks are commonly obtained when phosphoric acid is used for the sample decomposition, amounting in most cases to no more than 0.1 mg per hour. This reagent was recommended by Morgan(5) in place of hydrochloric acid, then in use in most laboratories. Where low blank values are important, as for example, where the rock material available is limited and where semi-micro working is necessary, the closed circuit gas recycling technique of Jeffery and Wilson(6) may be used.

For samples containing appreciable quantities of sulphide minerals, hydrochloric acid should be used, and mercuric chloride solution (20% w/v) added to the reaction vessel, as described by Corbett and Mizon.(7)

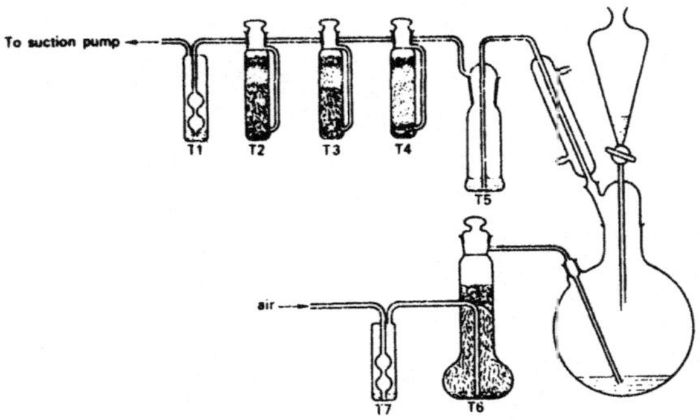

FIG.12 Apparatus for the gravimetric determination of carbon dioxide.

Method

Apparatus: The apparatus (Fig. 12) consists of a 150-ml, round-bottomed flask, tap funnel for the addition of diluted phosphoric acid, a water-cooled reflux condenser and a series of vessels T1 to T7. T2, 3 and 4 are Nesbitt bulbs, T1 and 7 are Arnold bulbs. T5 a 125-ml Drechsel bottle and T6 a Midvale bulb.

The Arnold bulbs contain syrupy phosphoric acid and are used to indicate the rate of flow of gas through the apparatus, the Drechsel bottle contains concentrated sulphuric acid to remove the bulk of the water from the gas, thus prolonging the life of the drying tube T4. The acid level in the Arnold bulbs and the Drechsel bottle should be such as just to cover the end of the delivery tube. T6, used to remove carbon dioxide from the incoming air, is packed with cotton wool to cover the end of the delivery tube and with soda asbestos to the level indicated in the figure. T2, whose increase in weight is used to indicate the weight of carbon dioxide absorbed, is packed with soda-asbestos, covered with a ¾-inch layer of anhydrous magnesium perchlorate. T3, used to remove hydrogen sulphide from the gas stream, is packed with copper phosphate covered with a layer of anhydrous magnesium perchlorate, and T4 contains magnesium perchlorate only. In each Nesbitt tube, the reagent is kept in place with the aid of a plug of cotton wool. The tubes are connected together with butt joints. A micro burner is used to heat the flask.

Reagents: Phosphoric acid, diluted 1 + 1, freshly boiled and allowed to cool.
Magnesium perchlorate, anhydrous, 14-22 mesh.
Soda-asbestos, 8-14 mesh.
Cupric phosphate, granular, specially prepared by pressing a stiff paste of the powdered reagent in 1 per cent starch solution through a sheet of perforated metal, and drying in an electric oven at 110°.

Procedure. With all the taps fully open, start the pump pulling air through the apparatus and observe that the two Arnold bulbs have air at about the same rate of 1 to 2 bubbles per second passing through them. Beginning at T6, close and then open each tap in turn and note that with each tap closed the current of air through the Arnold bulb T1 ceases to flow, indicating that each tap is operating correctly. Finally close all the taps and remove the Nesbitt Tube T2 to a balance case and allow to stand for 30 minutes. Carefully wipe the outside of the tube with a clean, dry, soft cloth , open and then close the tap to ensure atmospheric pressure within the tube, allow to stand in the balance case for a further 5 minutes and then weigh.

Weigh also a sample portion of from 1 to 5 g of the sample material, depending upon the anticipated carbon dioxide content and, using a little boiled water, rinse it into the flask. Transfer approximately 50 ml of diluted phosphoric acid to the tap funnel. Reassemble the apparatus as shown in the figure but with taps T1-7 open once more and with a clip between the suction pump and T1 restricting the flow of air through the apparatus fully closed.

Start the pump and open the screw clip to produce a steady stream of bubbles through the Arnold bulbs. Open the tap of the funnel to admit the phosphoric acid slowly to the flask, temporarily cutting off the current of air if a rapid evolution of carbon dioxide occurs. Close the tap of the tap funnel before all the acid has been added.

When all evolution of carbon dioxide has apparently ceased, bring the solution to the boil and boil gently whilst keeping the air current flowing for a further period of 45 minutes to sweep all the carbon dioxide out of the flask and into the Nesbitt tube. Finally turn off the burner, close all the open taps and open the tap of the tap funnel. Remove the Nesbitt tube T2 to a balance case, allow it to stand as before, wipe the tube and record the increase in weight due to the absorbed carbon dioxide.

Determine also a blank value by proceeding as above but omitting the sample material.

GRAVIMETRIC METHOD USING ABSORPTION TUBES AND THERMAL DECOMPOSITION

The difficulty of liberating all the carbon dioxide from some of the scapolite minerals by boiling with diluted mineral acid can be avoided by using thermal decomposition of the sample. The determination of carbon dioxide may then be combined with that of total water by measuring the increase in weight of separate tubes packed with anhydrous magnesium perchlorate and with soda-asbestos, as described in detail by Riley[8]. He attributed the high blanks that he first obtained for carbon dioxide to the formation of acidic oxides of nitrogen, and subsequently avoided this by replacing the current of air through the apparatus by a current of nitrogen gas from a cylinder supply, and by incorporating a short length of silica tube packed with copper wire and heated to a temperature of 700-750°. Interference from oxides of sulphur was prevented by the inclusion in this short length of silica tube of a layer of silver pumice, and by the use of a bubbler containing a saturated solution of chromium trioxide in phosphoric acid.

The main furnace is heated to a temperature of 1100-1200°, and the samples are inserted into the hot zone of the furnace by means of a stainless-steel push rod. Care must be taken not to push the samples directly into the very hottest part if they contain much water or easily decomposed carbonate minerals such as siderite or magnesite. Failure to observe this precaution can result in mechanical loss of the sample from the containing boat, and to loss of carbon dioxide by incomplete absorption following the too rapid evolution.

Most carbonate minerals readily lose carbon dioxide at 1100°, but strontianite, scapolite and witherite require to be heated to 1200°. The complete recovery of carbon dioxide from the latter may require up to 3 hours heating at this temperature for complete expulsion of all the carbon dioxide.

Method

Apparatus: For details of the construction of the two furnaces used by Riley[8] the original paper should be consulted.

High-temperature furnace (1100-1200°). The furnace tube consists of a 45-cm length of silica tube of internal diameter 1.8 cm, with an insertion device to enable the samples contained in 2-ml alumina boats to be pushed gradually into the heated zone of the furnace. For the majority of rock samples a temperature of 1100° should be used, and only if the sample contains minerals that are difficult to decompose (such as staurolite, cordierite, topaz or the carbonate minerals listed above) should the temperature be raised to 1200°. This is somewhat in excess of the recommended temperature for silica tubing, and with repeated use at this temperature, distortion will gradually occur. Each tube should last for at least three months with regular use at 1100°.

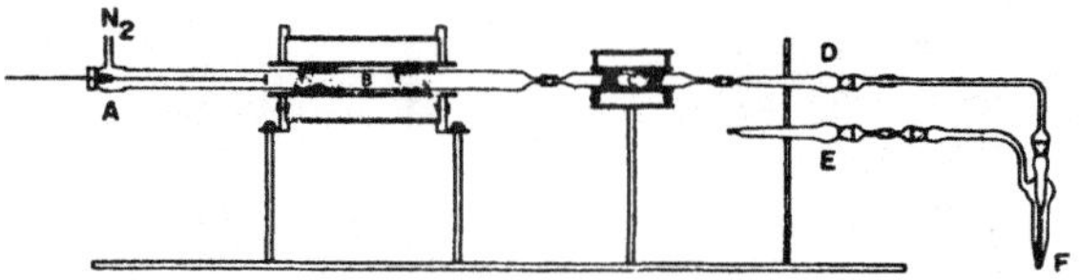

FIG.13 Apparatus for determining carbon dioxide and water. A-insertion device, B-silica combustion tube, C-silica tube containing copper wire and silver pumice, D-absorption tube for water, E-absorption tube for carbon dioxide, F-bubbler.

Low-temperature furnace (700-750°). The furnace tube consists of a 10-cm length of tubing, 1 cm internal diameter, packed with alternate layers of copper wire and silver pumice, held in place with plugs of asbestos wool. Both ends of the tube are fused to 3-cm lengths of silica tubing of 5 mm external diameter. At least once a week a current of coal gas should be passed through the heated tube to reduce any copper oxide to the metal. The life of the tube is about 3 months of continual use with rocks of low sulphur content.

Gas purification train. Riley has recommended using nitrogen of normal quality (ie not "white spot" - oxygen-free quality), from a cylinder fitted with a twin-stage regulator and a needle valve. A bubbler filled with sulphuric acid is used to indicate the flow of gas, regulated to about 3 litres per hour. The nitrogen is purified by passage through tubes containing soda lime, fused calcium chloride and finally anhydrous magnesium perchlorate.

Absorption tubes, those shown in Fig. 13 are those designed by Riley. They are rather smaller than many absorption tubes in common use, and should be replaced with larger tubes for samples containing much carbon dioxide or water. Tube D contains anhydrous magnesium perchlorate and tube E, soda asbestos.

Bubbler, with samples containing more than 0.5 per cent sulphur, the bubbler, filled with a saturated solution of chromium trioxide in syrupy phosphoric acid, should be interposed between the water absorption tube and the carbon dioxide absorption tube. Its side arm contains anhydrous magnesium perchlorate.

Reagents:

Anhydrous magnesium perchlorate, 14-22 mesh.
Soda-asbestos, 8-14 mesh.
Chromium trioxide.
Copper wire.
Silver pumice, prepared by evaporating 14-mesh pumice with concentrated silver nitrate solution and igniting strongly.
Magnesium oxide, ignited as required.

Procedure. Before use, check that the two furnaces are at their correct temperatures and adjust the flow of nitrogen to about 3 litres per hour. Allow the gas to pass through the apparatus for about 20 minutes, then remove the absorption tubes and allow to stand in a balance case for 30 minutes. Wipe them carefully with a clean, dry, soft cloth. Open and then close each tap, and after standing in a balance case for a further 5 minutes, weigh each tube separately.

Weigh a 0.5-1.5 g portion of the finely powdered rock material into a previously ignited 2-ml alumina boat lined with a piece of nickel foil. If the sample contains much fluorine or sulphur, cover the sample with a layer of freshly ignited magnesium oxide. Insert the boat into the end of the combustion tube (A), replace the insertion device and allow 5 minutes for the flow of nitrogen to sweep all introduced air out of the apparatus. Reconnect the absorption tubes and push the sample into the furnace using the push rod. Samples containing a lot of water or easily decomposed carbonate mineral should be pushed only into the cooler region outside the hot zone, and finally be inserted into the hottest part of the furnace tube when the decomposition is nearly complete. After a heating period of 30 to 40 minutes in the hottest zone, remove the absorption tubes, wipe them, allow to stand and weigh as before.

Carry out a blank determination in the same way, but without the sample material, both before and after the first determination, and after the end of each batch of samples. These reagent blanks should normally amount to no more than about 0.1 mg water per hour and 0.2 mg carbon dioxide per hour respectively. Higher blank values for carbon dioxide usually indicate that the packing in the low-temperature furnace tube is exhausted, which should then be regenerated as described.

TITRIMETRIC DETERMINATION OF CARBON DIOXIDE

In the gravimetric procedures described above, the blank values can be of the same order of magnitude as the carbon dioxide evolved from some samples. To avoid this difficulty Read[9] has described a titrimetric procedure based upon a non-aqueous titration. As with some other methods, the carbon dioxide is liberated by heating with diluted orthophosphoric acid. The liberated gas is carried in a stream of nitrogen through absorbents to remove water and hydrogen sulphide and is then absorbed in a 5 per cent solution of monoethanolamine in dimethylformamide containing thymolphthalein as indicator. The carbon dioxide is titrated directly with a solution of tetrabutylammonium hydroxide in toluene.

A comparable method was described by Sen Gupta[10] in which the absorption medium is a solution of monoethanolamine in acetone which contains a suitable excess of sodium methoxide together with phenolphthalein as indicator. The excess base is titrated with a standard solution of benzoic acid in methanol.

In a somewhat simpler titrimetric method by Bush[11], the carbon dioxide liberated with phosphoric acid was absorbed in 0.1 N barium hydroxide solution, precipitating the sparingly soluble barium carbonate. The excess barium hydroxide was titrated with standard hydrochloric acid using thymolphthalein as indicator. Lindner[12] has however commented on the incompleteness of absorption of carbon dioxide in baryta, whilst Belcher *et al*[13] noted significant errors in the titration of excess barium hydroxide.

DETERMINATION OF CARBON DIOXIDE BY GAS CHROMATOGRAPHY

Even with the more refined techniques using gas-absorption tubes, blank values of the order of 0.1-0.2 mg per hour can be obtained. These give rise to uncertainty where the rock material contains 0.02 per cent carbon dioxide or less. For concentrations in this range the procedure described by Jeffery and Kipping[1] using gas chromatography will give a more accurate and also positive result. The sample material is boiled under reflux with phosphoric acid and the liberated gases are transferred to a chromatograph in a stream of carrier gas, hydrogen being suggested. The separation of carbon dioxide from oxygen and nitrogen takes place on a column packed with silica gel, although activated charcoal could also be used. A thermal conductivity detector is used.

Some silicate rocks were found to contain 50 ppm carbon dioxide, and smaller quantities could undoubtedly be recovered and determined in this way. Jeffery and Kipping[1] suggested that as little as 5 ppm could be detected, but Carpenter[14] has given a lower limit of 0.2 ppm carbon dioxide. The original papers should be consulted for details.

Determination of Carbon ("non-carbonate carbon")

Graphitic carbon is sometimes determined by measuring the loss on ignition although this method has little to commend it except simplicity and speed. For graphite concentrates the loss on ignition is an important industrial parameter, but it should not be used for silicate rocks as a measure of the carbon content. The only procedures that can be recommended for silicates are those based upon total oxidation of the carbon followed by quantitative determination of the carbon dioxide produced. This may be carried out after decomposing any carbonate minerals present,[15] or at the same time as the decomposition of carbonates. In this latter case a separate carbonate determination must be made and the result calculated as carbon subtracted from the result for total carbon.

Oxidants suggested for the conversion of carbon and carbon compounds include chromic acid, potassium dichromate, sodium peroxide and oxygen. Oxidation with chromic acid[16] is not always complete, as volatile carbon compounds may distil from the acid mixture and give rise to low results. Some improvement can be obtained by using a "closed-circulation" system,[6] but for many samples an alternative procedure should be used. The same objection can be made to the use of potassium dichromate in sulphuric acid solution, which reacts as chromic acid.

Fusion with sodium peroxide in a closed bomb has been used[17] for determining the total carbon in marine shales. Samples containing more than 30 per cent organic carbon are easily oxidised, but those containing less generally require the addition of a combustion aid such as powdered aluminium or magnesium to provide the heat necessary for ensuring complete oxidation of the carbon. Any sulphides present will be oxidised to sulphates and do not interfere. Only small samples (0.4 g or less) can be taken, and as the sodium peroxide tends to absorb carbon dioxide from the air, a blank correction is necessary. This method is suitable only for those samples containing 0.2 per cent or more of carbon.

COMBUSTION TUBE METHODS

The apparatus shown in Fig. 13 using a combustion tube for the determination of carbon dioxide, can be used in a somewhat similar way for the determination of total carbon in rocks and minerals. By heating the sample material with a suitable flux, such as potassium dichromate, lead chromate or vanadium pentoxide, complete oxidation takes place and carbon dioxide from both carbonate minerals and non-carbonate carbon is evolved and collected. The carbonate content of the sample must be determined separately. Any sulphides present will be oxidised, and both sulphur dioxide and trioxide formed. These are removed prior to collection of the carbon dioxide in the weighed absorption train.

The difficulties that occur in the presence of much sulphide minerals (20-30% S as sulphides of zinc, lead or iron) are discussed by Corbett and Mizon.[7] These are not likely to be encountered in the analysis of silicate or carbonate rocks.

The determination of non-carbonate carbon in clay materials poses a number of special problems. These were investigated by Ferris and Jepson[18] who recommended a combustion tube method. The clay material is heated at a temperature of 900° in a stream of purified oxygen to convert all carbonaceous material to carbon dioxide, which is absorbed in a solution of monoethanolamine in dimethylformamide. The determination is completed by titration with a solution of tetra-butylammonium hydroxide in a mixture of toluene and methanol with phenolphthalein as indicator - a method similar to that described by Read.[9] Impurities in the gas stream include sulphur dioxide, silicon tetrafluoride and hydrogen fluoride. These interfere with the titration and must be removed.

Combustion in a stream of oxygen using a high-frequency induction furnace was described by Sen Gupta[10] for total carbon in rocks and meteorites. The determination of the evolved carbon dioxide was completed by non-aqueous titration as described earlier. A high-frequency induction furnace has also been used in combustion methods in which the volume of the liberated carbon dioxide was determined and in which the determination was completed by thermal conductivity measurement[19].

Determination of Organic Matter

The separation of organic material from both consolidated and unconsolidated sediments by extraction with organic solvents is an empirical procedure. Such factors as the weight of the sample in relation to the volume of solvent, the grain size to which the sample has been crushed or ground, the length of time during which the sample is extracted and the particular solvent used can all materially affect the results.[20] Extractants that have been used include chloroform, benzene and mixtures of benzene with methanol and acetone. The extracted material, after removal of the solvent, is often known as the "bituminous fraction" or more commonly bitumen. It may resemble petroleum and can usually be separated into its component hydrocarbons and other organic compounds, which are then determined by techniques familiar to the organic chemist, for example gas chromatography, infrared spectroscopy and mass spectrometry.

A further fraction can be separated by extraction of the residue with aqueous sodium hydroxide solution. This fraction is known as the "humic carbon content", "humic and fulvic acids", or simply as "humic acids". It may be determined by dichromate oxidation as described by Smeral.[21]

The greater part of the organic content of sedimentary rocks often remains with the mineral fraction after these extractions, possibly in an altered form. This solvent-insoluble fraction, or <u>kerogen</u>,[22] (also referred to as kerabitumen[23]) can be isolated by the stepwise removal of all the other constituents of the sediment. Carbonates are removed by dissolution in dilute hydrochloric acid, and quartz and silicates by treatment with hydrofluoric acid. Pyrite can be troublesome as it is frequently abundant in carbonaceous sediments. A separation with heavy media is sometimes effective, otherwise a zinc-hydrochloric acid reduction is used. The insoluble organic fraction remaining after this treatment may be contaminated with small amounts of zircon and other similar detrital minerals. The original papers should be studied for details of the analysis procedures adopted.[24,25,26] For review of the chemical methods in use, see eg Saxby.[23]

The separation and determination of organic compounds (hydrocarbons, fatty acids, amino acids, carbohydrates, pigments etc) is beyond the scope of this book. Standard texts[27,28] should be consulted.

References

1. JEFFERY P G and KIPPING P J., Analyst (1962) 87, 379
2. SHAPIRO L and BRANNOCK W W., Analyt. Chem. (1955) 27, 1796
3. FAHEY J J., US Geol. Surv. Bull. 950, p.139,1946
4. BORGSTROM L H., Zeit. Anal. Chem. (1914) 53, 685
5. MORGAN G T., J. Chem. Soc. (1904) 85, 1001
6. JEFFERY P G and WILSON A D., Analyst (1960) 85, 749
7. CORBETT J A and MIZON K J., Chem. Geol. (1976) 17, 155
8. RILEY J P., Analyst (1958) 83, 42
9. READ J I., Analyst (1972) 97, 134
10. SEN GUPTA J G., Anal. Chim. Acta (1970) 51, 437
11. BUSH P R., Chem. Geol. (1970) 6, 59
12. LINDNER J., Mikrochem. (1936) 20, 209
13. BELCHER R, THOMPSON J H and WEST T S., Anal. Chem. Acta (1958) 19, 309
14. CARPENTER F G., Analyt. Chem. (1962) 34, 66
15. ELLINGBOE J L and WILSON J E., Analyt. Chem. (1964) 36, 435
16. DIXON B E., Analyst (1934) 59, 739
17. FROST I C., US Geol. Surv. Prof. Paper 424-C, p.B-480, 1961
18. FERRIS A P and JEPSON W B., Analyst (1972) 97, 940
19. CAUWET G., Chem. Geol. (1975) 16, 59
20. TOURTELOT H Aand FROST I C., US Geol. Surv. Prof. Paper 525-D, p.D-73, 1966
21. SMERAL J., Pr. Vysk. Ustavu CS Naft. Dolu. Publ. (1965) 24, 27
22. FORSMAN J P and HUNT J M., Geochim. Cosmochim. Acta (1958) 15, 170
23. SAXBY J D., Chem. Geol. (1970) 6, 173
24. DOUGLAS A G, EGLINTON G and MAXWELL J R., Geochim. Cosmochim Acta (1969) 33, 579
25. NISSENBAUM A, BAEDECKER M J and KAPLAN I R., Geochim. Cosmochim. Acta (1972) 36, 709
26. BROWN F S, BAEDECKER M J, NISSENBAUM A and KAPLAN I R, Geochim. Cosmochim. Acta (1972) 36, 1185
27. "Organic Geochemistry". Editor E Ingerson, Monograph No 16, Earth Science Series, Pergamon/MacMillan, New York 1963
28. "Organic Geochemistry". Editor G Eglinton and M T J Murphy, Longman/Springer-Verlag, London etc. 1969.

CHAPTER 16

Chlorine, Bromine and Iodine

Gravimetric methods, based upon the precipitation of silver chloride, have long been used for the determination of chlorine. It is however difficult to collect and weigh the small quantity of silver chloride obtained from rocks containing 0.05 per cent or less of chlorine. For these rocks a titrimetric procedure can often be used but this is imprecise below about 0.01 per cent. A number of spectrophotometric procedures have been described, and these can be adapted for rocks containing small amounts of chlorine.

The introduction of ion-selective electrodes has given rise to new methods of analysis and adaptations of older ones for a number of elements. Chlorine is one of these. In the procedure by Haynes and Clark,[1] a chloride electrode is used to determine the chloride ion activity in a method similar to the classical technique involving the precipitation of silver chloride. The equivalence point in the silver nitrate titration is determined directly by measuring the electrode potentials of excess silver with a silver electrode.

The simple and rapid determination of both chlorine and fluorine in granitic rocks using ion-selective electrodes has been described by Haynes,[2] whilst Akaiwa et al[3] used a chloride electrode as a detector following an anion-exchange separation.

Determination of Water-soluble Chlorine

The presence of water-soluble chlorine in more than trace amounts is usually indicative of halite, kalsilite, carnalite or other evaporite mineral in the rock material. To determine the water-soluble chlorine, heat a suitable weight of the sample material to boiling with about 100 ml of water. Boil for 30 minutes, filter through a close-textured paper and wash the residue with warm water. Discard the residue. Add 5 ml of concentrated nitric acid to the filtrate and precipitate the chlorine with silver nitrate solution as described below for acid-soluble chlorine.

Determination of Acid-soluble Chlorine

Many of the chlorine minerals are completely decomposed by heating with dilute nitric acid. Unless the petrological examination has shown the presence of scapolite and similar minerals that are not decomposed in this way, this can be used in place of the more lengthy procedure described below for the determination of total chlorine. Loss of chlorine may occur if the acid strength is too high or if the heating is prolonged.

If the amount of acid-soluble chlorine in the rock is too small to permit the silver chloride to coagulate, it may be determined turbidimetrically with a spectrophotometer or by comparing the turbidity with that obtained by adding standard sodium chloride solution to an acid solution of silver nitrate.

Method

Reagents: Silver nitrate solution, dissolve 5 g of silver nitrate in 100 ml of water containing 1 ml of concentrated nitric acid.

Procedure. Accurately weigh 2 g or more of the finely powdered rock material into a 150-ml beaker and carefully add 2 ml of concentrated nitric acid and 40 ml of water. Cover the beaker with a clock glass and transfer to a steam bath. When any effervescence (if carbonate minerals are present) has subsided, boil for 2 minutes, but no longer, and allow to cool. Filter the liquid through a close-textured filter paper and wash the residue well with hot water. Combine the filtrate and washings in a 400-ml beaker and allow to cool to room temperature.

Shield the beaker from light either by wrapping it in stout brown paper, or by covering the outside with lacquer, and cover a clock glass in the same way. Now add a slight excess of silver nitrate solution and replace the beaker on the steam bath to coagulate the precipitated silver chloride. Transfer the beaker to a dark cupboard and allow to cool overnight. Collect the precipitate on a small, weighed, sintered-glass crucible of medium porosity, and wash with cold 0.16 N nitric acid. Finally wash twice with cold water, dry in an electric oven set at a temperature of about 150° and weigh as silver chloride (Note 1).

Note: 1. The sintered-glass crucibles may be cleaned by passing a small quantity of concentrated ammonia solution through them, followed by water to remove silver salts, nitric acid to remove silver stains and finally water again to remove nitric acid.

Determination of Total Chlorine

For this determination the sample material is fused with alkali carbonate and the chlorine precipitated as silver chloride from the acidified aqueous extract. The determination may be combined with those of sulphur, barium, zirconium, chromium and vanadium as described by Bennett and Pickup.[4]

The addition of potassium nitrate to the flux serves to increase the fluidity of the melt, and also to oxidise all ferrous iron present in the sample. Some part of the sodium carbonate can be replaced with potassium carbonate to reduce the temperature required for the fusion, and borax glass can be added to assist the decomposition of refractory oxide minerals.

Method

Reagents: Sodium carbonate wash solution, dissolve 10 g of anhydrous reagent in 500 ml of water.

Procedure. Accurately weigh approximately 2 g of the finely powdered rock material into a large platinum crucible and mix with 0.5 g of potassium nitrate and 10 g of anhydrous sodium carbonate. Fuse the contents of the crucible, first over a Bunsen burner and then over a Meker burner for a total of about 1 hour and then allow to cool. Extract the melt into hot water containing 2 drops of ethanol to reduce any manganate formed in the fusion. Allow to cool and collect the residue on an open-textured filter paper and wash well with warm sodium carbonate wash solution. Discard the residue. Combine the filtrate and washings in a 600-ml beaker and dilute to about 400 ml with water.

Add 2 or 3 drops of methyl red indicator solution, 2 N nitric acid until the red form of the indicator is obtained, and then add 1 ml in excess. Stir vigorously to remove the liberated carbon dioxide. Precipitate the chlorine present in the solution and complete the determination as described above for acid-soluble chlorine. Collect, dry and weigh the silver chloride.

Note: 1. Provided that the precipitation is made in a sufficiently large volume, and the solution is made only faintly acid, silica will not be precipitated. If silica does separate, it should be collected with the silver chloride. The latter can then be dissolved in a little hot dilute ammonia solution, filtered from silica and re-precipitated by adding dilute nitric acid.

Spectrophotometric Determination of Chlorine

Kuroda and Sandell[5] have described a procedure using the formation of a colloidal suspension of silver sulphide in aqueous ammoniacal solution. The chlorine present in the silicate rock is recovered by fusion with alkali carbonate, the melt extracted with water and the chlorine present in the filtrate precipitated as silver chloride by the procedure given in the preceding section. The silver chloride is collected on a small sintered-glass filter curcible, washed and dissolved in a small volume of aqueous ammonia. Sodium sulphide is added and the solution diluted to 10 ml for photometric measurement.

A photometric procedure is described by Bergmann and Sanik,[6] in which mercuric thiocyanate is added to the acidified filtrate from an alkaline fusion. Any chloride ions present displace thicyanate ions, which are then free to form the intense red-coloured ferric thiocyanate (Note 1). The procedure given below is based upon a modification of this described by Huang and Johns,[7] using the sample solution prepared for their determination of fluorine.

Method

Reagents: Ferric solution, dissolve 12.0 g of ferric ammonium sulphate, $FeNH_4(SO_4)_2.12H_2O$ in 100 ml of 9 M nitric acid.

Mercuric thiocyanate saturated solution, shake 0.35 g of mercuric thiocyanate $Hg(CNS)_2$ with 100 ml of 95 per cent ethanol and allow to stand overnight to allow the precipitate to settle.

Standard sodium chloride stock solution dissolve 0.824 g of dried sodium chloride in 500 ml of water. This solution contains 1 mg chlorine per ml.

Standard sodium chloride working solution, dilute a 5 ml aliquot of the stock solution to 500 ml with water. This solution contains 10 μg chlorine per ml.

Procedure. Decompose the rock material and obtain the chlorine in dilute nitric acid solution as described in the chapter dealing with fluorine. Transfer a 10- or 20-ml aliquot to a 25-ml volumetric flask and add 2 ml of the ferric ammonium sulphate solution and 2 ml of the mercuric thiocyanate solution. Mix the solution, dilute to volume and mix again. Measure the optical density of the solution against water in 1-cm cells with the spectrophotometer set at a wavelength of 460 nm. Measure also the optical density of a reagent blank solution prepared in the same way but omitting the sample solution.

Calibration. Transfer 2 to 10 ml aliquots of the standard sodium chloride solution containing 20 to 100 μg chlorine to separate 25-ml volumetric flasks, dilute each to

20 ml with water. Add 2 ml of ferric ammonium sulphate solution and 2 ml of the mercuric thiocyanate solution, dilute to volume and mix well. Measure the optical density of each solution as described above (Note 2).

Notes: 1. Bromide and iodide ions undergo a similar reaction with mercuric thiocyanate; they are however unlikely to be present in silicate rocks in amounts sufficient to interfere.

2. The calibration graph given by this procedure should be checked with each new solution of the ferric ammonium sulphate and mercuric thiocyanate reagent.

The spectrophotometric procedure described here has been combined with pyrohydrolysis by Farzaneh and Troll[8]. The dried rock material was mixed with silica gel in the ratio 1:1, then with 20 g of iron metal chips in a silica crucible and hydrolysed in an induction furnace under a flow of steam for 10 minutes. The hydrogen chloride liberated was collected in 0.02 M aqueous sodium hydroxide and diluted to volume prior to photometric determination.

Determination of Bromine

A procedure for the determination of bromine in silicate rocks has been given by Behne.[9] In this a 1-2 g portion of the rock material is fused with sodium hydroxide and the melt extracted with water. Insoluble material is filtered off and the filtrate evaporated to dryness after converting all excess hydroxide to sodium carbonate and bicarbonate with an excess of gaseous carbon dioxide. The dry salts are extracted with ethanol, when any sodium bromide and (iodide) passes into organic solution. The ethanolic solution is filtered from the bulk of the sodium carbonate and the solvent removed by evaporation. The dry residue contains sodium bromide which is converted to bromate by oxidation with hypochlorite:

$$Br^- + 3HOCl = BrO_3^- + 3H^+ + 3Cl^-$$

Potassium iodide is then added:

$$BrO_3^- + 6I^- + 6H^+ = Br^- + 3H_2O + 3I_2$$

liberating six equivalents of iodine for each equivalent of bromine (method of Van der Meulen[10]). The liberated iodine is titrated with a standard solution of sodium thiosulphate in the presence of ammonium molybdate as catalyst.

Sodium hydroxide is used for the decomposition in preference to potassium hydroxide, as sodium bromide is more soluble in ethanol (11.8 g per litre at 30°) than potassium bromide (3.2 g per litre at 30°).

Behne reported that only about 90 per cent of the bromine could be recovered from standard bromide solutions. As the Van der Meulen titration procedure is quantitative, it must be presumed that the loss of bromine occurs at the extraction stage, although this was not confirmed by Behne. Iodine reacts in the same way as bromine, and the results obtained will include any iodine present in the rock sample. The sensitivity of the method is about 1 μg Br, which is thus unsuitable for those silicates containing less than 1 ppm Br. For rocks containing less than this, a more sensitive procedure must be used - such as that described by Filby,[(11)] with a sensitivity of 0.001 ppm Br, using a neutron activation technique.

Determination of Iodine

Crouch[(12)] has described a procedure for determining iodine in silicate rocks based upon the blue colour given by free iodine with starch. The rock material is decomposed by fusion with sodium hydroxide and the iodine isolated by co-precipitation with silver chloride. Bromine is used to oxidise iodide ions to iodate, which is then reacted with cadmium iodide to give free iodine according to the equation:

$$IO_3^- + 5I^- + 6H^+ = 3I_2 + 3H_2O$$

The limit of detection is of the order of 0.15 μg I, indicating that large-sized samples will be required for most silicate rocks.

A method for iodine has been described by Schneider and Miller[(13)] using the catalytic action of iodine on the reaction between cerium (IV) and arsenic (III). Silica and manganese must be removed. The sensitivity is given as 0.2 ppm I.

A somewhat simpler method, based upon an extraction of elemental iodine into carbon tetrachloride, was reported by Grimaldi and Schnepfe.[(14)] The sample is decomposed by sintering with a mixture of sodium carbonate, potassium carbonate and magnesium oxide. Iodide in the aqueous extract is oxidised to iodate with alkaline permanganate and reduced to iodide with stannous sulphate. After adding sodium nitrite and urea, elemental iodine is extracted into carbon tetrachloride solution and the optical density measured at a wavelength of 517 nm. The lower limit of detection is of the order of 1 ppm, making the method suitable only for those rocks that are particularly rich in iodine.

This sample decomposition procedure described by Grimaldi and Schnepfe[(14)] can be used with an iodide-sensitive electrode as suggested by Ficklin[(15)]. Sulphide ion interferes and must be removed by oxidation. As the iodide-sensitive electrode responds also to silver ions, these must also be removed.

References

1. HAYNES S J and CLARK A H., Econ. Geol. (1972) 67, 378
2. HAYNES S J., Talanta (1978) 25, 85
3. AKAIWA H, KAWAMOTO H and HASEGAWA K., Talanta (1979) 26, 1027
4. BENNETT W H and PICKUP R., Colon. Geol. Min. Res. (1952) 3, 171
5. KURODA P K and SANDELL E B., Analyt. Chem. (1950) 22, 1144
6. BERGMANN J G and SANIK J Jr., Analyt. Chem. (1957) 29, 241
7. HUANG W H and JOHNS W D., Anal. Chim. Acta (1967) 37, 508
8. FARZANEH A and TROLL G., Z. Anal. Chem. (1978) 292, 293
9. BEHNE W., Geochim. Cosmochim. Acta (1953) 3, 186
10. VAN DER MEULEN J H., Chem. Weekblad (1931) 28, 82
11. FILBY R H., Anal. Chim. Acta (1964) 31, 434
12. CROUCH W H Jr., Analyt. Chem. (1962) 34, 1689
13. SCHNEIDER L A and MILLER A D., Zhur. Anal. Khin. (1965) 20, 92
14. GRIMALDI F S and SCHNEPFE M M., Anal. Chim. Acta (1971) 52, 181
15. FICKLIN W H., J. Res. U S Geol. Surv. (1975) 3 (6), 753

CHAPTER 17

Chromium

No single spectrophotometric method is appropriate for all silicate rocks, where the chromium content may range from less than 1 part per million to as much as 10 per cent or more. To cover this range three photometric methods are in common use, based upon the colour given with EDTA, the colour of alkali chromate and the colour given with diphenylcarbazide. The ranges for which these three methods are commonly used are given in Table 4.

TABLE 4 USEFUL RANGES OF PHOTOMETRIC METHODS FOR CHROMIUM

Method	Cr
Diphenylcarbazide	from 0.5 ppm to 200 ppm
Alkali chromate	from 10 ppm to 5 per cent
EDTA	from 0.5 per cent to 10 per cent

All three methods depend upon making an initial separation of chromium from other elements present in silicate rocks. This is usually done by fusing the sample material with a suitable alkaline flux, often containing an oxidising agent, and leaching the melt with water. Sodium carbonate, sodium carbonate with potassium nitrate, potassium chlorate, or magnesium oxide, potassium hydroxide and sodium peroxide have all been advocated for this decomposition. Some small part of the sample material may remain unattacked, and as this residue may contain an appreciable proportion of the chromium bearing minerals, it is advisable to re-treat the residue. Even when the sample decomposition is complete, some small part of the chromium may remain with the water-insoluble residue despite extensive washing of the residue with dilute sodium carbonate solution.

The determination of chromium as alkali chromate has the advantage that once the initial alkali separation has been made, no further chemical operation other than dilution to volume is required.

Chromium (VI) reacts with sym-diphenylcarbazide in dilute acid solution to give a reddish-violet colour that can be used for the photometric determination of chromium. The reaction is of some complexity, involving both an oxidation of the reagent and the formation of a complex molecule containing chromium.[1]

Chromium can readily be determined by atomic absorption spectrometry, although there can be interference from argon with the chromium lines at 357.9 and 360.5 nm when argon-filled hollow cathode lamps are used. The sensitivity to chromium depends upon the flame conditions, with highest sensitivity in fuel-rich, air-acetylene flames. Under these conditions interference from iron and other elements is enhanced and to minimise these effects, a lean-fuel flame has been recommended. A low sensitivity is obtained in a nitrous oxide-acetylene flame,[2] but interference from other elements is again reduced. A flameless technique using a graphite furnace has been described by Schweizer[3] for chromium and other elements in carbonate rock; it is adaptable also for silicate rocks.

Ethylenediaminetetraacetic acid (EDTA) reacts with chromium (III) to form a stable, violet-coloured complex that can be used for photometric measurement at a wavelength of 540 nm. This reaction is of particular value for the determination of chromium when present as a major constituent in such minerals as chromite, but has also been used with advantage for a number of rocks and silicate minerals containing 1 per cent or more.

Determination of Chromium with EDTA

Silicate rocks are decomposed by sinter or fusion with sodium peroxide, followed by extraction of the melt with water. (See below for recovery and further treatment of any residue remaining). On filtration the alkali chromate is separated from iron and from any cobalt, copper and nickel that also form coloured complexes with EDTA. Aluminium and other elements that form colourless complexes with EDTA do not interfere, provided always that excess complexone is present. Silica does not interfere, except by precipitation from solution, imposing a limit of about 0.3 g to the weight of rock material that can be taken for the analysis.

The alkali chromate is reduced to chromium (III) at a pH of between 4 and 5 with sodium sulphite, and the complex with EDTA is formed by boiling the solution with an excess of the reagent for a minimum period of 20 minutes. Once formed the complex is very stable, and the solutions obey the Beer-Lambert Law in concentrations of up to at least 12 mg Cr per 100 ml, or 120 ppm.

Method

Reagents: Sodium peroxide.

Buffer solution, dissolve 10.5 g of sodium acetate trihydrate and 10 ml of glacial acetic acid in water and dilute to 100 ml.

EDTA solution, dissolve 1 g of the di-sodium salt of ethylenediaminetetraacetic acid in 100 ml of water.

Sodium sulphite solution, dissolve 1 g of hydrated sodium sulphite in 100 ml of water. Prepare freshly each week.

Standard chromium stock solution, dissolve 0.373 g of dried potassium chromate in water and dilute to 100 ml in a volumetric flask. This solution contains 1 mg Cr per ml.

Standard chromium working solution, dilute 50 ml of the stock solution to 250 ml with water and transfer to a clean, dry polythene bottle. This solution contains 200 μg Cr per ml.

Procedure. Accurately weigh 0.2-0.3 g of the finely powdered silicate material into a small zirconium crucible, mix with 0.6-1 g of sodium peroxide and sinter at a temperature of 550° in an electric furnace or fuse over a gas burner. Place the crucible on its side in a small beaker, cover with a clock-glass, and add sufficient water to cover the melt. Warm gently until the melt has completely disintegrated, then rinse and remove the crucible. Boil the solution for about 10 minutes to decompose the excess peroxide and filter on to a small hardened filter paper. Wash the residue several times with small quantities of a dilute sodium carbonate solution and set the filtrate aside.

Rinse the residue back into the beaker with a jet of water, add more water to bring the total volume to about 50 ml and add concentrated hydrochloric acid, warming if necessary to bring all soluble material into solution. If the initial attack of the mineral was complete, no dark-coloured gritty particles should be visible in the acid solution, and this solution can then be discarded. If black grains are discernible, collect on a small filter, wash with water, ignite in the zirconium crucible previously used for the fusion, and fuse with a little more sodium peroxide, extracting with water, boiling and filtering as before. Combine the two alkaline filtrates and, if necessary, reduce the volume by evaporation to 75-80 ml.

Using a pH meter, or pH papers, adjust the pH of the solution with 6 N hydrochloric acid to a value in the range 4 to 5, add 5 ml of the acetate buffer solution, 5 ml of the EDTA solution and 5 ml of the sodium sulphite solution. Add a few fused alumina granules, cover the beaker with a clock-glass and gently boil for about 30 minutes. In the presence of chromium, the solution gradually becomes violet in colour, usually within the first 5 minutes of boiling. After 30 minutes, cool the solution and dilute to volume with water in a 100-ml volumetric flask. Prepare also a reagent blank solution in the same way, but omitting the sample material. Measure the optical density of the rock solution relative to water in 4-cm cells, with the spectrophotometer set at a wavelength of 540 nm. Measure also the optical density of a reagent blank solution prepared in the same way as the sample solution, but omitting the sample material. Calculate the chromium content of the sample material by reference to a calibration graph.

Calibration. To obtain the calibration graph, pipette aliquots of 5-30 ml of the standard solution containing 1-6 mg of chromium into separate 150-ml beakers. Dilute each to about 50 ml and add acetate buffer, EDTA and sodium sulphite solutions and continue as described in the method above. Plot the relation of optical density to chromium concentration.

Note: 1. Iron or zirconium metal crucibles can be used for the fusion; nickel crucibles should be avoided as they may contain appreciable amounts of chromium, some of which passes into solution giving high results, particularly if sodium peroxide has been used for the decomposition. Zirconium crucibles may also contain a little chromium, but the attack of the crucible is very much less, and the chromium contributed to the alkaline filtrate in this way can usually be ignored.

Determination of Chromium as Alkali Chromate

The maximum light absorption of chromate solutions occurs at a wavelength of 370 nm and for a concentration range of up to at least 12 ppm a straight line calibration is obtained. When a filter photometer or absorptiometer fitted with violet filters is used, the measurement can only be made at the edge of the absorption band, and a curved calibration is obtained. The sensitivity is also much reduced.

Similar yellow colours are given by uranium and cerium, but only in rare cases are these elements present in amounts sufficient to interfere. Iron in colloidal solution can also impart a yellow colour to the sample solution, but this is easily removed by digesting the sample solution on a steam bath, standing overnight and filtering in the cold. Platinum metal introduced into the solution from the crucible has also been suggested as a source of interference. This can best be limited by restricting the amount of oxidising agent added to the alkaline melt, and by using as low a temperature as possible for the decomposition. Any platinum that is brought into solution is removed on digesting the melt with water containing a little alcohol. The addition of a little borax to the melt serves to assist in the decomposition of the more refractory chromium-bearing minerals such as magnetite and chromite.

Some filter papers contain a small quantity of organic material that gives rise to yellow-coloured filtrates with the alkaline solutions used. Interference from this source can easily be avoided by washing each paper with dilute alkali carbonate solution before use.

The complete omission of oxidising agent may lead to low recoveries of chromium from some silicate rocks, particularly those that are rich in ferrous iron.(4) Potassium nitrate is commonly added to the flux, but alkali nitrite, formed during the fusion, absorbs strongly at the wavelength used and unless all traces of nitrite are destroyed high values of optical density will be obtained. This interference from nitrite can be avoided by measuring the solutions with an absorptiometer fitted with a violet gelatine filter (which has a maximum transmission at about 430 nm) as described by Bennett & Pickup.(4) To avoid the interference from nitrite Rader and Grimaldi(5) used potassium chlorate to complete the oxidation of the rock material. A variation of this procedure is described below.

Many methods for the determination of chromium described a decomposition procedure involving fusion over a gas burner. It has been noted, however, that platinum crucibles are appreciably porous to gases from the burner, and complete oxidation to chromate does not always occur. For this reason, the use of a small electric muffle furnace is recommended. Rather more care is necessary, particularly when using a carbonate flux, in the early stages of the attack when carbon dioxide is being liberated, to prevent too rapid a fusion with consequent loss of material and damage to the furnace. A good indication that a satisfactory, completely oxidised, fusion has been obtained is the green colour of the melt when cold (oxidation of some of the manganese to manganate). This is particularly evident with basic and other rocks containing much manganese, but can usually also be seen with granitic and other rocks. Where this green colour is not obtained, a further quantity of potassium chlorate can be added and fusion continued.

Method

Reagents: Potassium chlorate.
Anhydrous sodium carbonate.
Borax glass, powdered.

Sodium carbonate wash solution, dissolve 10 g of the anhydrous reagent in 500 ml of water.

Working standard chromium solution, dilute 20 ml of the stock solution (see above) to 1 litre in a volumetric flask and transfer immediately to a clean dry polythene bottle. This solution contains 20 μg Cr per ml.

Procedure. Accurately weigh 1 g of the finely powdered rock material into a platinum crucible and mix with 5 g of anhydrous sodium carbonate and 0.5 g of potassium chlorate. If the rock material contains much chromite or magnetite add also 0.5 g of borax glass. Cover the crucible with a platinum lid and transfer to a cold electric muffle furnace. Gradually raise the temperature to 900-950°, and maintain at this temperature for a full hour. Care must be taken to ensure that mechanical loss of the mixture does not occur in the initial stages of the fusion. Allow the melt to solidify around the sides of the crucible to give a thin layer, tinged green in colour. If this colour is not obtained, continue the fusion with an additional quantity of potassium chlorate. Extract the melt with water, add a few drops of ethyl alcohol and digest on a steam bath or hot plate to ensure that all soluble material has passed into solution and all manganate decomposed. Crush any lumps of solid material with the flattened end of a glass rod. Collect the residue on a hardened open-textured filter paper, that has previously been washed several times with hot sodium carbonate wash solution, and wash the residue several times with this wash solution. The residue should be examined as described below (Note 1). Collect the filtrate and washings in a beaker and again digest on a steam bath or hot plate for at least 1 hour, evaporate to a volume of 40-45 ml, and allow to cool, preferably overnight. When quite cold, filter the solution from any small insoluble material that may have separated into a 50-ml volumetric flask and dilute to volume with water. Discard the insoluble material. Prepare also a reagent blank solution in the same way but omitting the rock sample.

Measure the optical density of the solution, relative to water in 1-cm cells, using the spectrophotometer at a wavelength of 370 nm. Deduct the reagent blank and determine the chromium content of the sample by reference to the calibration graph.

Calibration. Transfer aliquots of 5-30 ml of the standard chromium solution containing 100-600 μg Cr, to separate 50-ml volumetric flasks. Add 5 g of anhydrous sodium carbonate to each flask, sufficient water to dissolve the solid material, and then dilute to volume with water. Measure the optical density of each solution in a 1-cm cell at a wavelength of 370 nm, using a similar solution without added chromium as the reference solution. Plot the relation of optical density to chromium concentration.

Note 1 : The residue may be used for the determination of barium or zirconium and should in any case be examined to ensure that the rock portion has been completely decomposed. If decomposition has been complete it will dissolve in hydrochloric acid to give a clear solution with no more than a small flocculent precipitate of silica, and no black or gritty particles. If any such particles of unattacked rock material remain, they should be collected, dried, ignited and fused with a little sodium peroxide in a zirconium crucible, the melt extracted with water and the solution obtained added to the main rock solution.

Determination of Chromium using sym-Diphenylcarbazide

The maximum absorption of the coloured solution given by chromium (VI) with sym-diphenylcarbazide is at a wavelength of 540 nm and the molar absorptivity has been calculated(6) as 31,400. The reaction is therefore one of high sensitivity, and can be used to determine the few parts per million chromium that occur in some silicate rocks. The Beer-Lambert Law is obeyed at the concentrations used.

Other elements that give colours with the reagent include molybdenum - which forms a similar violet colour, mercury - which gives a blue colour, and iron and vanadium - which both give brown colours. Of these, mercury is most unlikely to be encountered in silicate materials in amounts sufficient to interfere. The reaction of diphenylcarbazide with molybdenum is of a lower order of sensitivity to that with chromium and, as most silicate rocks contain considerably less molybdenum than chromium, interference from this source is therefore also unlikely. Iron is usually separated from chromium by alkaline fusion; in extracting the melt with water only traces of iron pass into the filtrate. These can be removed by passing the solution through a cation-exchange column, as described by Frohlich,(7) but this is not usually considered necessary. Any small amount not removed will react with the reagent, but as the absorption at 540 nm is small, it can be neglected.

A serious interference is that from vanadium, which is not separated from chromium in any of the procedures used to decompose the sample material. The brown-coloured complex formed with the reagent is unstable - the colour fades rapidly on standing - and nearly correct results can be obtained for chromium by measuring after 10-15 minutes. Partly for this reason and partly because the vanadium complex has a much smaller absorption than the chromium complex at 540 nm, the removal of vanadium is only necessary if its amount grossly exceeds that of the chromium. A procedure for this involving an extraction of the vanadium complex with 8-hydroxyquinoline is described by Sandell.(8)

The violet colour given by chromium with the reagent forms rapidly in 0.2 N sulphuric acid solution; it is unstable at higher acid concentrations, and does not develop completely at lower acidities. The correct acid concentration is often obtained by neutralising the solution and then adding a measured volume of sulphuric acid. Grogan et al[9] prefer to use a meter to bring the pH of the solution to a value of between 1.3 and 1.7.

The small loss of chromium that occurs when an alkaline fusion is used to decompose the silicate sample material, and to which reference is made above, can be avoided by using an acid decomposition. For this Spangenberg et al[10] describe the use of a potassium bifluoride fusion, although the more usual hydrofluoric-sulphuric acid evaporation can also be used. Iron can be removed by extraction with amyl acetate or other organic solvent from 10 N hydrochloric acid, chromium III is oxidised to chromate with ceric ammonium sulphate and the excess reagent destroyed with sodium azide, which also reduces any oxidised manganese. The violet-coloured complex with diphenylcarbazide is then formed and measured using a spectrophotometer as described below.

Other difficulties that occur include the tendency of chromium (VI) ions to be adsorbed on the walls of glass vessels, noted by Chuecas and Riley,[11] and the loss of chromium that occurs when silica precipitates from acid solutions. These difficulties may be avoided as described by Fuge,[12] by working in platinum and plastic ware apparatus wherever possible, and by eliminating most of the silica at the extraction stage. This is accomplished by adding magnesium oxide to the sodium carbonate flux used for the decomposition of the silicate rock material forming the insoluble magnesium silicate which is retained in the insoluble residue. The procedure given below is based upon those described by Fuge[12] and others.

Method

Reagents:

Sodium carbonate-magnesium oxide mixture, mix intimately 12 g of anhydrous sodium carbonate and 3 g of magnesium oxide. This provides sufficient flux for twenty-five determinations.

Sodium carbonate wash solution, dissolve 5 g of anhydrous sodium carbonate in 25 ml of water.

sym-Diphenylcarbazide solution, dissolve 0.25 g of a good quality reagent in 100 ml of acetone. This solution deteriorates slowly on standing, and should be replaced when appreciably brown in colour.

Standard chromium working solution, dilute 20 ml of the freshly prepared standard chromium solution containing 20 µg per ml (see above) to 1 litre with water and transfer immediately to a clean, dry polythene bottle. This solution contains 0.4 μg chromium per ml.

Procedure. Accurately weigh approximately 0.1 g of the finely powdered silicate rock material into a small platinum crucible and mix intimately with approximately 0.6 g of the sodium carbonate-magnesium oxide mixture. Cover the crucible with a platinum lid and transfer to an electric muffle furnace set at a temperature of about 900°, and maintain at this temperature for 45 minutes. Remove the crucible from the furnace and allow to cool. Add a few millilitres of water, together with 2 drops of ethanol, and heat on a hot plate to disintegrate the melt. Break up any lumps of solid material with the aid of a stout polythene rod, and replace water lost by evaporation as necessary. Digest on the hot plate for 1 hour, then remove the crucible and allow to cool.

Collect the residue on a 9-cm close-textured filter paper supported in a polyethylene funnel, and wash with successive small volumes of the sodium carbonate wash solution. Collect the filtrate and washings in a small polypropylene beaker and dilute to about 20 ml with water. Using a pH meter, bring the pH of the solution to a value in the range 1.3-1.7 by adding 10 N sulphuric acid, swirling to remove as much of the liberated carbon dioxide as possible. Add 1 ml of the diphenylcarbazide solution, transfer to a 25-ml volumetric flask, shake to remove any residual carbon dioxide, and dilute to volume with water.

Prepare also a reagent blank solution in the same way as the sample solution, but omitting the sample material. Allow the coloured solutions to stand for 15 minutes and then measure the optical density in 4-cm cells relative to water using a spectrophotometer set at a wavelength of 540 nm. Measure also the optical density of the reagent blank solution.

Calibration. Transfer aliquots of 0-20 ml of the standard chromium solution containing 0-8 μg Cr to separate 25-ml volumetric flasks. Dilute each solution to about 20 ml with water and add 1 ml of 6 N sulphuric acid and 1 ml of the diphenyl-carbazide solution to each. Dilute to volume, allow to stand for 15 minutes and measure the optical density of each solution in 4-cm cells at a wavelength of 540 nm as described above. Plot the relation of optical density to chromium concentration.

Determination of Chromium by Atomic Absorption Spectroscopy

Chromium can be determined by this technique, and a number of authors have given procedures[2,3,13,14,15] with varying flame and other conditions. The determination of chromium may be combined with those of copper, cobalt, vanadium, nickel and barium.[15] In this procedure the rock material is decomposed with a mixture of hydrofluoric and perchloric acids in a pressure decomposition vessel, and boric acid added to dissolve any precipitated fluorides. For the measurement of chromium, an air-acetylene flame is used and the spectrometer set to a wavelength of 357.9 nm. A high background is obtained with this flame, and a correction to the absorption should be made for it. This correction can be obtained by similar measurement of a 'background' or 'blank' solution prepared in the usual way. A detection limit of 2ppm (2 μg/g) is claimed.

References

1. BOSE M., Anal. Chim. Acta (1954) 10, 201 and 209
2. BECCALUVA L and VENTURELLI G., Atom. Absorp. Newsl. (1971) 10, 50
3. SCHWEIZER V B., Atom. Absorp. Newsl. (1975) 14, 137
4. BENNETT W H and PICKUP R., Colon. Geol. Min. Res. (1952) 3, 171
5. RADER L F and GRIMALDI F S., US Geol. Surv. Prof. Paper 391-A, p.A-10 (1961)
6. ROWLAND G P Jr., Ind. Eng. Chem. Anal. Ed. (1939) 11, 442
7. FROHLICH F., Z. Anal. Chem. (1959) 170, 383
8. SANDELL E B., Ind. Eng. Chem. Anal. Ed. (1936) 8, 336
9. GROGAN C H, CAHNMANN H J and LETHCO E., Anal. Chem. (1955) 27, 983
10. SPANGENBERG J D, RUSSEL B F and STEEL T W., Nat. Inst. Metall. Johannesburg S.A. Rept. No. 265, (1967)
11. CHUECAS L and RILEY J P., Anal. Chim. Acta (1966) 35, 240
12. FUGE R., Chem. Geol. (1967) 2, 289
13. RUBESKA I and MIKSOVSKY M., Colln. Czech. Chem. Comm. (1972) 37, 440
14. CHOWDHURY A N and DAS A K., Z. Anal. Chem. (1972) 261, 126
15. WARREN J and CARTER D., Canad. J. Spectrosc. (1975) 20, 1

CHAPTER 18

Cobalt

The earliest attempts to determine cobalt in silicate and other rocks used a separation based upon the insolubility of cobalt sulphide in alkaline solution. Cobalt sulphide was precipitated together with the sulphides of nickel, manganese and zinc. In view of the small quantities of cobalt in silicate rocks, it is doubtful if accurate results can be obtained using this separation.

Little difficulty exists in applying colorimetric methods[1,2] to the determination of cobalt in basalts and other similar rocks. Care is however required in their application to granitic rocks, particularly those containing less than 1 ppm of cobalt, for which the more sensitive methods are advised.

Cobalt forms a red-violet coloured complex with dithizone that can be extracted into carbon tetrachloride or chloroform solution for photometric measurement. More frequently, this procedure is used to concentrate the cobalt into a small volume and to separate it from certain other metals. Both Carmichael and McDonald[1] and Rader and Grimaldi[3] have used this technique, combining it with photometric measurement with a separate colour-forming reagent. It has been reported[4] that an excessive concentration of hydroxyammonium chloride in the aqueous solution impedes the extraction of cobalt with dithizone, giving somewhat low recoveries. Only a few sensitive but selective photometric reagents are known for cobalt. One that has been applied to the analysis of silicate rocks is nitroso-R-salt, and a procedure using this reagent following a dithizone extraction is given below. Other reagents proposed for the photometric determination of cobalt in silicates include CyDTA, 2-nitroso-1-naphthol[5,6] and thiocyanate with tri-n-butylamine.[7] Of considerable interest is the reaction of cobalt with 4-(3.5-dibromo-2-pyridylazo)-1.3-diaminobenzene. This reagent prepared by Kiss[8] and used by him[9] to determine cobalt in silicates and meteorites, has a molar extinction coefficient of 123×10^3. It may be used in aqueous fluoride solution, combining simplicity and rapidity with both selectivity and sensitivity to cobalt.

Cobalt is commonly determined by atomic absorption spectroscopy using a wavelength of 240.7 nm. However departure from linearity has been observed, attributed to the presence of a non-absorbing line, to line-broadening at high hollow-cathode lamp current and to flame composition. Lines at 241.2, 242.5, 243.6 and 252.1 nm all give reasonable sensitivity and can be used for analytical measurement. Interferences can be minimised by using dilute solutions, making background corrections[10] and by using a lean nitrous oxide-acetylene flame[11]. A statistical treatment of the interferences has been given by Thompson et al.[12]

In view of the ease and convenience of atomic absorption methods for the determination of cobalt, it is not surprising that this technique has been widely recommended[13,14] for application to silicate rocks. In the method referred to in the chapter on Ni, due to Warren and Carter,[15] the determination of cobalt is combined with that of copper, vanadium, chromium nickel and barium. Flameless (carbon furnace) techniques can also be used.[16,17]

Spectrophotometric Determination using Nitroso-R-salt

Nitroso-R-salt (the di-sodium salt of 1-nitroso-2-hydroxynaphthalene-3:6-disulphonic acid) reacts with cobalt in acid solution to give a stable red-coloured complex, soluble in aqueous solution. The reagent itself is highly coloured and gives an appreciable background absorption. Procedures have been devised for reducing this background absorption by bleaching the excess reagent with nitric acid or bromine, but acceptable results for silicate rocks can be obtained by working at a wavelength where the background absorption is much reduced. See Sandell[18] for a discussion of the most appropriate wavelength to use.

In the procedure described below[6] the cobalt complex of the reagent is developed in a boiling citrate-phosphate-borate buffer solution as described by Marston and Dewey.[19] Any nickel or copper extracted from the rock solution in a prior dithizone separation step (Note 1) will form coloured complexes with the reagent, but these complexes are decomposed by boiling with hot dilute nitric acid.

Method

Reagents:

Citric acid solution, dissolve 50 g of purified citric acid monohydrate in water and dilute to 100 ml.

Citric acid solution, 0.20 M solution, dissolve 4.2 g of the purified reagent (monohydrate) in water and dilute to 100 ml.

Dithizone solution, dissolve 0.05 g of analytical grade diphenyl-thiocarbazone in 100 ml of carbon tetrachloride. Store in a refrigerator.

Nitroso-R-salt solution, dissolve 0.05 g of the reagent in 100 ml of water.

Buffer solution, dissolve 6.2 g of boric acid, 35.6 g of disodium hydrogen phosphate heptahydrate and 20 g of sodium hydroxide in water and dilute to 1 litre.

Standard cobalt stock solution, dissolve 0.807 g of cobaltous chloride hexahydrate in water containing 2 ml of concentrated hydrochloric acid and dilute to volume in a 1 litre volumetric flask. This solution contains 200 μg cobalt per ml.

Standard cobalt working solution, dilute 5 ml of this stock solution to 1 litre with water. This solution contains 1 μg cobalt per ml.

Procedure. Accurately weigh approximately 1 g of the finely powdered silicate rock into a platinum dish (Note 2), moisten with water and add 10 ml of hydrofluoric acid, 10 ml of nitric acid and 4 ml of perchloric acid. Cover the dish with a platinum or polypropylene cover and heat on a steam bath for 30 minutes. Rinse and remove the cover, transfer the dish to a hot plate and evaporate to fumes of perchloric acid. Allow to cool, rinse down the sides of the dish with a little water and again evaporate to fumes of perchloric acid. Again rinse down the sides of the dish and evaporate, this time almost, but not quite, to dryness. Avoid overheating the moist salts. Add 2 to 4 ml of 6 N hydrochloric acid to the residue followed by 10 ml of water. Digest on a steam bath for a minute or two until all soluble salts have dissolved. If any residue remains, this should be collected and decomposed as described in Note 3.

Evaporate the solution if necessary so that it may be diluted to volume in a 25-ml volumetric flask. At this stage some sodium chloride or potassium perchlorate may crystallise out. Allow any precipitated salts to settle, pipette a 10-ml aliquot of the solution into a small beaker and add 5 ml of the concentrated citric acid solution. Using a pH meter, adjust the pH of the solution to 9 by adding aqueous ammonia. Transfer the solution to a 60-ml separating funnel and add 5 ml of dithizone solution, stopper the funnel and shake vigorously to extract the cobalt complex into the carbon tetrachloride solution. Run off and retain the lower organic layer. Extract the aqueous phase with further 5-ml portions of the dithizone reagent solution until the green colour of the organic phase persists after shaking for at least 1 minute. (Three or four 5-ml portions are usually required for this). Discard the aqueous layer and combine the organic extracts.

Return the organic solution to the separating funnel and add 5 ml of dilute aqueous ammonia (1 + 99). Shake vigorously for 1 minute, allow the layers to separate and transfer the organic solution to a 50-ml beaker. Discard the aqueous fraction. Evaporate the carbon tetrachloride solvent on a water bath. Add 0.5 ml of 20 N sulphuric acid and between 0.25 ml and 0.5 of perchloric acid to the residue and evaporate on a hot plate until a colourless liquid is obtained, then evaporate the excess sulphuric and perchloric acids, including any drops that have condensed on the sides of the beaker.

Add 1 ml of 6 N hydrochloric acid, and using this acid, wet the inside walls and bottom of the beaker. Rinse down the walls with a very small quantity of water and evaporate the solution to dryness. Complete the drying by removing the last traces of water and acid in an electric oven set at 140°. Add 1.0 ml of 0.2 M citric acid solution to the beaker followed by 1.2 ml of the buffer solution. Stir the solution whilst adding exactly 2 ml of the nitroso-R-salt reagent solution. Boil for 1 minute and then allow to cool.

Add 1.0 ml of nitric acid to the solution and again boil for 1 minute. Cool the solution, transfer to a 10-ml volumetric flask and dilute to volume with water. Measure the optical density of this solution against water using a spectrophotometer set at a wavelength of 475 nm. The reagent blank solution is prepared by adding 1.0 ml of 0.2 M citric acid solution to 1.2 ml of the buffer solution, adding exactly 2 ml of the nitroso-R-salt reagent and boiling for 1 minute. This solution is then cooled, boiled for a further minute with nitric acid, cooled and diluted to 10 ml in a volumetric flask as described for the sample solution.

Calibration. Transfer 0-30 ml aliquots of the standard cobalt solution containing 0-30 μg cobalt to 50-ml beakers. Evaporate to dryness on a hot plate to remove all traces of acid and proceed as described in the method above beginning with the addition of 1.0 ml of 0.2 M citric acid and 1.2 ml of buffer solution. Plot the relation of optical density of these solutions to their cobalt content.

Notes: 1. Although the prior separation stage with dithizone forms an integral part of this procedure, acceptable results have been obtained by a somewhat shorter method involving acid attack, neutralisation of an aliquot with sodium carbonate to a pH between 5.5 and 8, addition of the buffer solution, nitroso-R-salt and concentrated nitric acid. The solution was allowed to cool in the dark, diluted to volume and the optical density measured as soon as possible.

2. If the sample contains any organic material it should be ignited for a few minutes at a temperature of 700-800°C before moistening with water.

3. Any residue should be carefully collected, washed with a little water, dried, ignited and fused with the minimum quantity of anhydrous sodium carbonate. Dissolve the melt in a little 6 N hydrochloric acid and add to the main rock solution.

Determination by Atomic Absorption Spectroscopy.

The selected procedure for the determination of cobalt by this technique is based upon that described by Warren and Carter.[15] This determination may readily be combined with those of copper, vanadium, chromium, nickel and barium. The rock matrix is decomposed with a mixture of hydrofluoric and perchloric acids in a pressure decomposition vessel, and boric acid added to dissolve any precipitated fluorides. For the measurement of cobalt, an air-acetylene flame is used and the cobalt absorption measured at 241.2 or 240.7 nm with background correction. This procedure is given in greater detail in the chapter dealing with nickel.

References

1. CARMICHAEL I and McDONALD A J., Geochim. Cosmochim. Acta (1961) 22, 87
2. LODOCHNIKOVA N V., Informatsionnyi sbornik VSEGEI (1956) No. 3
3. RADER L F and GRIMALDI F S., US Geol. Surv. Prof. Paper, 391A, 1961
4. CARMICHAEL I and McDONALD A J., Geochim. Cosmochim. Acta (1961) 25, 189
5. CLARK L J., Analyt. Chem. (1958) 30, 1153
6. BODART D E., Chem. Geol. (1970) 6, 133
7. STANTON R E, McDONALD A J and CARMICHAEL I., Analyst (1962) 87, 134
8. KISS E., Anal. Chim. Acta (1973) 66, 385
9. KISS E., Anal. Chim. Acta (1975) 77, 320
10. GOUETT G J S and WHITEHEAD R E., J. Geochem. Explor. (1973) 2, 121
11. PEARTON D C G and MALLETT R C., Nat. Inst. Metall. Rept. No 1435, (1972)
12. THOMPSON M, WALTON S J and WOOD S J., Analyst (1979) 104, 299
13. ARMANNSSON H., Anal. Chim. Acta (1977) 88, 89
14. CHAO T T and SANZOLONE R F., J. Res. U S Geol. Surv. (1973) 1, 681
15. WARREN J and CARTER D., Canad. J. Spectrosc. (1975) 20, 1
16. SCHWEIZER V B., Atom. Absorb. Newsl. (1975) 14, 137
17. CRUZ R B and VAN LOON J C., Anal. Chim. Acta (1974) 72, 231
18. SANDELL E B., Colorimetric Determination of Traces of Metals, Interscience, (3rd ed.), 1959, p.420
19. MARSTON H R and DEWEY D W., Austral. J. Expl. Biol. Med. Sci. (1940) 18, 343

CHAPTER 19

Copper

A number of analytical procedures for the determination of copper in soils and silicate rocks have been designed - mostly based upon the need to provide rapid, easy and cheap methods of the geochemical prospecting for this element in remote areas. Such procedures invariably determine only that part of the copper present in a readily accessible form; for example, that part that can be extracted into dilute sulphuric acid[1] or into aqueous solution after fusion with potassium pyrosulphate.[2] Although they may be ideal for their purpose, these methods do not give a measure of the total copper content when applied to granitic and similar rocks. For these materials it is preferable to decompose the silicate minerals present, using hydrofluoric acid in combination with other mineral acid in the usual way.[3]

Except in rare cases, the amount of copper present in silicate rocks is insufficient for either gravimetric or volumetric determination, and photometric methods are in more general use. The most sensitive of reagents for copper is dithizone, but this reacts also with many other metal ions. Less sensitive than dithizone are the dithiocarbamate reagents. The diethyl compound is the best known, although the dibenzyl compound has been recommended.[4] These reagents react also with other metallic ions. Although considerably less sensitive than the other two, (molar extinction coefficient in only 5500), diquinolyl("cuproine") is very much less subject to interference from other metallic ions. It can be used directly for copper in silicate rocks, as described by Riley and Sinhaseni,[3] or after a preliminary isolation of copper, zinc and lead with dithizone, as described by Rader <u>et al</u>.[5]

Diquinolyl is one of a number of specific reagents for copper (I) that contain the "ferron" structure but are sterically inhibited from reacting with iron (II) by the presence of ortho substituents. 2,9-Dimethyl-1,10-phenanthroline and 2,9-dimethyl-4,7-diphenyl-1,10-phenanthroline, although slightly more sensitive than diquinolyl as reagents for copper, are considerably more expensive. Both have been used for the determination of copper in silicate rocks by Bodart.[6] The reaction with diquinolyl was first used for the determination of copper by Breckenridge <u>et al</u>[7] and, because of its specificity, has been extensively used for the examination of a wide range of materials and products. The pink complex formed by the cuprous ion is extracted into a suitable alcoholic solvent at a pH in the range 4.5-7.5.[8] Hoste <u>et al</u>[9] has suggested that complete extraction can be obtained in the range 2-9, but Riley and Sinhaseni[3] have found that practically no copper could be extracted at a pH below 2.7, and recommended working in the range 4.2-5.8.

Both normal and isoamyl alcohols have been used for this extraction. They are however appreciably soluble in water; Riley and Sinhaseni[3] noted that the less soluble n-hexanol could also be used. A purification procedure is usually necessary for all suggested solvents. Riley and Sinhaseni[3] have recommended making three extractions of the copper solution, with intervening addition of hydroxyammonium chloride to ensure that all the remaining copper is in the cuprous state. Fading of the colour of the cuprous-diquinolyl extracts has been reported by several workers, this can be prevented by adding a little hydroquinone to the organic extracts.

The cuprous-diquinolyl extracts have a maximum absorption at a wavelength of 540-545 nm and the Beer-Lambert Law is obeyed up to at least 10 ppm in the organic extract.

Atomic absorption spectroscopy provides an alternative technique for the determination of copper in silicate rocks.[10] Measurement may be made at 324.7 nm or the slightly less sensitive 327.4 nm. This method has the advantage of simplicity, rapidity and relative freedom from interference from other elements present in the rock material. The determination of copper can be combined with that of a number of other elements. The procedure given in the chapter on nickel is based upon that described by Warren and Carter,[11] which provides for the atomic absorption determination of copper, vanadium, chromium, nickel, cobalt and barium. An alternative procedure by Armannsson[12] involving a prior extraction of copper and other elements with dithizone does not appear to offer any advantage over this somewhat simpler method.

Flameless atomic absorption using a graphite furnace, has been described by Schweizer[13] for the determination of copper and other elements in carbonate rocks. By extension this technique could be applied also to silicate rocks.

Spectrophotometric Determination of Copper with 2,2'-Diquinolyl

Method

Reagents: Diquinolyl reagent solution, dissolve 0.03 g of 2,2'-diquinolyl in 100 ml of n-hexanol that has been freshly distilled from solid sodium hydroxide.

Hydroxyammonium chloride solution, dissolve 25 g of analytical grade hydroxyammonium chloride in about 80 ml of water and dilute to 100 ml with water. If any appreciable amounts of copper are present in the reagent solution, extract with successive 10-ml portions of a 0.01 per cent solution of dithizone in carbon tetrachloride until there is no change in the green colour of the dithizone solution. Then extract the solution with carbon tetrachloride until all colour has been removed from the aqueous solution.

Sodium acetate buffer solution, dissolve 136 g of sodium acetate trihydrate in water and dilute to 1 litre. If the reagent contains more than a trace of copper, purify this solution by extraction with 0.01 per cent dithizone solution in carbon tetrachloride as for the hydroxyammonium chloride solution.

Hydroquinone solution, dissolve 1 g of reagent in 100 ml of redistilled ethanol as required.

Standard copper stock solution, dissolve an accurately weighed 0.1 g portion of pure copper (Note 1) in 3 ml of concentrated nitric acid, add 1 ml of 20 N sulphuric acid and evaporate to fumes of sulphuric acid. Allow to cool, dissolve the residue in distilled water and dilute to 500 ml. This solution contains 200 μg copper per ml.

Standard copper working solution, dilute 5 ml of the stock solution to 250 ml with water. This solution contains 4 μg Cu per ml, and should be used for the calibration of 1-cm spectrophotometer cells. For the calibration of 4-cm cells, dilute 5 ml of the stock solution to 1 litre with water, to give a solution containing 1 μg Cu per ml.

Procedure: Accurately weigh approximately 1 g of the finely powdered rock material into a platinum crucible and add 2 ml of concentrated nitric acid and 15 ml of concentrated hydrofluoric acid. Set the covered crucible aside overnight on a water bath, and then evaporate to dryness. Fuse the dry residue with from 1.5 to 2 g of potassium pyrosulphate at dull red heat for 5 minutes, taking care not to lose any sample material in the early stages of the ignition when excessive effervescence may occur. Dissolve the fused cake by warming on a water bath with 100 ml of water containing 1.5 ml of concentrated hydrochloric acid. When cold, transfer the solution to a 250-ml volumetric flask and dilute to volume with water.

Transfer a 100-ml aliquot containing not more than 80 μg of copper to a 250-ml separating funnel and add 2.5 ml of hydroxyammonium chloride solution and 25 ml of sodium acetate buffer solution. Shake with 6 ml of diquinolyl reagent solution for 5 minutes and allow the phases to separate. Run the lower aqueous layer into another separating funnel, add 2 ml of hydroxyammonium chloride solution and extract again with 2.5 ml of diquinolyl reagent solution. Separate the phases and again extract the aqueous layer with 2 ml of the diquinolyl reagent solution. Combine the three organic extracts in a 10-ml volumetric flask containing 0.5 ml of hydroquinone solution and dilute the solution to volume with n-hexanol. Measure the optical density of the solution against n-hexanol in 1- or 4-cm cells with the spectrophotometer set at a wavelength of 540 nm. Measure also the optical density of a reagent blank solution prepared in the same way as the sample solution but omitting the rock material.

Calibration. Use the standard solution containing 4 µg Cu per ml to calibrate the 1-cm cells and the standard containing 1 µg Cu per ml for the 4-cm cells. Transfer aliquots of 0-25 ml to 250-ml separating funnels, add 1.5 ml of concentrated hydrochloric acid and dilute each solution to 100 ml with water. Add hydroxylamine and buffer solution, extract the copper with diquinolyl reagent solution and measure the optical densities as described above. Plot the relation of optical density to copper concentration for the ranges 0-25 µg Cu (4-cm cells) or 0-100 µg Cu (1-cm cells). A solution containing 25 µg Cu in 10 ml of organic extract should have an optical density of about 0.984 in 4-cm cells, or 0.246 in 1-cm cells.

Notes: 1. The metallic copper should be cleaned with hydrochloric acid to remove any oxide layer.

2. This method can be used to determine copper in carbonate rocks or minerals as follows. Dissolve 5 g of the carbonate by gradual addition of 30 ml of 4 N nitric acid. (If foam tends to rise to the top of the flask it can be broken by the addition of a drop of octyl alcohol). Cautiously evaporate the solution to dryness on a hot plate. If organic matter is present, add 10-15 ml of concentrated nitric acid and repeat the evaporation. Evaporate the residue twice to dryness with 15 ml of concentrated hydrochloric acid to remove the nitric acid. Dissolve the residue in 100 ml of distilled water containing 1.5 ml of concentrated hydrochloric acid and dilute to 250 ml in a volumetric flask. Determine the copper in a 100-ml aliquot of this solution as described above.

Determination by Atomic Absorption Spectroscopy

The selected procedure for the determination of copper by this technique is based upon that described by Warren and Carter.(11) This determination may readily be combined with those of cobalt, vanadium, chromium, nickel and barium. The rock matrix is decomposed with a mixture of hydrofluoric and perchloric acids in a pressure decomposition vessel, and boric acid added to dissolve any precipitated fluorides.

For measurement of the copper absorption, the solution is aspirated into an air-acetylene flame and the spectrometer set at a wavelength of 324.7 nm. The detection limit is said to be 2 ppm (2 μg/g), which is fully adequate for silicate and many other rocks. This procedure is given in greater detail in the chapter dealing with nickel.

References

1. ALMOND H and MORRIS H T., Econ. Geol. (1951) 46, 608
2. ALMOND H., U S Geol. Surv. Bull. 1036-A, 1955
3. RILEY J P and SINHASENI P., Analyst (1958) 83, 299
4. MARTENS R I and GITHENS R E Jr., Analyt. Chem. (1952) 24, 991
5. RADER L F and GRIMALDI F S., U S Geol. Surv. Prof. Paper 391-A, 1961
6. BODART D E., Chem. Geol. (1970) 6, 133
7. BRECKENRIDGE J G, LEWIS R W and QUICK L., Can. J. Res. (1939) B17, 258
8. GUEST R J., Analyt. Chem. (1953) 25, 1484
9. HOSTE J, EECKHOUT J and GILLIS J., Anal. Chim. Acta (1953) 9, 263
10. SANZOLONE R F and CHAO T T., Anal. Chim. Acta (1976) 86, 163
11. WARREN J and CARTER D., Canad. J. Spectrosc. (1975) 20, 1
12. ARMANNSSON H., Anal. Chim. Acta (1977) 88, 89
13. SCHWEIZER V B., Atom. Absorp. Newsl. (1975) 14, 137

CHAPTER 20

Fluorine

The oldest reliable method for determining fluorine appears to be that of Berzelius[1] involving the decomposition of the silicate rock by fusion with alkali carbonate, removal of silica and alumina by precipitation with ammonium and zinc carbonates, separation of phosphate and chromate by precipitation with silver nitrate solution, and finally precipitation of the fluorine as calcium fluoride from an acetate solution. Good results could be obtained only by a skilled analyst with the expenditure of much time and patience. However, the method had two considerable advantages: firstly, there was no other reliable method available, and secondly, it was possible to recover silica from the ammonium and zinc carbonate precipitates, giving a method - the only method then available - for the determination of silica in materials containing fluorine. The limit of detection appears to be about 0.05 per cent F.[2]

A considerable improvement was the distillation from sulphuric or perchloric acid solution described by Willard and Winter.[3] Unfortunately the presence of much silica or aluminium hinders the volatilisation of fluorine, so that small amounts of this element are difficult to recover, making the ammonium and zinc carbonate separations essential preliminary stages of the method. The fluorine in the distillate was determined titrimetrically using thorium nitrate as titrant and the pink-coloured zirconium-alizarin lake as indicator. The pink colour is bleached by the fluorine in the distillate, but returns at the end-point of the thorium nitrate titration. The thorium nitrate solution was standardised by titration with known amounts of fluorine as sodium fluoride. The end-point is poor, necessitating colour-matching and careful lighting when more than a few milligrams of fluorine is present.

Gravimetric and titrimetric procedures based upon precipitation of lead chlorofluoride[4] have found some use in the analysis of samples containing appreciable quantities of fluorite, but are not applicable to normal silicate and carbonate rocks. The useful range of the methods is from 1 to about 20 per cent fluorine. A method involving the precipitation of lead bromofluoride has also been described.[5]

One of the earliest colorimetric procedures to be used for fluorine was that of Steiger[6] involving the bleaching action of fluorine on the yellow colour produced by titanium with hydrogen peroxide. The modification of Steiger's method by Merwin[7] gave a procedure more sensitive than Berzelius's gravimetric method, but still "not delicate enough to indicate with certainty the presence of 0.01 per cent of fluorine". It was still necessary to remove silica and aluminium by precipitation with ammonium carbonate.

A series of spectrophotometric methods have been based upon the bleaching of coloured zirconium and thorium complexes. One such method, devised by Megregian[8] and studied in further detail by Sarma,[9] is described below following the work of Evans and Sergeant.[10] Other spectrophotometric methods have been based upon the reaction of fluorine with alizarin complexan and either cerous or lanthanum solution.[11,12] This reaction, the first direct positive colour-forming reaction of the fluoride ion, cannot be used without prior separation from aluminium and iron.[13,14] A separation has been described based upon the extraction of a boron trifluoride-isopropanol complex after prior precipitation of interfering elements with ammonia.[15]

When more than about 0.5 per cent of fluorine is present (as in certain pegmatites and mineral concentrates), the technique of pyrohydrolysis can be used with advantage.[16,17,18] This is simple and rapid, no prior separations are required, and a relatively simple titrimetric procedure can be used to determine fluorine in the distillate. Details of this procedure are given below.

Fluorine may also be determined in silicate rocks using an ion-selective electrode[18,19,20,21,22]; a procedure for this is also given below.

Spectrophotometric Determination of Fluorine

The procedure given here is based upon that described by Evans and Sergeant[10] - the choice has been made largely because a preliminary separation of silica and alumina is not required with the small portion of sample used. A "Willard and Winter" separation by distillation of hydrofluorosilicic acid is used to recover the fluorine from about 0.2 g of rock material. As with other procedures involving distillation, a fluorine recovery of about 95 per cent is achieved. The distillate passes directly to a column of anion exchange resin, which serves to concentrate the fluorine, and from which it is eluted with a small volume of ammonium acetate solution. The final determination is made spectrophotometrically by an eriochrome cyanine R method.

This photometric method is based upon the bleaching action of fluorine on the coloured complex formed between zirconium and eriochrome cyanine R. The conditions for this reaction were studied in some detail by Sarma[9] but have been modified for this particular application. In the range 2-54 μg F per 100 ml of solution, a negative linear calibration is obtained. The reagent blank amounts to about 3 μg F, corresponding to about 15 ppm.

Method

<u>Apparatus</u>. The distillation apparatus and ion-exchange column are shown in Fig. 14. The ion-exchange resin used - De Acidite FF - is converted to the hydroxyl form by treatment with 5 ml of N sodium hydroxide solution, followed by washing free from excess alkali with about 50 ml of water. After use the resin may be regenerated by further treatment in the same way. A column containing about 750 mg of resin is sufficient for about forty determinations before this regeneration becomes necessary.

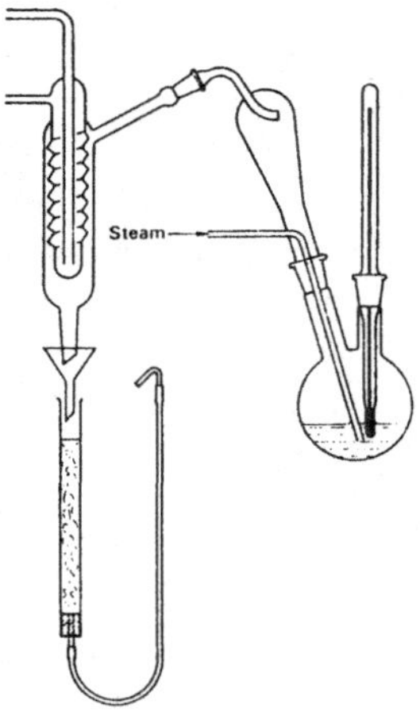

FIG. 14 Apparatus for the separation and recovery of fluorine.

<u>Reagents</u>: Sodium acetate solution, 0.1 M

Zirconyl chloride solution, dissolve 0.1325 g of zirconyl chloride $ZrOCl_2.8H_2O$ in a little water, add 600 ml of concentrated hydrochloric acid and dilute to 1 litre with water.

Eriochrome cyanine R solution, dissolve 0.8 g of the solid reagent in 1 litre of water. Different batches of this reagent have been found to contain varying amounts of sodium sulphate. This may be removed by dissolving the dye in methanol, filtering and evaporating the solution to dryness under reduced pressure. Store the purified reagent in a desiccator.

Standard fluorine stock solution, dissolve 0. 111 g of sodium fluoride in water and dilute to 500 ml. This solution contains 100 μg F per ml. Store in a polythene bottle.

Standard fluorine working solution, dilute 10 ml of the stock solution to 500 ml with water. This solution contains 2 μg F per ml. Store in a polythene bottle.

Procedure. Accurately weigh 0.2 g of the finely powdered silicate rock into a platinum crucible, add 1 g of anhydrous sodium carbonate and fuse over a Meker burner for 15 to 20 minutes. Allow to cool, add 10 ml of 2 N sulphuric acid slowly and allow the mixture to digest for 30 minutes, then transfer the contents of the crucible to the distilling flask. Rinse the crucible and lid with 45 ml of 18 N sulphuric acid and transfer the acid to the distilling flask. Set up the apparatus as shown in the figure. Pass steam into the flask when the temperature has reached 125°, (Note 1). The distillation temperature should be kept in the range 145-150° with a distillation rate of 6-8 ml per minute (Note 2).

Allow between 300 and 400 ml of distillate to pass through the ion-exchange column, and then elute the absorbed fluorine from the column with 25 ml of 0.1 M sodium acetate solution, collecting the eluate in a 100-ml volumetric flask. Dilute to volume with water.

Pipette an aliquot of this solution containing not more than 50 μg of fluorine into a clean 100-ml volumetric flask and dilute to about 70 ml with water. Add 10 ml of the zirconyl chloride solution, followed by 10 ml of the eriochrome cyanine R solution, dilute to volume with water and mix well. Allow the solution to stand for 30 minutes then measure the optical density against water in 1-cm cells with a spectrophotometer set at a wavelength of 525 nm. The reference solution is prepared from 70 ml of water, 6 ml of concentrated hydrochloric acid and 10 ml of eriochrome cyanine R solution followed by dilution to 100 ml as for the sample solution.

Calibration. Transfer 5-25 ml aliquots of the standard fluorine solution containing 10-50 μg F to separate 100-ml volumetric flasks and dilute each solution to 70 ml with water. Add zirconyl chloride and eriochrome cyanine R solutions as described above, dilute to volume with water and measure the optical density against a reference solution as for the sample solution. Plot the relation of optical density to fluorine concentration.

Notes: 1. Electric heating mantles can conveniently be used for both the distilling flask and the steam generator.

2. This can be done by appropriate adjustment to the heater and supply of steam. In the original paper a contact thermometer was used to control the heater.

Spectrophotometric Determination without Distillation

Huang and Johns[24] also used a spectrophotometric procedure based upon the work of Megregian[8] and Sarma,[9] but by incorporating zinc oxide into the alkaline flux used for sample decomposition, found it possible to dispense entirely with the distillation stage. Iron and phosphate remain entirely with the residue after aqueous extraction. Aluminium is partly extracted, but by making the solution alkaline and allowing the coloured solution to stand for 60 to 90 minutes before photometric measurement, the aluminium content can be tolerated. Sulphates react similarly to fluorine, but the sulphate level in most rocks is too low to give serious interference.

Procedure. Mix 0.5 g of the finely ground rock material, 3.5 g of anhydrous sodium carbonate and 0.6 g of zinc oxide in a large platinum crucible, and fuse in an electric muffle furnace at a temperature of about 900° for 20 to 25 minutes. Allow to cool, add 10 ml of water and 3 drops of ethanol to reduce any manganate formed in the fusion, and heat on a hot plate to disintegrate the melt. Boil for 1 minute and allow to cool. Filter the solution through a close-textured filter paper, collecting the filtrate in a 100-ml polyethylene beaker. Wash the residue five times by decantation with 2-ml portions of hot water and discard it.

Add 4.1 ml of concentrated nitric acid, which should leave the solution faintly acid, and stir to promote the release of most of the carbon dioxide. Transfer the solution to a 50-ml volumetric flask and dilute to volume with water (Note 1). Use 5 ml aliquots of this solution for the determination of fluorine. Add 6 drops of 6 M sodium hydroxide solution, mix well and add 3 ml of eriochrome cyanine R solution (Note 2) followed by 3 ml of zirconyl chloride solution, dilute to volume with water, mix well and allow to stand for 60 to 90 minutes before photometric measurement.

Notes: 1. This solution can be used also for the determination of chlorine.

2. The zirconyl chloride and eriochrome cyanine R solutions should contain twice the weight of reagent as the solutions prepared for the procedure described above.

Pyrohydrolytic Determination of Fluorine

When fluorine containing minerals are heated in a stream of moist air hydrolysis occurs and the fluorine is evolved - probably as hydrogen fluoride. Complete evolution of fluorine can be achieved by the use of a suitable flux and an accelerator. The flux suggested is a mixture of the oxides of bismuth and vanadium with the addition of tungstic oxide as accelerator.(16) The use of this mixture enables a relatively low temperature of 655-665° to be used. The liberated hydrogen fluoride is collected in sodium hydroxide solution and determined titrimetrically by a cerous-EDTA method.

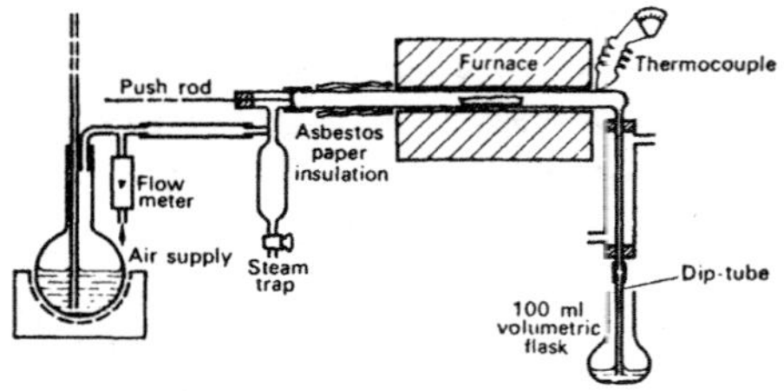

FIG 15 Apparatus for the recovery of fluorine by pyrohydrolysis

Method

Apparatus. This is shown in Fig. 15. It consists of a horizontal tube furnace with a chamber 30 cm in length by 4 cm in diameter, together with a reaction tube fabricated from silica with a narrow diameter side tube, water cooled. Silica combustion boats are used, of dimensions 75 x 15 x 12 mm. One of the commonest causes of incomplete recovery of fluorine is too low a temperature, and the value indicated should be checked from time to time by observing the melting point of silver sulphate (m.p. 652°) in a boat at the hottest part of the furnace.

Reagents:

Vanadium pentoxide.

Bismuth trioxide

Sodium bismuthate (Note 1)

Tungstic oxide

Sodium hydroxide solution, dissolve 50 g of reagent in 1 litre of water.

p-Nitrophenol indicator solution, dissolve 0.3 g of the solid indicator in water and dilute to 100 ml.

Chloroacetate buffer solution, pH 2.7, dissolve 9.45 g of monochloroacetic acid in 50 ml of water. Using a pH meter adjust the pH of the solution to 2.7 by adding sodium hydroxide solution, and then dilute to 100 ml with water.

Xylenol orange indicator solution, dissolve 0.067 g of the solid reagent in a mixture of 25 ml of water with 25 ml of ethanol, dilute to 100 ml with water and filter if necessary. Prepare a fresh solution every few days.

Gelatin, powdered.

Acetate buffer solution, pH 6, dissolve 46 g of ammonium acetate and 18 g of sodium acetate trihydrate in a litre of water. Using a pH meter adjust the pH to a value of 6 by adding glacial acetic acid.

Cerous nitrate 0.05 M solution, dissolve 21.7 g of cerous nitrate hexahydrate in water and dilute to 1 litre.

Arsenazo III indicator solution, dissolve 0.02 g of the solid reagent in 100 ml of water containing 2 drops of sodium hydroxide solution.

Sodium fluoride, pure.

Ethylenediaminetetraacetic acid, 0.025 M solution, dissolve 9.31 g of the disodium salt of EDTA dihydrate in water and dilute to 1 litre. Standardise by weighing out 0.1 g of the pure sodium fluoride, transfer to a 100-ml volumetric flask and dissolve in about 30 ml of water. Add 10 ml of sodium hydroxide solution and follow the procedure given below, beginning with the addition of p-nitrophenol indicator.

Procedure. Grind together in an agate mortar 1.5 g of vanadium pentoxide, 0.5 g of bismuth trioxide and 0.2 g of tungstic oxide. Add a weighed portion of the sample material containing from 1 to 30 mg fluorine, mix thoroughly and grind together. Brush the mixture into the silica combustion boat and press down with a spatula.

Check that the furnace temperature is between 655° and 665°. Adjust the air flow to between 100 and 110 litres per hour and the steam rate to condense at about 0.5 ml per minute (Note 1). Remove the air/steam inlet tube from the reaction tube and wipe the inlet dry. Run 10 ml of sodium hydroxide solution into a 100-ml volumetric flask and connect to the condenser as shown in Fig. 15. Insert the boat containing the sample into the end of the reaction tube, replace the air/steam inlet and push the boat just inside the furnace. Advance the boat about 4 cm every minute until it is just in the hot zone of the furnace, withdrawing the push rod each time to cool part of the tube. Allow the pyrohydrolysis to continue for 30 minutes (Note 2) after the furnace has regained its original temperature.

Disconnect the dip-tube and rinse down into the flask using not more than 5 ml of water. Add 1 drop of p-nitrophenol indicator solution and neutralise with 5 N nitric acid. Add 0.5 ml of chloroacetate buffer and two or three drops of xylenol orange indicator solution. If the colour is red, titrate just to yellow with EDTA solution to complex any traces of metal compounds that may have distilled or been carried over. Add 5-10 mg of powdered gelatine and exactly 20 ml of cerous nitrate solution and warm to about 40° in a water bath for 5 minutes. Add 10 ml of approximately 0.1 N nitric acid, dilute to volume with water and filter through a <u>dry</u> open-textured filter paper. Pipette 50 ml of the filtrate into a 250-ml conical flask, add 10 ml of acetate buffer solution and 1-2 ml of arsenazo III indicator solution. Titrate to a red end-point with the EDTA solution (Note 3).

Carry out a blank determination, beginning with the addition of 10 ml of sodium hydroxide to the volumetric flask.

<u>Notes</u>: 1. The apparatus should be steamed through at the beginning of each batch of samples. Blank determinations should be made at the beginning and end of each day's run.

2. In the presence of appreciable amounts of lead as galena, or barium as barytes, it may be necessary to increase the time of pyrohydrolysis to 40 minutes to complete the evolution of fluorine. In these cases, and also for other sulphide minerals, 0.3 g of the bismuth trioxide should be replaced with an equal weight of sodium bismuthate. This prevents the tendency for sulphur compounds to distil.

3. 1 ml of 0.05 M cerous nitrate solution is equivalent to 2.85 mg of fluorine. As only one-half of the hydrolysate is titrated, each ml of 0.025 M EDTA titre is equivalent to 2.85 mg fluorine.

A much higher temperature, obtained by the use of an induction furnace, is described by Farzaneh and Troll.(18) The rock material is mixed with silica gel, ground with lead dioxide, mixed with metallic copper and pyrolysed for about 10 minutes in an induction furnace giving a temperature of about 1200°. The fluorine determinations were completed using a fluoride-selective electrode. This combination of pyrohydrolysis with fluoride-selective electrode has also been used by Hutchison(19) for the determination of fluorine.

Determination of Fluorine using an Ion-selective Electrode

Fluoride-selective electrodes are now widely used for the determination of fluorine in silicate rocks. They are used in conjunction with reference electrodes, which may be of the double junction calomel type, in which the outer chamber is filled with sodium hydroxide-citrate buffer at pH 6.0.

A variety of procedures have been described using rock decomposition with sodium carbonate and zinc oxide,[20,23] a mixture of sodium and potassium carbonates,[21] lithium metaborate[22] and sodium carbonate with potassium nitrate.[23] The procedure given below is based upon that of Troll et al.[21]

Method

Apparatus. A fluoride-selective electrode, reference electrode and pH meter suitable for measuring or recording solution potentials.

Reagents: Buffer and complexing solution, dissolve 58 g of sodium chloride and 294 g of sodium citrate dihydrate in 500 ml of water. Add a solution of 18.2 g of 1:2-cyclohexanediaminotetraacetic acid and 4 g of sodium hydroxide in 100 ml of water. Adjust pH to 6.0 with either hydrochloric acid or aqueous sodium hydroxide and dilute to 1 litre with water.

Procedure. Accurately weigh from 0.05 to 1.0 g (depending on the expected fluorine content) of the finely ground silicate material into a platinum crucible already containing 4 g of a sodium carbonate-potassium carbonate mixture. Mix carefully without entraining the layer of carbonate adjacent to the walls of the crucible and cover with a further 4 g of carbonate mixture.

Place the crucible in a pre-heated furnace set at a temperature of 500° and gradually raise the temperature to 950°. After an hour raise the temperature to 1000°C for a further period of 30 minutes. Allow to cool and extract the melt in a polyethylene beaker containing 20 ml of M hydrochloric acid and 60 ml of water. A further small quantity of hydrochloric acid may be added if needed to assist the decomposition of the melt and ensure a slightly acid solution. Dilute this solution to volume with water in a 250-ml volumetric flask, ignoring any residue that may be present. Transfer to a polyethylene bottle for storage

Transfer by pipette 5 ml of this rock solution to a 100-ml polyethylene beaker, add 50 ml of the buffer and complexing solution and measure the solution potential using the fluoride selective electrode.

The initial potential reading, taken after one hour and measured while the solution is being stirred will give a measure of the fluorine concentration by reference to a calibration curve prepared from standard fluoride solutions. Alternatively, additions of standard fluoride solution may be made to the rock solution to effect the calibration. An elegant procedure for this has been described by Troll et al.[21] Attention has however been drawn[25] to a revision of the derived formula for concentration by the addition technique of Durst.[26]

References

1. BERZELIUS J J., reported in HILLEBRAND W F. (ref. 2)
2. HILLEBRAND W F, The Analysis of Silicate and Carbonate Rocks, U S Geol. Surv. Bull. 700, p.225, 1919
3. WILLARD H H and WINTER O B., Ind. Eng. Chem. Anal. Ed. (1933) 5, 7
4. HOFFMAN J I and LUNDELL G E F., Bur. Stds, J. Res. (1929) 3, 581
5. CHEBURKOVA E E., Zavod. Lab. (1950) 16, 1009
6. STEIGER G., J. Amer. Chem. Soc. (1908) 30, 219
7. MERWIN H E., Amer J. Sci. 4th Ser. (1908) 28, 119
8. MEGREGIAN S., Analyt. Chem. (1954) 26, 1161
9. SARMA P L., Analyt. Chem. (1964) 36, 1684
10. EVANS W H and SERGEANT G A., Analyst (1967) 92, 690
11. BELCHER R, LEONARD M A and WEST T S., J. Chem. Soc. (1959) 2577
12. BELCHER R, LEONARD M A and WEST T S., Talanta (1959) 2, 92
13. JEFFERY P G and WILLIAMS D., Analyst (1961) 86, 590
14. JEFFERY P G., Geochim. Cosmochim. Acta (1962) 26, 1355
15. SHKROBOT E P and TOLMACHEVA N S., Zhur. Anal. Khim. (1976) 31, 1491
16. KERR G O., Unpublished work.
17. CLEMENTS R L, SERGEANT G A and WEBB P J., Analyst (1971) 96, 51
18. FARZANEH A and TROLL G., Geochem. J. (1977) 11, 177
19. HUTCHISON D, Unpublished work.
20. INGRAM B L., Analyt. Chem. (1970) 42, 1825
21. TROLL G, FARZANEH A and CAMMANN K., Chem. Geol. (1977) 20, 295
22. BODKIN J B., Analyst (1977) 102, 409
23. HAYNES S J., Talanta (1978) 25, 85
24. HUANG W H and JOHNS W D., Anal. Chem. Acta (1967) 37, 508
25. TROLL G., personal communication.
26. DURST R A., Nat. Bur. Stds. Publ. 314, p.385, Washington DC, USA (1969)

CHAPTER 21

Gallium

Spectrophotometric Determination of Gallium

The most frequently used reagent for the determination of gallium is the dyestuff rhodamine B. Kuznetzov and Bol'shakova[1] have, however, noted the unfavourable effect of the carboxyl group of this compound upon the solubility of the rhodamine complex and recommended the use of the butyl ester under the name butylrhodamine. This latter reagent was used by Skrebkova[2] for the determination of gallium in unroasted lead dust, bauxites, copper-zinc ores and quartz-topaz greisens.

In the procedures based upon rhodamine B, the red-coloured compound formed with the chlorogallate anion is extracted into organic solution prior to measurement of the optical density. Onishi and Sandell[3] have shown that the most favourable extraction into benzene is achieved from a 6 N hydrochloric acid solution containing 0.036 per cent rhodamine B. Under these conditions the extraction coefficient $Ga_{benzene}/Ga_{water}$ has a value of 0.57 at about 25°. This extraction can be increased by adding acetone to the solution as described by Kuznetsov and Tananaev,[4] by replacing the benzene with a mixture of chlorobenzene and carbon tetrachloride (Culkin and Riley[5]), or by a combination of both these methods.

Straight-line calibrations are obtained at the concentrations used, and the maximum absorption is at a wavelength of 562 nm.

Under the conditions used to extract the chlorogallate, coloured organic extracts are also given by rhodamine B with antimony (III), gold (III), thallium (III) and iron (III). Kuznetsov and Tananaev[4] reduced all these elements to lower valency states with titanous chloride before extracting gallium. Onishi and Sandell[3] separated gallium from these interfering elements by a prior extraction into di-isopropyl ether. This separation was used also by Burton et al[6] following the recovery of germanium with carbon tetrachloride, in a procedure for the determination of these two elements in the same sample portion. The procedure below describes the separation with di-isopropyl ether, followed by the extraction of the rhodamine B chlorogallate into chlorobenzene-carbon tetrachloride in the presence of acetone.

Method

Reagents: Titanous chloride solution, approximately 15 per cent.

Di-isopropyl ether, freshly distilled from solid sodium hydroxide, b.p. 68°. As with diethyl ether, peroxide formation can give unstable, explosive mixtures.

Rhodamine B, 0.5 g dissolved in 100 ml of water.

Acetone.

Chlorobenzene-carbon tetrachloride solvent, add 125 ml of carbon tetrachloride to 375 ml of monochlorobenzene (Note 1).

Standard gallium stock solution, dissolve 0.5 g pure gallium in 50 ml of 6 N hydrochloric acid. Transfer to a 1-litre volumetric flask, add 60 ml of concentrated hydrochloric acid and dilute to volume with water. This solution contains 0.5 mg Ga per ml in N hydrochloric acid.

Standard gallium working solution, prepare a dilute solution containing 1 μg Ga per ml by serial dilution of the stock solution with N hydrochloric acid.

Procedure. Accurately weigh about 0.2 g of the finely powdered rock material into a small platinum crucible and add 2 ml of 20 N sulphuric acid, 0.5 ml of concentrated nitric acid, and 5 ml of hydrofluoric acid. Heat the crucible on a hot plate or sand bath until copious fumes of sulphuric acid are evolved. Allow to cool, rinse down the sides of the crucible with a little water and again evaporate, this time to dryness. Dissolve the residue in approximately 25 ml of 6 N hydrochloric acid and transfer the solution to a 100-ml separating funnel.

Add a 15 per cent titanous chloride solution drop by drop until the yellow colour of the ferric ion is completely discharged, and then add a few drops in excess. Add 10 ml of di-isopropyl ether and shake for 20 to 30 seconds. Run the aqueous phase into a second separating funnel and shake again with 5 ml of di-isopropyl ether. Repeat the extraction with a further 5 ml portion of ether, discard the aqueous layer and combine the organic extracts in a 50-ml beaker.

Evaporate the ether in a fume cupboard and add 15 ml of 6 N hydrochloric acid to the dry residue. Warm gently and then transfer the solution to a 100-ml flask. Rinse the beaker with a little 6 N hydrochloric acid, add the washings to the flask and dilute to a volume of 25 ml, also with 6 N hydrochloric acid. Add titanous chloride solution drop by drop until a faint violet colour persists, indicating that an excess is present, and allow the solution to stand for 10 minutes. Add 2 ml of the aqueous rhodamine B solution followed by 5 ml of acetone, mix well and allow to stand for a further period of one hour. Using a safety pipette, transfer 10 ml of the chlorobenzene-carbon tetrachloride solvent and shake vigorously to extract the chlorogallate into the organic phase.

Allow the phases to separate. Transfer the lower, organic layer to a 1-cm spectrophotometer cell and measure the optical density against the pure organic solvent at a wavelength of 562 nm. Measure also the optical density of a reagent blank solution prepared in the same way as the rock solution but omitting the sample material.

Calibration. Transfer 0-10 ml aliquots of the standard gallium solution containing 0-10 μg gallium to separate 100-ml flasks and dilute each to 25 ml with 6 N hydrochloric acid. Extract the gallium as chlorogallate using the procedure described above and measure the optical densities in the same way. Plot the relation of optical density to gallium concentration.

Note: 1. Where reagent grade solvents were used, the chlorogallate extracts were found to fade appreciably. This was avoided by allowing the solvent mixture to stand over concentrated sulphuric acid for a few days, separating the organic layer, shaking with aqueous sodium hydroxide, and finally shaking with water.

Determination of Gallium by Atomic Absorption Spectroscopy

The level at which gallium occurs in most silicate rocks and minerals is too low for direct determination by this technique using a typical rock solution directly aspirated into a suitable flame, or when using a solvent extraction technique to effect a concentration increase.(7) Such a procedure has been described by Lypka and Chow(8) for the determination of gallium in ores using a 4 to 5 g sample portion, evaporation with sulphuric and hydrofluoric acids, extraction from chloride solution into isopropyl ether, back extraction into hydrochloric acid and direct aspiration into a lean air-acetylene flame. A considerable increase in sensitivity was obtained by Chow and Lipinsky(9) using a nitrous oxide-acetylene flame, which had the added advantage of eliminating both matrix and background effects that were noted with the lower temperature flame.

Direct volatilisation from the solid state using a graphite furnace was described by Langmyhr and Rasmussen(10) as giving a detection limit of 2.5 ng gallium - equivalent to 5 or 0.25 ppm when using the 0.5 and 10 mg sample weights suggested.

The inherent higher sensitivity obtainable using flameless atomic absorption can be combined with a solvent extraction concentration step which serves also to reduce matrix interferences. Warren and Harrison,[11] reporting on the routine application of flame and flameless atomic absorption spectroscopy to rock analysis, used this procedure for the determination of gallium. Methylisobutylketone was used to extract gallium from the rock solution adjusted to 6M in hydrochloric acid and containing ascorbic acid to minimise the co-extraction of iron.

References

1. KUZNETZOV V I and BOL'SHAKOVA L I., Zhur. Anal. Khim (1960) 15, 523
2. SKREBKOVA L M., Zhur. Anal. Khim (1961) 16, 422
3. ONISHI H and SANDELL E B., Anal. Chim. Acta (1955) 13, 159
4. KUZNETSOV V K and TANANAEV N A., Ivz. Vyssh. Uchebn. Zavedenii, Khim.i Khim. Tekhnol. (1959) 2, 840
5. CULKIN F and RILEY J P., Analyst (1958) 83, 208
6. BURTON J D , CULKIN F and RILEY J P., Geochim. Cosmochim. Acta (1959) 16, 151
7. CRESSER M S and TORRENT-CASTELLET J., Talanta (1972) 19, 1478
8. LYPKA G N and CHOW A., Anal. Chim. Acta (1972) 60, 65
9. CHOW A and LIPINSKY W., Anal. Chim. Acta (1975) 75, 87
10. LANGMYHR F J and RASMUSSEN S., Anal. Chim. Acta (1974) 72, 79
11. WARREN J and HARRISON M P., Proc. Anal. Div. Chem. Soc. (1976) Sept, 290

CHAPTER 22

Germanium

Spectrophotometric Determination of Germanium

A number of reagents have been described for the spectrophotometric determination of germanium, of which the most popular is phenylfluorone (9-phenyl-2,3,7-trihydroxy-6-fluorone or 2,3,7-trihydroxy-9-phenylxanthen-6-one). This reagent introduced by Gillis, Hoste and Claeys[1] as a spot test reagent for germanium, has been studied by Cluley[2,3] for the determination of germanium in the presence of a wide variety of other elements. The complex formed by germanium with phenylfluorone has a maximum optical density at a wavelength of 504 nm, and the Beer-Lambert Law is obeyed at the concentrations encountered in silicate rocks.

Although it is suggested[4] that prior separation of the germanium is not necessary, most authors have recommended either distilling or extracting germanium from hydrochloric acid as a means of separation from other elements. The distillation serves also to concentrate the germanium in a small volume, free from all elements except some of the sulphur, arsenic, antimony and tin present in the original rock. In the amounts in which they are normally present in silicate rock, these elements do not interfere.

The extraction procedure, described by Schneider and Sandell,[5] is also usually applied from a hydrochloric acid solution. For the amounts of germanium normally encountered, a single extraction with an equal volume of carbon tetrachloride was regarded as sufficient. Burton et al[6] however suggest that only 85-90 per cent of the germanium is extracted in a single operation, and recommend three successive extractions. The addition of EDTA has been recommended[7] as a means of preventing the extraction of other elements into the carbon tetrachloride solution.

Silicate rocks and minerals can be decomposed by evaporation with a mixture of sulphuric, nitric and hydrofluoric acids. After the removal of excess sulphuric acid the residue can be dissolved in hydrochloric acid. Hybbinette and Sandell[8] have shown that this method of attack can be used for samples containing even as much as 0.05 per cent chlorine, without loss of germanium by volatilisation of germanium tetrachloride. This is presumably due to the formation of the involatile fluogermanic acid.

Method

Reagents:

Carbon tetrachloride

Ethylenediaminetetraacetic acid solution, dissolve 1 g of the disodium salt in water and dilute to 200 ml.

Gum arabic solution, dissolve 0.1 g in about 80 ml of boiling water, filter and dilute to 100 ml with water.

Phenylfluorone solution, dissolve 0.05 g of the solid reagent in 75 ml of ethanol and 5 ml of 5 N sulphuric acid by warming on a steam bath. When cold, dilute to 100 ml with water.

Standard germanium stock solution, weigh 0.036 g of pure germanium oxide into a small beaker, add 0.5 g of sodium hydroxide and 20 ml of water. When solution is complete transfer to a 250-ml volumetric flask and dilute to volume with water. This solution contains 100 μg Ge per ml.

Standard germanium working solution, dilute 10 ml of the stock solution to 1 litre with water as required. This solution contains 1 μg Ge per ml.

Procedure. Accurately weigh approximately 0.5 g of the finely powdered silicate rock material into a platinum dish and decompose by evaporation to fumes with 3 ml of 20 N sulphuric acid, 0.5 ml of concentrated nitric acid and 5 ml of concentrated hydrofluoric acid. Allow to cool, rinse down the sides of the dish with a little water and again evaporate to fumes. In each case avoid prolonged or strong fuming with sulphuric acid. When cold, add 5 ml of water and heat to near boiling for a few minutes to disintegrate the residue. Cool and transfer the mixture to a separating funnel, completing the transfer with the aid of 5 ml of concentrated hydrochloric acid. Stopper the funnel and allow to stand with occasional shaking until all solid material has dissolved, or for 30 minutes if it does not all dissolve. If the sample material is rich in calcium or barium, some insoluble sulphate will remain undissolved, but this does not interfere.

Add sufficient concentrated hydrochloric acid to give an acid concentration of 9 molar, mix well and cool to room temperature. Now add 10 ml of carbon tetrachloride and shake vigorously for 2 minutes. Run the organic layer off into a small beaker and repeat the extraction with two successive 10 ml portions of carbon tetrachloride. Combine the organic extracts in a separating funnel, add exactly 5 ml of water containing 1-2 drops of 0.01 M sodium hydroxide solution and shake for 2 minutes. Discard the organic phase and transfer the aqueous solution to a 10-ml volumetric flask. Add 1 ml of 10 N sulphuric acid, 0.4 ml of EDTA solution, 2.4 ml of ethanol, 0.4 ml of gum arabic solution and 0.6 ml of the phenylfluorone solution. Dilute the solution to volume with water, mix and allow to stand for 1 hour at room temperature. Measure the optical density of the solution relative to water in 1-cm cells, with the spectrophotometer set at a wavelength of 504 nm. Measure also the optical density of a reagent blank solution prepared in the same way as the sample solution, but omitting the rock material.

Calibration. Transfer 1-5 ml aliquots of the standard germanium solution containing 1-5 μg Ge to a series of 10-ml volumetric flasks. Add to each 10 N sulphuric acid, EDTA solution, ethanol, gum arabic and phenylfluorone solution as described above. Dilute each solution to volume with water, allow to stand and measure the optical density at 504 nm in the same way as the sample solution. Plot the relation of optical density to germanium concentration.

Determination of Germanium by Atomic Absorption Spectroscopy

The determination of germanium by atomic absorption spectroscopy is difficult by direct solution methods. Even using acetone-water solvent systems and nitrous oxide-acetylene flames the sensitivity is too low for determination in silicate rocks.(9) The use of a graphite tube atomiser by Johnson *et al*(10) has the advantage that only very small quantities of solution are required. The possibility of using the hydride of germanium generated by reaction with sodium borohydride was examined by Smith.(11) None of these methods appears to have been so far adopted for the determination of germanium in rock material.

References

1. GILLIS J, HOSTE J and CLAEYS A., *Anal. Chim. Acta* (1947) 1, 302
2. CLULEY H J., *Analyst* (1951) 76, 523
3. CLULEY H J., *Ibid.* (1951) 76, 530
4. DEKHTRIKYAN S A., *Izv. Akad. Nauk Armyan S S R Nauka Zemle* (1966) 19, 97
5. SCHNEIDER W A and SANDELL E B., *Mikrochim. Acta* (1954) 263
6. BURTON J D, CULKIN F and RILEY J P., *Geochim. Cosmochim. Acta* (1959) 16, 151
7. BURTON J D and RILEY J P., *Mikrochim. Acta* (1959) 586
8. HYBBINETTE A G and SANDELL E B., *Ind. Eng. Chem. Anal. Ed.* (1942) 14, 715
9. POPHAM R E and SCHRENK W G., *Spectrochim. Acta* (1968) 23B, 543
10. JOHNSON D J, WEST T S and DAGNALL R M., *Anal. Chim. Acta* (1973) 67, 79
11. SMITH A E., *Analyst* (1975) 100, 300

CHAPTER 23

Hydrogen

Hydrogen occurs in a number of roles in silicate rocks, including free or elemental hydrogen, organically bound hydrogen, adsorbed moisture, water of crystallisation, "combined water", and as hydroxyl groups present within the lattice structure of a number of minerals. It is not always possible to differentiate between some of these forms, particularly where some of the silicate minerals have been altered and are represented by later, more hydrous species.

Most silicate rocks release considerable volumes of hydrogen on ignition[1]. Tokhtuyev and Frantsuzova[2] have demonstrated that free or elemental hydrogen occurs in metamorphic rocks, by recovering the gases evolved during the powdering of specimens under vacuum, without heating - a technique developed earlier by Elinson.[3] Shorokhov[4] has also reported the wide occurrence of free hydrogen in sedimentary rocks.

Organic material is an essential component of many silicate rocks, particularly those of sedimentary origin. The separation and determination of organic material is beyond the scope of this book.

Water Evolved at 105°

"Hygroscopic water", "moisture", "water evolved at 105°", and "non-essential water" are all terms that have been used to describe the loss in weight experienced by a powdered rock specimen on heating to constant weight at 105°. The results of this determination are used to give a measure of the "free" as distinct from the "combined" water. In this respect it is an unsatisfactory measure in that it includes a certain amount of the essential water of some easily decomposed minerals (such as zeolites) whilst at the same time not including some of the non-essential water of many other minerals. This determination is, however, always included in the list of thirteen constituents accepted as the minimum required for a full analysis of any silicate rock.

It is the custom in some laboratories to remove this "moisture" or "hygroscopic water" by drying the ground rock powder at a temperature in the region 105-110°. This procedure should not be used where the presence of much chlorite would indicate that the dried material is likely to be extremely hygroscopic. The day-to-day variation of the weight of such samples is often considerable, even for samples that have not previously been dried. This is shown in Fig. 16; the weight changes of a chloritised

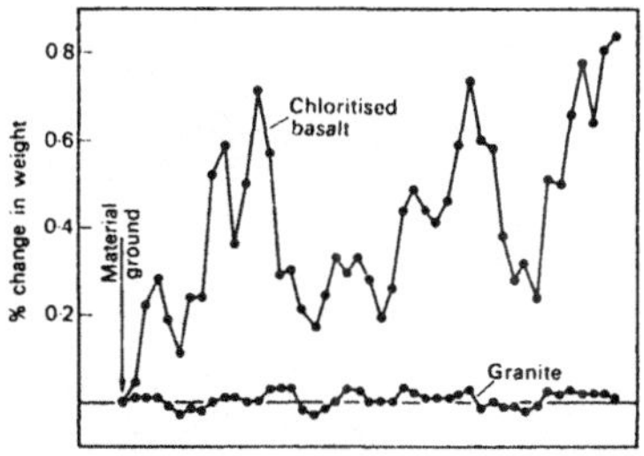

FIG 16 Weight changes of two ground silicate rocks on exposure to air

basalt were as much as 0.73 per cent of the initial weight of the material. In contrast, a granite, ground and exposed to the atmosphere at the same time as the altered basalt, changed weight by only 0.03 per cent during the same period. Where such chloritised rocks are being examined, it is advisable to weigh out at one time all the several portions required for analysis, including that for moisture and combined water.

The water content of the silicate rock sample will vary with the extent of grinding of the sample. Hillebrand,[5] for example, shows that for a number of rocks excessive grinding results in an increase of the water content of the sample, and that in the presence of certain minerals, gypsum for example, the reverse may occur, and water be lost by overgrinding.

The simplest way of determining "water evolved at 105°", is to measure the loss in weight of about 2 g of the powdered sample when heated in electric oven for 2 to 3 hours or even longer, until a constant weight is obtained. The former practice of using toluene baths now appears to be obsolete. Certain rocks containing much ferrous iron may not reach constant weight, due to the slow continual oxidation that takes place at this temperature. Many rocks are hygroscopic when completely dry; for this reason a squat weighing bottle with a closely fitting lid, as in Fig.17 is commonly used for the determination.

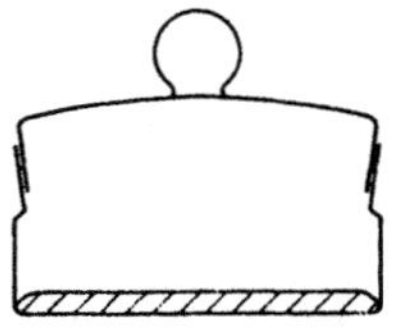

FIG 17 Apparatus for determining loss in weight by heating at 105°

The difficulties associated with the weighing of a very hygroscopic material and with the presence of much easily-oxidised ferrous iron can be avoided by making a direct determination of the water evolved. An apparatus for this, described by Jeffery and Wilson,[6] is shown in Fig. 18. It consists of a closed-circulation system comprising a small electric pump, a heating chamber maintained at a temperature of 104-105° by means of boiling isobutyl alcohol (a thermostatically controlled heating block set at 105° may be more convenient), two absorption tubes and a bubbler containing orthophosphoric acid to indicate the rate of flow of air through the apparatus. Before inserting the sample, tube A is replaced by a short piece of glass tubing, and the air in the system is dried by operating the pump for about 30 minutes. Tube A is weighed and replaced, the sample boat inserted, and the circulation of air continued until all the water evolved from the sample has passed into the absorption tube. In practice the apparatus is kept running for about 2 hours, which is sufficient to drive over all the water, including any that may condense on the cooler parts of the heating chamber.

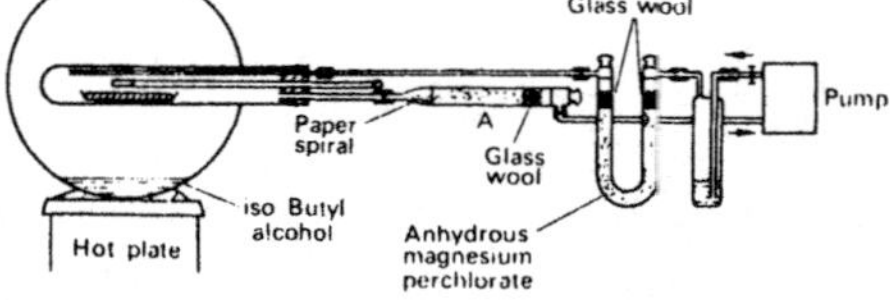

FIG 18 Apparatus for the direct determination of water evolved by heating at 105°

The blank value, ie the increase in weight of the absorption tube A when the determination is carried out without the insertion of a sample, is less than 0.1 mg. Tube A, which is used in place of a more conventional U-tube, contains a paper spiral that prevents clogging of the magnesium perchlorate when large amounts of water are absorbed. The open end of the tube, fitted with a B7 ground glass joint, is closed with a cap during weighing. When fully packed, the tube and cap together weigh 20-25 g.

METHOD OF DEAN AND STARK

Shales and certain altered igneous rocks that contain several per cent of moisture, often lose part of this water content in the process of sample preparation. For these materials and for clays that cannot be prepared for chemical analysis without drying, an approximate moisture content can be obtained by the method of Dean and Stark,[7] in which the roughly crushed or broken material, 20-50 g in weight, is boiled with toluene under reflux apparatus (Fig. 19). Moisture evolved at 105° collects in the side arm, which is calibrated directly in millilitres.

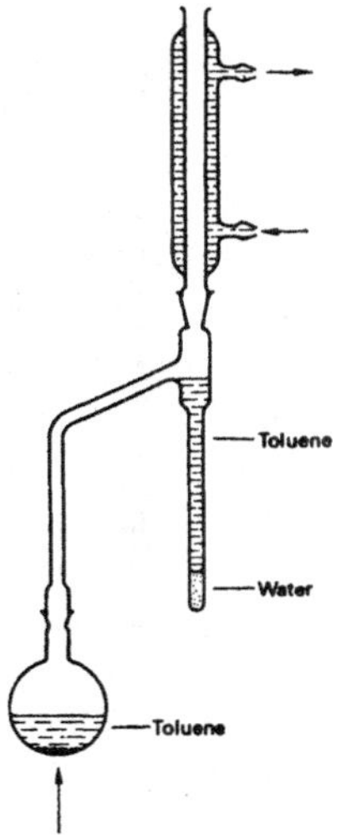

FIG. 19 Dean and Stark apparatus for the determination of water

Total Water

The methods used for determining total water rely upon the direct collection and weighing of all the water than can be expelled from the sample by heating to a high temperature. One of the earliest methods was that of Penfield[8] in which the sample, contained in a bulb blown at the end of a hard-glass tube, was heated at a temperature just below fusion point of the glass. The water released condensed on the cooler parts of the tube, from which the heated part could be drawn away. The water content of the sample was then obtained by weighing the cooler part before and after drying. The method suffers from the difficulty of ensuring that all of the water does condense, particularly under conditions of low humidity, and also of ensuring that all the water present in the sample had been liberated at the particular temperature used.

The Penfield method does have the advantage of being both simple and rapid, and for this reason a number of modifications to it have been described, notably those of Courville,[9] Shapiro and Brannock[10] and Harvey.[11]

COURVILLE-PENFIELD METHOD

This method is close to the original Penfield method, in that after heating the sample, the containing bulb is removed and the remaining part of the glass tube is weighed before and after drying. In order to increase the efficiency of liberation of water from the sample, it is mixed with lead oxide (litharge) before transfer to the ignition tube. The mixture fuses on heating and the tube can then be gently rotated. Additional precautions are necessary when certain volatiles, such as fluorine, are present in the sample.

A similar method for use on a micro scale was described earlier by Sandell,[12] 20-30 mg of sample were used with litharge if fluorine or other volatile matter was present, or ignited lime if sulphur trioxide or hydrogen chloride were likely to be evolved on heating.

SHAPIRO AND BRANNOCK-PENFIELD METHOD[10]

This method has been used as part of a scheme for the rapid, complete analysis of silicate rocks. The sample is heated with sodium tungstate in a pyrex glass boiling tube, and the water evolved is collected in a weighed strip of filter paper. Not all the water is retained by the filter paper, and an empirical correction is necessary; this is in the form of an addition of 10 per cent of the recorded weight if less than 20 mg are obtained, or 2 mg if more than 20 mg are obtained. It is difficult to recommend this method except when a rapid, approximate result is required.

HARVEY-PENFIELD METHOD[11]

In this method the sample is contained in a silica tube, and the water that is evolved on heating is absorbed in anhydrous calcium chloride contained in a tube attached to the heating chamber. Improvements to this early procedure of Harvey were described by Wilson,[13] who used anhydrous magnesium perchlorate and the apparatus shown in Fig. 20 , in which the ground glass joints were lubricated with PTFE sleeves. These sleeves permit the use of higher working temperatures. A heating period of 1 hour was used, although 10 minutes were stated to be sufficient in some cases. Before and after the ignition, the absorption tube is sealed with a cap and plug for weighing.

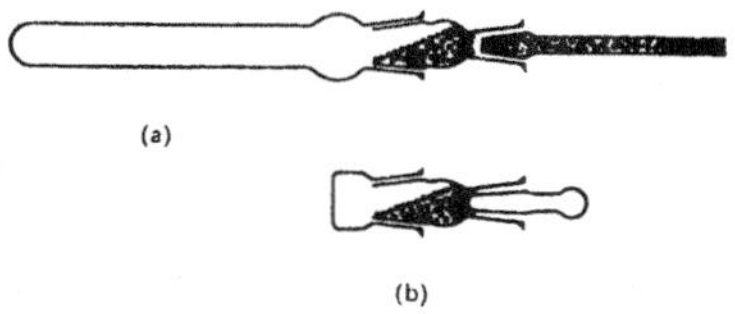

FIG 20 Wilson's apparatus for determining the total water content of silicate rocks, (a) assembled for use, (b) capped and plugged for weighing.

The Penfield method, and the modifications described, suffer from the disadvantage that many minerals are not completely decomposed at the temperature used for the ignition, and that further quantities of water can be obtained by heating these minerals to a higher temperature. Minerals that behave in this way include talc, topaz, staurolite, cordierite and epidote. Other disadvantages of the original method and some of the modifications are that the collection of the water evolved is not always complete, and that other volatile constituents (sulphur and fluorine have been noted) are sometimes collected and weighed as water. The alternative to the Penfield method is to collect the liberated water in an absorption tube that can be weighed before and after the ignition. The sample material, contained in an alumina or platinum boat, can be heated to a temperature of 1000° in a silica tube furnace heated by either gas burners or electricity, and the water evolved can be collected using magnesium perchlorate or other suitable absorbent. This method, described in detail by Groves,(14) can incorporate provision for retaining gases and vapours such as fluorine, sulphur and oxides of sulphur. Using a platinum boat and a flux, such as sodium tungstate, even refractory minerals such as staurolite can be decomposed in a reasonable time. Sodium carbonate and metafluoborate have also been used as fluxes in this way. Blank values are always rather high when fluxes are used, and are by no means negligible with simple ignition.

In an effort to reduce these blank values Jeffery and Wilson(6) used the "closed-circulation system" as a modification to both the ignition method as described by Groves, and a simple fusion using a mixture of equal weights of sodium tungstate and borax glass. The blank value for either determination should be no more than 0.1 mg if the air is recycled within the apparatus as described. If the air is drawn through the apparatus as described by Groves,(14) a blank value of several milligrams will probably be obtained when the apparatus is first set up, but this should fall to a constant value of about 0.5-1 mg after several days operation.

Procedure. The apparatus is shown in Fig. 21 , it consists of a silica tube containing the sample in an alumina or unglazed porcelain boat that had previously been ignited to 1000°. A current of air from the small circulating electric pump is passed into a U-tube containing magnesium perchlorate and soda asbestos, over the sample, through a basic lead chromate packing and then to a weighed absorption tube containing anhydrous magnesium perchlorate. The circulation is completed through a bubbler containing orthophosphoric acid, which indicates the rate of flow of air through the apparatus, and finally back to the pump. A packing of basic lead chromate, retained in place by copper spirals, and kept at a temperature of 300-400°, serves to retain oxides of sulphur. Copper wire and silver pumice at a temperature of 700-750° have also been used for this purpose. The wide end of the silica tube is closed with a silicone rubber bung.

The absorption tube is replaced with a short length of glass tubing, and the air within the apparatus dried by operating the pump for 1 hour. During this time the packing heater is allowed to reach its operating temperature. The tube is then weighed and replaced. The weighed sample, contained in an inert boat is inserted into the furnace, which is then allowed to reach its maximum temperature, and is

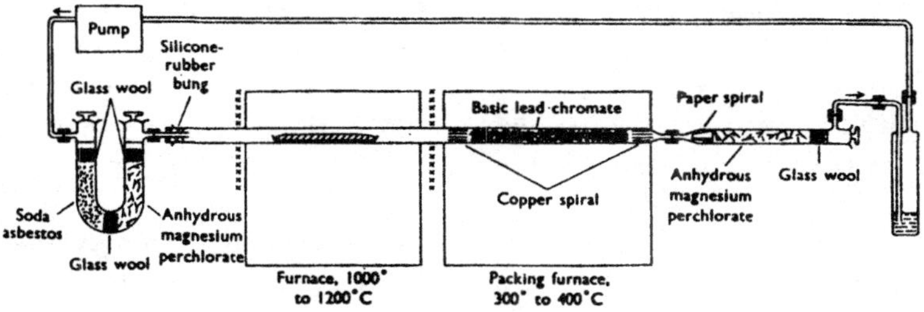

FIG 21 Closed-circulation apparatus for the determination of total water.

maintained at this temperature for 1 hour. The absorption tube is then removed from the apparatus, allowed to stand in a balance case for 30 minutes and is then reweighed to give the weight of total water released.

The chief disadvantage of the methods similar to that described by Groves is that no more than three determination can be completed in any one day, as it is necessary to allow the furnace tube to cool before the next sample can be inserted. In order to increase this number Riley[15] has described an apparatus (see Fig. 13, in chapter on carbon), in which the sample is pushed gradually into the hot zone of a furnace maintained at a sufficiently high temperature to release all the water within about 20 minutes. This enables twelve determinations to be made in a single working day. For most rocks a temperature of 1100° was found to be sufficient, but for refractory minerals, topaz, epidote, staurolite etc, 1200° was required. At this higher temperature it was found possible to determine also any carbon dioxide present in the material by including an absorption tube packed with soda-asbestos in the absorption train. Interference from sulphur compounds was prevented by including a layer of silver pumice and a bubbler containing a saturated solution of chromium trioxide in orthophosphoric acid in the train. A current of nitrogen gas, obtained from a gas cylinder is used, and blank values of 0.1 mg per hour of water and 0.2 mg per hour of carbon dioxide were reported. Details of the procedure using this apparatus for the determination of carbon dioxide and water are given in the chapter on carbon.

GOOCH-SERGEANT METHOD

The problems associated with the releasing of water from refractory materials do not arise when using a technique based upon fusion with anhydrous sodium carbonate, due originally to Gooch.[16] In this method, the use of an alkaline flux ensures the retention of both sulphur and fluorine. Sergeant[17] has reported the adaptation in which the rock sample is heated with the flux in a platinum or other refractory metal crucible supported inside a silica vessel using a high-frequency induction furnace. Nitrogen is used as carrier gas and sodium chlorate is added to avoid possible loss of water by interaction with any ferrous iron present in the silicate rock.

The Karl Fischer reagent (a solution of iodine and sulphur dioxide in a mixture of methanol and pyridine) can be used for the determination of evolved water. The sample material is mixed with lead oxide[18] or lead chromate[19] and heated in a tube furnace. The water evolved is collected in dry methanol or a mixture of methanol and pyridine for the titration. The reaction of the reagent with water proceeds according to the stoicheiometry of the equation -

$$I_2 + SO_2 + 2H_2O = H_2SO_4 + 2HI$$

References

1. JEFFERY P G and KIPPING P J., Analyst (1963) 88, 266
2. TOKHTUYEV G V and FRANTSUZOVA T A., Geokhimiya (1963) 961
3. ELINSON M M., Izv. Acad. Nauk. SSR., Otd. teknich, nauk. No. 2, 1949
4. SHOROKHOVA N R., Trudy Soyzn. Geologporsk. Kontura Gluagaza pri Sov. Min. USSR. 1960, 64
5. HILLEBRAND W F., US Geol. Surv. Bull. 700, 1919
6. JEFFERY P G and WILSON A D., Analyst (1960) 85, 749
7. As in, eg, BSS 756:1952
8. PENFIELD S L., Am. J. Sci. 3rd Ser. (1894) 48, 31
9. COURVILLE S., Canad. Mineral (1962) 7, 326
10. SHAPIRO L and BRANNOCK W W., Analyt. Chem. (1955) 27, 560
11. HARVEY C O., Bull. Geol. Surv. Gt. Brit. (1939) (1), 8
12. SANDELL E B., Mikrochim. Acta (1951) 38, 487
13. WILSON A D., Analyst (1962) 87, 598
14. GROVES A W., Silicate Analysis, Allen & Unwin, 2nd ed., 1951, p.95
15. RILEY J P., Analyst (1958) 83, 42
16. GOOCH F A., Amer. Chem. J. (1880) 2, 247
17. SERGEANT G A., Rept. Govt. Chem. Lond. (1971) p.111
18. TUREK A, RIDDLE C, COZENS B J and TETLEY N W., Chem. Geol. (1976) 17, 261
19. FARZANEH A and TROLL G., Zeit. Anal. Chem. (1977) 287, 43

CHAPTER 24

Indium

At the ppb levels encountered in silicate rocks, the sensitivity required is difficult to obtain by chemical or spectrochemical methods. For this reason neutron activation analysis[1] is of particular value. Attempts to use atomic absorption spectrometry[2] have not been entirely successful in this application.

A spectrophotometric procedure based upon the use of large samples was used by Rozbianskaya, and described briefly by Ivanov.[3] A 5-g portion of the silicate material was dissolved in a mixture of sulphuric and hydrofluoric acids, and the insoluble material treated by fusion with potassium bisulphate. Iron, titanium and other elements including indium were precipitated with ammonia, and after solution in 5 N hydrobromic acid the indium was recovered by extraction into ether. The indium was returned to the aqueous phase by back-extraction into 6 N hydrochloric acid containing hydrogen peroxide. The complex with rhodamine C was then formed in 2 N hydrobromic acid containing ascorbic acid, and extracted into a 3:1 mixture of benzene and ether. The determinations were completed by a visual comparison with similar extracts prepared from a standard indium solution. For indium contents exceeding 2 μg per ml, the solutions were viewed in transmitted light, for contents less than 2 μg per ml the solutions were examined under ultraviolet light and the fluorescence compared.

References

1. SMALES A A, SMIT J van R and IRVING H M, Analyst (1957) 82, 539
2. LANGMYHR F J and RASMUSSEN S, Anal. Chim. Acta (1974) 72, 79
3. IVANOV V V, Geokhimiya (1963) (12), 1101

CHAPTER 25

Iron

Iron is one of the commonest of elements, occurring to an extent of about 5 per cent by weight of the Earth's crust, and fourth in the list of abundances after oxygen, silicon and aluminium. Native iron, although rare, has been reported from a number of localities. Silicate minerals vary considerably in iron content, and this variation is reflected in the iron content of silicate rocks. Hortonolite and other similar basic rocks may contain 35-40 per cent iron (calculated as FeO), whilst many acidic rocks contain no more than 1 per cent total ferric and ferrous oxides.

Sulphide minerals containing iron are of considerable economic importance. These minerals occur in both igneous and sedimentary rocks and, if present in more than trace amounts, can introduce difficulty in the determination of ferrous iron. Somewhat similar difficulties are encountered in the presence of native or tramp iron and of carbonaceous material - particularly common in mudstones, marls and shales.

Determination of Metallic Iron

Native iron occurs in both massive form and as disseminated grains. The method of analysis given below has been adapted from that described by Easton and Lovering[1] for the metallic phase of chondritic meteorites. It can be used for silicate rocks provided that the sample material has been ground sufficiently finely to liberate all the metallic phase - usually to pass a 200-mesh sieve.

The procedure is based upon an extraction of the metal by heating the sample with an aqueous solution of mercuric and ammonium chlorides as proposed by Friedheim.[2] An anion-exchange resin is used to separate iron from the excess mercuric salt.

Method

Apparatus. An anion-exchange column, with a resin bed 2 cm diameter and 12 cm in length is packed with Dowex 1 x 4, 100-150 mesh (or other similar strongly basic anion exchange) resin. Before use, wash with 100 ml of 6 N hydrochloric acid, 150 ml of 0.5 N hydrochloric acid and again with 100 ml of 6 N hydrochloric acid.

Reagents:

Mercuric chloride

Ammonium chloride

Hydroxyammonium chloride solution, dissolve 10 g of the reagent in 100 ml of water.

1,10-Phenanthroline solution, dissolve 0.2 g of the solid reagent in 200 ml of water.

Sodium citrate solution, dissolve 10 g of the dihydrate in 100 ml of water.

Standard iron stock solution, dissolve 0.100 g of pure iron wire in 10 ml of 6 N hydrochloric acid and dilute to 500 ml with water. This solution contains 0.2 mg Fe per ml.

Standard iron working solution, dilute 25 ml of the stock iron solution to 250 ml with water. This solution contains 20 μg Fe per ml.

Procedure. Accurately weigh about 0.5 g of the finely powdered (200-mesh) sample material into a stoppered 100-ml volumetric flask, add 0.6 g of mercuric chloride, 0.6 g of ammonium chloride and about 70 ml of warm water. Swirl the flask to ensure thorough mixing, wrap in foil to exclude light and then stand on a heated water bath for 4 to 5 days. Collect the residue on an open-textured filter paper and wash several times with water to ensure the complete recovery of all soluble salts.

Evaporate the combined filtrate and washings to dryness on a steam bath, and dissolve the resultant residue in the minimum quantity of 6 N hydrochloric acid. Transfer this solution to the anion exchange column and elute with three column volumes of 6 N acid. Discard the eluate. Elute the iron from the column with four column volumes of 0.5 N hydrochloric acid. The excess mercuric chloride is retained on the anion exchange resin, which can then be discarded. Transfer the solution containing the eluted iron to a 250-ml volumetric flask and dilute to volume with water.

Transfer an aliquot of this solution, containing not more than 500 μg of iron, to a 100-ml volumetric flask, dilute to a volume of about 25 ml, add 5 ml of hydroxyammonium chloride solution and allow to stand for a period of 5 minutes. Now add 10 ml of 1,10-phenanthroline solution and 10 ml of sodium citrate solution and dilute to volume with water. Mix well, allow to stand for 2 hours and measure the optical density against water in a 1-cm cell with a spectrophotometer set at a wavelength of 508 nm. Determine also the optical density of a reagent blank solution similarly prepared, but omitting the sample material. Use the calibration graph to obtain the concentration of iron in the solution and hence calculate the metallic iron content of the sample.

Calibration. Transfer aliquots of 0-25 ml of the standard iron solution containing 0-500 μg of iron, to separate 100-ml volumetric flasks and dilute each solution to 25 ml with water. Add hydroxyammonium chloride, 1,10-phenanthroline and sodium citrate solutions and measure the optical densities as described above. Plot the relation of optical density to concentration of metallic iron. A straight-line graph should be obtained.

Determination of Ferrous Iron

The ease with which ferrous iron is converted into the ferric state by atmospheric oxidation is well known and, during the course of the preparation of the rock sample and at every subsequent stage of the determination, care must be taken to ensure that such oxidation is kept to the minimum. Hillebrand has shown that partial oxidation of the ferrous iron in the rock may occur during the grinding of the sample, and for that reason recommended that samples rich in ferrous iron should be ground only until they passed a 70-mesh sieve. The introduction of mechanical mortars has increased the tendency to over-grind rock samples, and frequent sievings are necessary if serious oxidation is to be avoided. In special cases the grinding may be completed under absolute alcohol.

Most of the extensive literature describing the oxidation of ferrous iron that occurs on grinding, relates to grinding by hand or in mechanically driven agate mortars. The general conclusions reached are not generally applicable to grinding in modern high-speed mechanical equipment such as disc mills with tungsten carbide grinding barrels. These produce finely milled material in a very short time, but in doing so, subject the sample material to temperatures sufficiently elevated to cause extensive oxidation of ferrous iron. Fitton and Gill[3] have shown that in using such a mill, enhanced oxidation occurs within a few minutes. As would be expected the degree of oxidation is strongly dependent upon mineralogical composition, with hydrous silicates being very susceptible. With samples rich in ferrous iron the limiting factor in the oxidation appears to be the quantity of oxygen enclosed within the mill. Samples ground in this way should not be used for the determination of ferrous iron.

The presence of an appreciable amount of "acid decomposable sulphide" invalidates the ferrous iron determination, although the approximate figures can often be quoted. Pyrite is not appreciably attacked by the mixture of hydrochloric acid and sulphuric acid used to decompose silicate rocks, but other sulphides such as pyrrhotite are more extensively decomposed, liberating hydrogen sulphide which results in the reduction of some of the ferric iron of the rock. It is possible to correct the ferrous and ferric iron figures of the rock for the pyrrhotite content but such corrections are seldom valid as some sulphur is usually lost and the attack of the mineral is often not complete.

Other constituents of silicate rocks that affect the ferrous iron determination are manganese dioxide, which occurs in some sedimentary rocks, vanadium, which is oxidised to V(V) during the titration (a correction can be applied if the V(III) content of the rock is known), and organic matter. Graphite itself does not appear to have any effect upon the ferrous iron determination, but other forms of organic matter may completely invalidate the determination.

The method commonly used for the determination of ferrous iron is known in some laboratories as the "Pratt method". Although the form in use differs somewhat from that originally described by Pratt,[4] the principle of the method is unchanged. A tall platinum or platinum-iridium alloy crucible of approximately 80-ml capacity is used for this determination. Any "acid decomposable sulphide" is first removed by heating with diluted sulphuric acid.

The rock sample is then decomposed in the crucible by adding hydrofluoric acid. The crucible and contents are plunged into boric acid solution and the ferrous iron liberated during the acid decomposition titrated with an oxidising agent. Originally potassium permanganate was used, but in the presence of hydrofluoric acid a very poor and transient end-point is obtained. A more satisfactory end point is obtained with potassium dichromate using barium diphenylamine sulphonate as indicator.

Improvements to this method have included a number of devices to exclude air from the sample during the decomposition - such as the copper apparatus described by Harris,[5] and the lead box described by Treadwell,[6] in which atmospheres of carbon dioxide are maintained.

In a modification by Lo-Sun Jen,[7] the platinum crucible is replaced by a 400-ml PTFE beaker with a well-fitting lid. The boric and phosphoric acids are added to the beaker after the decomposition of the rock material is complete.

An alternative procedure for the titrimetric determination of ferrous iron in silicate minerals and rocks is that due to Wilson,[8,9] who described the decomposition of the rock material at room temperature in the presence of an excess of vanadate solution. The excess of oxidising agent is determined by titration with ferrous iron. Vanadium as V(III) in the rock does not interfere but organic matter, manganese dioxide and "acid-decomposable sulphide" will invalidate the determination.

Reichen and Fahey[10] have also used the presence of an added oxidant - in this case potassium dichromate - to prevent the atmospheric oxidation of ferrous iron liberated by sulphuric and hydrofluoric acids. However, there is some reaction between the dichromate and hydrofluoric acid, the extent of which appears to be proportional to the amount of excess dichromate. For this reason Wilson's procedure based upon the use of ammonium vanadate is preferred.

When refractory minerals containing ferrous iron - such as chromite or staurolite - are present in the rock, the acid decomposition can be replaced by a fusion procedure such as that described by Rowledge,[11] involving fusion with sodium fluoride and boric acid in a sealed glass tube. Modifications to this have been described by Vincent,[12] Groves[13] and Meyrowitz.[14] Donaldson[15] has however shown that when applied to normal silicate rocks this fusion results in erratic results that are not in agreement with results obtained by other methods. This was attributed to reduction of iron oxides in the fusion stage.

Colorimetric procedures for the determination of ferrous iron in small sample weights have been described by Shapiro[16] and Wilson[9] based upon 1,10-phenanthroline and 2,2'-dipyridyl respectively. Wilson's method involves the use of beryllium solutions to complex fluoride ions; the procedure described below is based upon that of Shapiro, who used boric acid for this purpose.

The difficulties of the Pratt method were ventilated by French and Adams,[17] who recommended the use of wide-mouthed, screw-capped polypropylene bottles and a small domestic pressure cooker for the sample decomposition. These authors also recommended the use of a mixture of hot sulphuric and hydrofluoric acids. Although this may be ideal for some rocks, it produces a too-violent reaction with others and should therefore be used with caution.

THE "PRATT METHOD"

<u>Reagents</u>:

Boric acid, saturated solution.

Barium diphenylamine sulphonate solution, dissolve 0.15 g of reagent in 50 ml of water.

Potassium dichromate standard solution, dissolve 3.268 g of pure dry potassium dichromate in 2 litres of water. For most purposes this solution can be considered as standard N/30, but for accurate work it should be standardised by titration against an iron solution prepared from pure iron wire.

Procedure. Accurately weigh 0.5 g of the finely powdered rock material into a Pratt platinum crucible, moisten with a little recently boiled distilled water and add 10 ml of 20 N sulphuric acid followed by freshly boiled distilled water until the crucible is approximately half full. Cover the crucible with its platinum lid. Support the crucible on a triangle over a low flame protected from draughts and rapidly bring the contents of the crucible to the boiling point. Displace the lid slightly and add 10 ml of concentrated hydrofluoric acid, replace the lid and rapidly bring to the boil. Continue to boil for 7-10 minutes. During this boiling a constant jet of steam should issue from beneath the lid of the crucible, ensuring the complete exclusion of atmospheric oxygen. If the sample is appreciably decomposed by sulphuric acid alone, then the crucible should be filled with carbon dioxide from a small generator before the addition of the acid.

Remove the flame and plunge the crucible and lid below the surface of a boric acid solution prepared by adding 30 ml of saturated boric acid solution to 150 ml of cold, recently boiled distilled water. Rinse the crucible and lid and remove them. Add 10 ml of syrupy phosphoric acid and 5 drops of barium diphenylamine sulphonate solution, and titrate the ferrous iron immediately with potassium dichromate solution. After completing the titration, allow the beaker to stand and then decant off the aqueous solution. If any gritty or dark-coloured particles remain, indicating an incomplete attack of the rock powder, they should be collected and subjected again to the procedure described. If an appreciable quantity of the rock material remains unattacked after prolonged boiling, the rock powder should be ground more finely before repeating the decomposition.

The indicator does not change colour in the absence of ferrous iron, and a zero "blank" determination will not therefore give an end-point. A small quantity of a standard ferrous ammonium sulphate solution should therefore be added to the reagent "blank" solution before titrating, and the appropriate correction applied to the volume of potassium dichromate solution used.

THE "WILSON METHOD"

Apparatus. A vessel of 75-100 ml capacity, with tightly fitting lid, made of polyethylene, polypropylene or polycarbonate material, is used.

Reagents: Ammonium vanadate solution, dissolve 5 g of ammonium metavanadate and 2 g of sodium hydroxide in 100 ml of water.

Procedure. Accurately weigh approximately 0.5 g of the finely powdered rock material into the polyethylene vessel, add by pipette 2 ml of the vanadate solution and with a measuring cylinder 10 ml of concentrated hydrofluoric acid. Close the vessel with its tightly fitting cap and set it aside. Allow the mixture to stand until the decomposition of the rock powder is complete, as indicated by the absence of gritty particles. These should not be confused with the white fluoride precipitate that usually separates on standing. For most rocks the decomposition will be complete after standing overnight or at the most over two nights. In rare cases a more prolonged digestion may be required.

When the decomposition is complete, add 30 ml of 10 N sulphuric acid to the vessel and rinse the contents into a 800-ml beaker containing 250 ml of saturated boric acid solution. Add 5 drops of barium diphenylamine sulphonate solution, stir until the fluoride precipitate has largely dissolved and titrate the solution with a standard ferrous solution. Titrate also a "reagent blank" similarly prepared but omitting the rock sample, and also a 2 ml aliquot of the ammonium vanadate solution. This quantity of ammonium vanadate (0.1 g) is sufficient for rocks containing up to 12 per cent ferrous iron and should be increased or decreased if the rock material contains much or little ferrous iron.

SPECTROPHOTOMETRIC DETERMINATION OF FERROUS IRON

In the titrimetric procedure devised by Wilson and described above, the ferrous iron liberated on dissolving the silicate rock in hydrofluoric acid is immediately removed from the solution by reaction with ammonium vanadate. An alternative method of removing the free ferrous iron is by combining with a suitable chelating reagent. 1,10-Phenanthroline, 2,2'-dipyridyl or any of a number of similar reagents can be used for this, with the added advantage that the extent of ferrous chelation (and hence of ferrous iron present in the rock sample) can be observed by measuring the optical density of the coloured solution. The procedure described below is based upon that of Shapiro,[16] who prefers this procedure for rocks containing oxidisable material such as organic material or sulphide minerals for which reliable results are not otherwise obtained

On decomposing most rocks a cloudy solution is obtained due to the presence of insoluble fluorides and sulphates. The extent to which this cloudiness obscures the light path of the spectrophotometer is determined by measuring the optical density at a second, higher wavelength. The value obtained is subtracted from the optical density recorded at the lower wavelength.

Under the steaming conditions used the ferrous-phenanthroline complex is not completely stable, the colour decreasing by about 1 per cent for 3 minutes of standing. Too short a steaming period may result in incomplete attack of the sample, too long a period in a greatly reduced sensitivity. A 30-minute period was selected as a reasonable compromise. The error introduced by the bleaching action is limited by using a silicate rock of known ferrous iron content to calibrate the procedure.

Method

Apparatus. Polythene bottle, about 25-ml capacity (about 1 oz)

Reagents:

1,10-Phenanthroline, powdered solid reagent.

Boric acid, dissolve 5 g of the reagent in 100 ml of hot water and allow to cool.

Sodium citrate solution, dissolve 50 g of the dihydrate in 500 ml of water and filter if necessary.

Reference sample of silicate rock, choose a silicate rock of known ferrous iron content, similar in composition to the samples being examined.

Procedure. Accurately weigh about 10 mg of the finely powdered silicate material into a dry polyethylene bottle and a similar amount of the reference sample into a second bottle. Use a third bottle for the reference blank solution. Add approximately 20 mg of the 1,10-phenanthroline to each followed by 3 ml of 4 N sulphuric acid and 0.5 ml of concentrated hydrofluoric acid. Transfer the bottles to a steam bath in a fixed order, so that the first one on will be the first off, and leave for 30 minutes.

While the bottles are on the steam bath, transfer 5 ml of the boric acid solution to each of a series of 100-ml volumetric flasks. Using the same order as previously, remove the bottles from the steam bath and, as rapidly as possible, add 20 ml of sodium citrate solution to each bottle. As far as possible each bottle should be left on the steam bath for the same length of time. Transfer the contents of each bottle to one of the 100-ml volumetric flasks, using water to rinse the bottles. Cool to room temperature, dilute each solution to volume with water and mix well.

Using a spectrophotometer set at wavelengths of 555 nm and 640 nm, measure the optical densities of the solutions using the blank as the reference solution. For each solution, subtract the optical density at 640 nm from that obtained at 555 nm, and determine the ferrous iron content of the sample by reference to the optical density difference obtained for the silicate rock of known ferrous iron content.

Determination of Ferrous Iron in Carbonaceous Shales

The importance of knowing the ferrous iron content of carbonaceous shales has led Nicholls[18] to propose a method of determination in which up to 4 per cent of carbon can be tolerated. The shale is decomposed by heating with hydrofluoric and sulphuric acids, and the excess fluoride complexed with boric acid in the usual way. The ferrous iron is allowed to react with iodine monochloride, and the liberated iodine titrated with standard potassium iodate solution. This method of determination was proposed by Heisig[19] in 1928, and was used at a later date by Hey[20] for the determination of ferrous iron in silicate rocks. The procedure is described in detail below. As with other methods for the determination of ferrous iron following sample decomposition with hydrofluoric and sulphuric acids, some interference may be encountered from the presence of sulphide minerals.

Method

Reagents: Carbon tetrachloride.

Iodine monochloride solution, dissolve 10 g of potassium iodide and 6.44 g of potassium iodate in 150 ml of 6 N hydrochloric acid.

Potassium iodate standard solution, 0.1 N (M/40) dissolve 5.350 g of dried pure potassium iodate in water and dilute to volume in a 1 litre flask.

Procedure. Accurately weigh about 0.5 g of the finely powdered shale sample into a platinum crucible, moisten with water and add 10 ml of 20 N sulphuric acid and 5 ml of concentrated hydrofluoric acid. Cover the vessel with a suitable lid and simmer gently for a period of 5 minutes, or until the decomposition is judged to be complete. Pour the contents of the vessel into a 100-ml beaker containing 2 g of solid boric acid and rinse the platinum vessel and lid with cold, boiled distilled water and add these washings to the beaker to give a solution approximately 50 ml in volume.

Pour this solution into a 200-250 ml bottle containing 75 ml of concentrated hydrochloric acid and 6 ml of the prepared iodine monochloride solution. Rinse the beaker, adding the washings to the bottle and dilute the solution to about 150 ml. Now add 10 ml of carbon tetrachloride, stopper the bottle and shake for 20 seconds.

At this stage the carbon tetrachloride layer should be coloured deep purple with the liberated iodine extracted from the aqueous phase. The colour may be partially obscured by the presence of carbonaceous matter which tends to gather at the interface. Titrate the solution by adding standard potassium iodate solution from a burette, with intermittent shaking of the bottle, until the purple colour has almost disappeared from the organic layer. Now add a further 10 ml of carbon tetrachloride to form a zone free from carbonaceous material at the bottom of the bottle in which the end-point (the complete disappearance of the purple colour) can be detected. Towards the end of the titration shake the contents of the bottle after the addition of each drop of potassium iodate solution.

Determination of Ferric Iron

An adequate measure of the ferric iron content of silicate rocks can usually be obtained by deducting the ferrous iron from the separately determined total iron - both expressed as Fe_2O_3. As noted in the previous section, the presence of sulphide minerals or carbonaceous material may, by the reduction of trivalent iron, give rise to appreciable errors in the ferrous iron content which in turn are reflected in the calculated value for the ferric iron. An assumption is sometimes made that where sulphide minerals are present they are insoluble in the hydrofluoric-sulphuric acid mixture used for the determination of ferrous iron, and therefore do not affect this determination. This assumption is largely but not completely true for pyrite, Fe,S_2, the sulphide mineral occurring most frequently in silicate rocks. Where the rock sample contains more than a trace of pyrite, the sulphide iron, calculated as Fe_2O_3, is added to the ferrous iron (also calculated as Fe_2O_3) before the calculation of ferric iron. Pyrrhotite and certain other sulphide minerals are appreciably soluble in the acid mixture used for the ferrous iron determination.

When required, a direct determination of ferric iron can be made by a direct titration of the rock solution without the addition of nitric acid or other oxidising agent, using titanous chloride or sulphate solution as titrant. This method should in principle, provide a more accurate value for ferric iron than that derived by difference, especially when the ratio of ferric to ferrous iron is low. The method described below, due to Murphy et al[21] involving titration of ferric iron in solution with ferrocene (dicyclopentadienyliron) is based upon the reaction

$$Fe(C_5H_5)_2 + Fe^{3+} = Fe^{2+} + Fe(C_5H_5)_2^{+}.$$

In the presence of thiocyanate ions, the end point is indicated by the disappearance of the red colour of ferric thiocyanate and its replacement by the blue colour of the ferricenium ion.

Method

Reagents:

Ammonium thiocyanate solution, dissolve 5 g in 100 ml water.

Lissapol NDB solution, dissolve 5 ml of Lissapol NDB detergent liquid (available from Hopkin & Williams Ltd) in 95 ml of 95 per cent ethanol.

Ferrocene solution, dissolve 0.583 g of ferrocene (dicyclopentadienyliron) in 2-methoxyethanol and dilute with this solvent to 500 ml. Store in a glass stoppered bottle.

Boric acid solution, saturated aqueous.

Potassium permanganate solution, 0.1 per cent w/v aqueous.

Standard ferric iron solution, dissolve 0.140 g of pure iron wire in about 25 ml of water containing 10 ml of 20 N sulphuric acid and add a small excess of bromine water. Evaporate the solution to fumes of sulphuric acid, cool, cautiously add about 100 ml of water. Heat until a clear solution is obtained, then transfer to a 500-ml volumetric flask, cool to room temperature and dilute to volume with water. Prepare by dilution with water a solution containing 0.2 mg Fe_2O_3 per ml, to give the iron spike solution.

Procedure. To a clear plastic beaker of about 250 ml capacity, add 100 ml of saturated boric acid and to a similar 100-ml beaker add 50 ml of the same solution. Set these aside in readiness for the next stage.

To 0.5 g of the finely powdered rock material in a large platinum crucible ("Pratt crucible", see above), add a few ml of water and 10 ml of 20 N sulphuric acid. Add more water to about half fill the crucible and heat to boiling over a small flame protected from draughts. After a few seconds boiling add 10 ml of concentrated hydrofluoric acid without interrupting the heating, replace the lid immediately and note the time when boiling recommences. After 10 minutes boiling remove the flame and quickly rinse the lid into the crucible with boric acid solution from the 100 ml beaker. Add further boric acid solution almost to fill the crucible and without delay tip the contents into the 100 ml of boric acid solution in the 250 ml beaker. Rinse the crucible into the beaker with the boric acid solution remaining in the 100 ml beaker and then transfer the whole contents to a 200-ml volumetric flask and dilute to the mark. Mix well.

Transfer a 20 ml aliquot of this solution to a 150 ml glass beaker and dilute to about 75 ml with boiled water. Stir the solution and, after adding 10 ml of ammonium thiocyanate solution and 2 ml of the Lissapol NDB solution, titrate with ferrocene solution. The end point is indicated in good white light by the appearance of a blue colour without any trace of red ferric thiocyanate colour. Add 1 ml of the ferric iron spike solution and again titrate to the same end point, noting the total volume of titrant added. Carry out the above titration procedure using 1 ml of the ferric iron spike solution to which has been added 1 ml of 20 N sulphuric acid before diluting to 75 ml, and note the volume of titrant required.

To standardise the ferrocene solution, transfer 10 ml of the standard ferric iron solution by pipette to a 150 ml beaker and add potassium permanganate solution a drop at a time until there is a small visible excess, then dilute to 75 ml and proceed as described above, including the addition of the 1 ml of the ferric iron spike solution.

Determination of Total Iron

There is a wide selection of procedures for the determination of the total iron content of silicate and carbonate rocks. Photometric methods are of greatest use for those sample materials that contain only small amounts of iron, but as these procedures are some of the most precise of their kind, they have been widely adopted for the determination of iron even when present as a major component. Mercy and Saunders[22] have, however, shown that titrimetric procedures are more precise than photometric ones, although the differences they obtained were quite small. In a comparison of two titrimetric and three photometric procedures, the titrimetric methods were found to be consistent with each other, but there were some differences in the results from the three photometric methods, although all five methods gave approximately the same average values.

A number of separation procedures are available for the recovery of iron from solution. The simplest is precipitation with aqueous ammonia, followed by filtration and dissolution of the residue in dilute hydrochloric acid. Titanium, vanadium, chromium, phosphorus, most of the aluminium and some of the manganese will accompany the iron. A double precipitation is necessary to remove all the calcium and magnesium from the iron. Ion-exchange procedures have been developed with both cation and anion exchange resins. The anion-exchange procedure is particularly useful for the separation of iron from aluminium, titanium, manganese and other metallic elements.

Solvent extraction procedures have rarely been applied to the determination of total iron in silicate rocks, but Kiss[23] in a scheme for the determination of a number of major components utilises a well-known separation based upon the extraction of the chloroferrate ion from seven molar hydrochloric acid solutions with methylisobutylketone. The iron in the organic extract is recovered by shaking the extract with water.

Atomic absorption spectroscopy is now widely used for the determination of total iron in silicate rocks, see chapter 4. The absorption spectrum of iron is complex and a considerable number of lines have been suggested for analytical use. The most sensitive line is at 248.3 nm. Lines at 248.8, 252.3, 271.9 and 301.0 nm are all of sufficient sensitivity for most applications.

Interference from the presence of other elements has been noted when cool flames (eg air-coal gas or air-propane) are used, but only silicon interferes significantly when an air-acetylene flame is used. This is not likely to give rise to problems in rock analysis where the removal of silicon is the first step in the preparation of solutions for iron determination. For some observations on the determination of iron by this technique, see Thompson and Wagstaff.[24]

SPECTROPHOTOMETRIC DETERMINATION OF TOTAL IRON

There is no lack of colour-forming reagents for the determination of iron in silicate or carbonate rocks. Tiron (disodium salt of catechol-3,5-disulphonic acid) has been proposed for the determination of both iron and titanium in the same solution.[25] The iron complex is violet coloured with an absorption maximum at a wavelength of 560 nm. This solution can be decolorised by reduction of the iron using sodium dithionite, leaving the yellow-coloured titanium complex in solution. This has an absorption maximum at 430 nm and does not absorb at the wavelength of 560 nm, used for the iron determination. There is disagreement as to the best pH to use for this determination.[26,27] This, together with the recorded instability of the reducing agent leading to the formation of collodial sulphur, suggests that little advantage is to be gained in the use of tiron for the determination of iron, particularly as superior reagents are available for titanium.

Both phenyl-2-pyridylketoxime[28,29] and 4,7-dihydroxy-1,10-phenanthroline[30] (Snyder's reagent) react with ferrous iron in alkaline solution to give coloured products that have been used for the determination of iron in silicate materials. The latter reagent is extremely expensive and appears to offer little if any advantage over the former. Unpublished work[31] has shown that any time gained by working with silicate rocks in alkaline solution was more than offset by the time spent in removing the iron alloyed with the platinum of the crucibles used. A considerable loss of iron can occur in this way, although the amounts are rather variable.

1,10-Phenanthroline and 2,2'-dipyridyl are two of the most widely used reagents for iron. Both reagents readily form red-coloured complexes with ferrous iron that are stable over wide ranges of salt concentration and temperature. Precise control of pH is not necessary and interference from other elements is negligible. Both reagents are completely stable in the solid state. The ferrous complexes form rapidly, are completely stable in aqueous solution, and such solutions obey the Beer-Lambert Law. The procedure described below is based upon the use of 1,10-phenanthroline, although dipyridyl can be used in the same way.

Ammonium thiocyanate, NH_4CNS, has in the past been extensively used for the photometric determination of iron. It suffers from a number of disadvantages, particularly when compared with the previous two reagents. The optical densities of ferric thiocyanate solutions depend upon the conditions used for the reaction (temperature, acidity, excess of reagent), the solutions may suffer from some measure of fading, and do not completely follow the Beer-Lambert Law. The departures from this law are not regarded as very serious, and ammonium thiocyanate is still used in a number of laboratories. Other reagents that have been used for the photometric determination of iron in silicate materials include salicylic acid, EDTA and hydrogen peroxide, acetylacetone and sulpho-salicylic acid.

Some workers have recommended using a mixture of perchloric and hydrofluoric acids for the decomposition of silicate rocks and minerals and for the volatilisation of silica. This use of perchloric acid is ideal where the sample contains only easily decomposable minerals such as felspars, but the somewhat higher temperatures that can be obtained with sulphuric acid are preferred for general silicate rock analysis where a variety of accessory minerals are likely to be present. Even using sulphuric acid, a small unattacked residue is often obtained consisting of such minerals as tourmaline, zircon, ilmenite and rutile, together with barium precipitated as sulphate. Any iron present in this residue is recovered following fusion with sodium carbonate, and is added to the main rock solution. If it is required, the barium content of the rock sample can be determined in this insoluble residue.

The iron present in a suitable aliquot of the rock solution is reduced to the ferrous state using hydroxyammonium chloride, and the red colour given with 1,10-phenanthroline is formed in the solution buffered with ammonium tartrate. The absorption maximum occurs at a wavelength of 508 nm.

The reagents used for this determination all contain small amounts of iron and it is therefore particularly important that the reagent blank solution should be carefully prepared. Particular attention should be paid to the cleanliness of the apparatus used, especially the platinum crucibles and the spectrophotometer cells.

Method

Reagents:

Tartaric acid solution, dissolve 10 g of the solid reagent in 100 ml of water.

p-Nitrophenol indicator solution, dissolve 1 g of the solid reagent in 100 ml of water.

Hydroxyammonium chloride solution, dissolve 10 g of the solid reagent in 100 ml of water.

1,10-Phenanthroline solution, dissolve 0.1 g of the reagent in 100 ml of water.

Standard iron stock solution, accurately weigh 0.112 g of pure iron wire into a small beaker, add 20 ml of water and 5 ml of 20 N sulphuric acid. Warm gently until the metal has completely dissolved and then dilute to 1 litre with water. This solution contains 160 µg Fe_2O_3 per ml.

Standard iron working solution, pipette 25 ml of the stock solution into a 100-ml volumetric flask and dilute to volume with water. This solution contains 40 µg Fe_2O_3 per ml.

Procedure. Accurately weigh approximately 1 g of the finely ground rock material into a clean platinum dish, and add 10 ml of 20 N sulphuric acid, 2 ml of concentrated nitric acid and 10 ml of hydrofluoric acid. Transfer to a hot plate and evaporate just to fumes of sulphuric acid. Remove the dish, allow to cool, add a further 5 ml of hydrofluoric acid and again evaporate to fumes of sulphuric acid. Allow to cool once again, rinse down the sides of the dish with a little water, add a further 2 ml of water, break up any solid cake that has formed using a small platinum rod and again evaporate on the hot plate, this time to copious fumes of sulphuric acid. Allow the dish to cool.

Rinse down the sides of the dish and add 15 ml of water. Allow to stand on a hot plate for a few minutes for the solid residue to disintegrate and pass partly into solution. Rinse the solution and any remaining residue into a 250-ml beaker and heat the solution until all soluble material has dissolved. Collect any remaining insoluble material on a small, close-textured filter paper, wash well with very dilute sulphuric acid and transfer the paper and residue to a platinum crucible. Collect the filtrate and washings in a 250-ml volumetric flask and set it aside.

Dry and ignite the paper, fuse any remaining residue with a small quantity of anhydrous sodium carbonate for a period of at least 30 minutes and then allow to cool. Extract the melt with water and rinse the solution and residue into a small beaker. Warm the crucible with a little dilute sulphuric acid, and rinse this acid solution into the beaker. At this stage no unattacked particles of the original silicate rock should remain, and the only precipitate should be barium sulphate, which will form if the sample material contains an appreciable amount of barium. Collect this precipitate on a small close-textured filter paper, wash with very dilute sulphuric acid and discard it. Add the sulphuric acid solution and the washings of the barium sulphate precipitate to the solution contained in the 250-ml volumetric flask and dilute to volume with water.

Pipette a suitable aliquot of this solution into a 100-ml volumetric flask, add 10 ml of tartaric acid solution and a drop of p-nitrophenol indicator solution, followed by concentrated aqueous ammonia until the solution becomes pure yellow in colour. Add dilute hydrochloric acid drop-by-drop until this colour is just discharged. Cool the solution to room temperature, add 2 ml of hydroxyammonium chloride solution and 10 ml of 1,10-phenanthroline solution and dilute to volume with water. Mix well, allow to stand for an hour and measure the optical density against water at a wavelength of 508 nm. Measure also the optical density of a reagent blank solution, prepared in a similar way but without the sample material.

Calibration. For the preparation of the calibration graph for 1 cm spectrophotometer cells, pipette aliquots of 0-25 ml of the standard solution containing 0-1 mg Fe into separate 100-ml volumetric flasks and add to each 10 ml of tartaric acid and 1 drop of p-nitrophenol indicator solutions. Adjust the pH by adding aqueous ammonia and hydrochloric acid as described above, finally adding hydroxyammonium chloride and 1,10-phenanthroline solutions and diluting to volume with water. After standing, measure the optical densities of these solutions at 508 nm and plot the values obtained against the concentration of iron.

TITRIMETRIC DETERMINATION OF TOTAL IRON

Titration methods for the determination of total iron are based upon the prior conversion of all the iron present into one valency state, which is then converted in the course of the titration into the other. The titration of ferric to ferrous iron has never been popular and the most widely used procedures are those based upon the titration of ferrous to ferric.

Total iron by titration as ferrous. Only a small number of oxidants have been successfully used for this determination, and either potassium dichromate or ceric sulphate is recommended. Potassium permanganate may oxidise chloride ion and should be avoided for use with hydrochloric acid solutions.

A variety of reducing agents have been suggested for the reduction of ferric iron, although some of them are far from ideal. Sulphur dioxide and hydrogen sulphide for example, although effective reducing agents, must be added in excess, and the excess is then difficult to remove. Prolonged boiling is recommended for this, but some trace of sulphur compounds usually remains and oxidation of Fe^{2+} to Fe^{3+} is a real danger. Titanous sulphate and chloride have been recommended, but these solutions are unstable and the solutions cannot be stored for long periods. They offer no advantage over the more conventional reductant-stannous chloride solution described below. A great excess of stannous chloride should be avoided, otherwise the mercuric chloride which is added to remove the excess of stannous chloride, will be reduced to the black metallic state.

An alternative procedure for the stannous chloride reduction has been described by Hume and Kolthoff,[(32)] in which cacotheline is added to indicate the presence of a slight excess of stannous chloride. This slight excess is then titrated with ceric sulphate solution before the titration of the ferrous iron with the same reagent.

Although a number of metals have been suggested as suitable for reducing ferric to ferrous iron, only two, zinc and silver, have found extensive use in the analysis of silicate rocks. A Jones reductor of amalgamated zinc is less suited to this application as there is considerable interference from titanium which is reduced to Ti^{3+}. Chromium and vanadium which may be present to a minor extent, are reduced to Cr^{2+} and V^{2+} respectively. Many other elements are reduced by passage through the reductor, but are unlikely to be present in normal silicate rocks in amounts sufficient to interfere.

Neither titanium nor chromium (Cr^{3+}) is reduced by passage through a silver reductor, although vanadium (V^{5+}) is reduced to V^{4+}. The formation of hydrogen peroxide in silver reductors has been reported by Miller and Chalmers,[(33)] and this prevents the complete reduction of ferric iron. This problem has been overcome by the use of solutions saturated with carbon dioxide. Platinum, derived from the apparatus used, is reduced (Pt^{4+} to Pt^{2+}) in a silver reductor, and may also interfere with the titration of iron by catalysing the reduction of titanium. The introduction of platinum can be avoided by conducting the fusion in crucibles of gold or silver.

Method

Reagents:

Stannous chloride solution, dissolve 7.5 g of stannous chloride dihydrate in 100 ml of 6 N hydrochloric acid. Prepare freshly at frequent intervals.

Ceric sulphate 0.3 N standard solution, dissolve 10 g of ceric sulphate in 500 ml of water containing 50 ml of 20 N sulphuric acid and dilute to volume in a 1 litre volumetric flask. Standardise by titration against pure iron solution, or preferably against pure arsenious oxide in the presence of osmium tetroxide as catalyst.

Cacotheline reagent, grind together 0.5 g of the solid reagent and 0.5 ml of water. Add 50 ml of water. Shake before using.

Ferroin indicator solution, dissolve 0.742 g of 1,10-phenanthroline monohydrate in 50 ml of a solution containing 6.95 g of ferrous sulphate heptahydrate per litre.

Mercuric chloride solution, saturated.

Potassium dichromate 0.03 N standard solution, dry a small amount of the pure reagent at a temperature of 150° for 4 hours, allow to cool and then dissolve 1.471 g in water and dilute to volume in a 1 litre volumetric flask.

Diphenylamine indicator solution, dissolve 10 mg of barium diphenylamine sulphonate in a mixture of 50 ml of water and 50 ml of syrupy phosphoric acid.

Procedure. Accurately weigh approximately 1 g (Note 1) of the finely powdered silicate rock material into a platinum dish, moisten with water and add 5 ml of concentrated perchloric acid and 10 ml of concentrated hydrofluoric acid. Transfer the dish to a hot plate and evaporate first to fumes of perchloric acid and then to complete dryness. Allow to cool, moisten the dry residue with a little perchloric acid and again evaporate to dryness on a hot plate. Add 5 ml of water and 5 ml of concentrated hydrochloric acid to the residue and warm to bring all soluble material into solution. With some silicate rocks, complete solution will be obtained at this stage. With others a small residue, insoluble in hydrochloric acid, will remain. Collect this residue on a small filter, wash with water, dry and ignite in a small silica crucible. Add a small amount of potassium pyrosulphate to the residue and fuse at a dull red heat until a quiescent melt is obtained. Allow to cool, extract the melt into a little dilute hydrochloric acid and add to the main rock solution (Note 2).

Dilute the solution to about 75 ml, add 5 ml of concentrated hydrochloric acid, heat almost to boiling and add stannous chloride solution dry by drop until the yellow colour of the solution is completely discharged. Add 1 drop of stannous chloride solution in excess. Allow to cool, and add 5 drops of the cacotheline suspension which turns the solution deep violet in colour. Titrate the solution, drop by drop, with ceric sulphate solution until the violet colour fades, giving first a brown colour, then a pure yellow. Near the end-point it may be necessary to add 2 or 3 further drops of the cacotheline indicator suspension to observe the end-point.

Add 125 ml of water, 6 ml of 20 N sulphuric acid and 3 drops of ferroin indicator solution, and titrate with the standard ceric sulphate solution to the disappearance of the red colour given by the indicator with ferrous iron. There is a small indicator and reagent blank value which should be determined and subtracted from the sample titration value before calculation of the total iron content.

The excess of stannous chloride can also be removed by cooling and adding 5 ml of mercuric chloride solution and allowing the solution to stand - preferably under a blanket of inert gas such as nitrogen or carbon dioxide, for 5 minutes. Then add 50 ml of water and 10 ml of the diphenylamine indicator solution and titrate with **0.03 N potassium dichromate solution until a faint purple colour is obtained.**

The total iron content of carbonate rocks can be determined in the same way as that of silicate rocks, except that for the initial stage of the decomposition it is sufficient to evaporate the sample material to dryness with hydrochloric acid. After dissolution of the residue in dilute hydrochloric acid, any insoluble material can be recovered and decomposed by fusion with a little anhydrous sodium carbonate.

Notes: 1. This weight of sample material should give a titration of about 25 ml of 0.03 N oxidant with samples containing about 6 per cent total iron as Fe_2O_3. For basic and other rocks rich in iron, the sample weight should be reduced. Alternatively, 0.1 N oxidant solution can be used.

2. Most iron-containing minerals will be decomposed by this treatment. If any residue remains it should be collected and fused with a little sodium hydroxide, the melt extracted with water containing a little hydrochloric acid, and the solution added to the main rock solution.

<u>Total iron by titration as ferric</u>. The difficulties that arise during the reduction of ferric iron in the rock solution can be avoided by titrating not the ferrous to the ferric but the ferric to ferrous. Thornton and Chapman(34) have described a procedure in which potassium permanganate is used to ensure that all the iron present is in the higher valency state, and they then titrated the ferric iron with a standard solution of titanous sulphate. The end-point was indicated by the disappearance of the red colour given by ferric iron with thiocyanate. Some experience is necessary to judge the exact point at which this occurs, as the reaction is rather slow near the end-point.

Nitric acid, free hydrofluoric acid, vanadium, molybdenum and a number of other metallic elements interfere with the determination. The reagent can be prepared by dissolving pure titanium sponge in sulphuric acid. Once prepared it should be stored under a layer of inert hydrocarbon and away from direct sunlight. Titanous sulphate solutions are very readily oxidised, and special precautions are necessary for the delivery of the solution in the course of the analysis. Experience with titanous sulphate has shown the necessity of standardisation at frequent intervals, if accurate results are to be obtained.

Mercurous nitrate can also be used for this titration and possibly also a number of other titrants such as vanadous sulphate or chromous sulphate solutions. These reagents would also need to be protected from atmospheric oxidation, and are unlikely to offer any significant advantage over titanous sulphate solutions.

Ferric iron can also be titrated with solutions of EDTA. Neither ammonium thiocyanate nor salicylic acid is suitable for indicating the end points of the reaction, and a redox indicator such as variamine blue B is preferred.(35) An alternative procedure is the indirect titration suggested by Pribil and Vesely,(36) involving the addition of an excess of EDTA solution and a titration of this excess with bismuth, thorium or lead solution using xylenol orange as indicator.

References

1. EASTON A J and LOVERING J F., Geochim. Cosmochim. Acta (1963) 27, 753
2. FRIEDHEIM C., S B Akad. Wiss. Berlin (1888) 345
3. FITTON J G and GILL R C O., Geochim. Cosmochim. Acta (1970) 34, 518
4. PRATT J H., Amer. J. Sci. (1894) 48, 149
5. HARRIS F R., Analyst (1950) 75, 496
6. TREADWELL F P., Kurzes Lehrbush der Analytischen Chemie (1913) 2, p.425, (6th ed)
7. LO-SUN JEN, Anal. Chim. Acta (1973) 66, 315
8. WILSON A D., Bull. Geol. Surv. Gt. Brit. (1955) (9), 56
9. WILSON A D., Analyst (1960) 85, 823
10. REICHEN L E and FAHEY J J., U S Geol. Surv. Bull. 1144-B, 1962
11. ROWLEDGE H P., J. Roy. Soc. W. Aust. (1934) 20, 165
12. VINCENT E A., Geol. Mag. (1937) 35, 86
13. GROVES A W., Silicate Analysis, Allen & Unwin, London 1951 (2nd ed.)
14. MEYROWITZ R., Analyt. Chem. (1970) 42, 1110
15. DONALDSON E M., Analyt. Chem. (1969) 41, 501
16. SHAPIRO L., U S Geol. Surv. Research 1960, B-496 (1961)
17. FRENCH W J and ADAMS S J., Analyst (1972) 97,828
18. NICHOLLS G D., J. Sed. Petrol. (1960) 30, 603
19. HEISIG G B., J. Amer. Chem. Soc. (1928) 50, 1687
20. HEY M H., Amer. Mineral. (1949) 34, 769
21. MURPHY J M, READ J I and SERGEANT G A., Analyst (1974) 99, 273
22. MERCY E L P and SAUNDERS M J., Earth, Planet. Sci. Lett. (1966) 1, 169
23. KISS E., Anal. Chim. Acta (1967) 39, 223
24. THOMPSON K C and WAGSTAFF K., Analyst (1980) 105, 641
25. YOE J H and ARMSTRONG A R., Analyt. Chem. (1947) 19, 100
26. RIGG T and WAGENBAUER H A., Analyt. Chem. (1961) 33, 1347
27. ARCHER K, FLINT D and JORDAN J., Fuel, London (1958) 37, 421
28. TRUSELL F and DIEHL H., Analyt. Chem. (1959) 31, 1979
29. CLULEY H J and NEWMAN E J., Analyst (1963) 88, 3
30. SCHILT A A, SMITH G F and HEIMBUCH A., Analyt. Chem. (1956) 28, 809
31. RICHARDSON J and JEFFERY P G., Unpubli. work, 1962
32. HUME D S and KOLTOFF I M., Analyt. Chem. (1957) 16, 415
33. MILLER C C and CHALMERS R A., Analyst (1952) 77, 2
34. THORNTON W M Jr. and CHAPMAN J E., J. Amer. Chem. Soc. (1921) 43, 91
35. FLASCHKA H., Mikrochim.Acta (1954) 361
36. PRIBIL R and VESELY V., Talanta (1963) 10, 361

CHAPTER 26

Lead

In the determination of lead in silicate rocks, mixtures of hydrofluoric acid with either perchloric or nitric acid are commonly used to remove silica. In general sulphuric acid has been avoided, probably because of the possibility that lead sulphate may be co-precipitated with barium or calcium from rocks rich in these elements.

Procedures based upon ion-exchange separation have not been extensively used for the separation of lead in silicate rocks, although for the determination of lead in marine sediments Korkisch and Feik[1] have described the use of an anion exchange separation, based upon elution with tetrahydrofuran-nitric acid mixtures from a strongly basic resin (Dowex 1x 8).

Many analysts prefer to make this initial separation of lead from other elements present in silicate rocks by solvent extraction. Dithizone has been described for this,[2] but difficulties have been experienced with rocks rich in iron[3]. Diethyldithiocarbamate has been proposed[4] for this purpose and its use was considered to be satisfactory by Baskova,[5] who compared a number of methods for the separation and determination of lead in silicate rocks. Trioctylphosphine oxide in methyliso-butyl ketone has been recommended,[6] giving an extract which can be aspirated directly into the flame of an atomic absorption spectrometer.

Traces of lead can be precipitated with suitable metal carriers, eg as sulphide with mercury or zinc and as sulphate with barium. None of these methods gives a good recovery of lead[5] at the concentration levels encountered in some silicate rocks.

One procedure for separating lead that does not require either the removal of silica or an acid dissolution of the rock material, is based upon distillation or sublimation from the powdered material. The method as described by Iordanov and Kocheva[7] involved sublimation in vacuo, but a simpler procedure by Marshall and Hess[8] used sublimation in a stream of nitrogen at a temperature of 1400°. As little as 0.01 ppm of lead can be determined in this way.

The reagent most commonly used for the photometric determination of lead is dithizone. The reagent itself is dark green, almost black, in colour and gives green solutions in chloroform and carbon tetrachloride that deteriorate slowly with time. It reacts readily with a large number of metallic ions in solution to give highly coloured - mostly brown, orange or red - complexes that are soluble in organic solvents. In the presence of cyanide ions, only lead, bismuth, thallium, tin (II) and possibly indium are extracted as dithizonates. Bismuth, thallium, tin and indium are present in only very small amounts in silicate rocks, and are not likely to interfere. All four elements are, however, separated from lead in a preliminary concentration procedure involving extraction of the lead complex with diethyldithiocarbamate into organic solution.

Solutions of lead dithizonate in carbon tetrachloride (the solvent most frequently used) have a maximum absorption at a wavelength of 520 nm and obey the Beer-Lambert Law up to about 3 ppm Pb, although at this concentration the solutions are probably supersaturated. 0 to 1 ppm Pb is a convenient concentration range to use for the determination in silicate rocks.

In this procedure based upon that described by Korkisch and Gross,[9] (a useful paper giving 138 references to the determination of lead), the silicate material is evaporated with perchloric and hydrofluoric acids, and the lead separated on a column of anion-exchange resin in the bromide form. After elution with hydrochloric acid, lead is determined by atomic absorption spectrometry.

Spectrophotometric Determination of Lead

This method is based upon methods given in outline by Baskova[5] and in greater detail by Gage.[10] A mixture of hydrofluoric and nitric acid is used to remove the silica and to obtain the lead and other metallic constituents in solution. The initial separation from iron and certain other metals is by extraction of the lead complex of diethyldithiocarbamate into an organic solvent consisting of a mixture of pentanol and toluene. The lead is transferred to aqueous solution by shaking with dilute hydrochloric acid, and is added to an ammoniacal solution of dithizone containing potassium cyanide and sodium metabisulphite. The red-coloured lead dithizonate is extracted into carbon tetrachloride and the optical density of the extract measured at 520 nm.

Method

Reagents: Pentanol-toluene mixture, mix equal volumes of pentanol and toluene. Some batches of this mixture have given low recoveries of lead, even when "sulphur-free" solvents have been used.

These mixtures can be purified by treating with sufficient bromine to give a deep yellow colour, allowing to stand for 30 minutes, decolorising with bisulphite solution and washing with water.

Citrate-bicarbonate solution, dissolve 25 g of sodium citrate dihydrate and 4 g of sodium bicarbonate in water and dilute to 100 ml.

Sodium diethyldithiocarbamate.

Ammonia-cyanide-sulphite solution, mix 95 ml of 10 M aqueous ammonia with 5 ml of a solution of potassium cyanide (CARE!) containing 10 g per 100 ml, and dissolve 5 g of sodium metabisulphite in the mixture.

Dithizone solution, dissolve 2 mg of the solid reagent in 100 ml of pure carbon tetrachloride. Store in a refrigerator.

Standard lead stock solution, dissolve 0.160 g of the lead nitrate in water and dilute to 1 litre. This solution contains 100 µg Pb per ml.

Standard lead working solution, dilute 10 ml of stock solution to 1 litre with water. This solution contains 1 µg Pb per ml.

Procedure. Accurately weigh from 0.5 to 1 g of the finely powdered silicate rock material into a small platinum basin and evaporate to dryness with 10 ml of concentrated hydrofluoric acid and 5 ml of concentrated nitric acid. Add 5 ml of hydrofluoric acid and 5 ml of nitric acid to the dry residue and again evaporate to dryness. Moisten the residue with 2 ml of concentrated nitric acid and evaporate to dryness, and repeat this evaporation with two further portions of nitric acid to expel hydrofluoric acid. Moisten the dry residue with nitric acid and rinse the solution into a small beaker (Note 1). Evaporate the solution to dryness twice with concentrated hydrochloric acid to convert nitrates to chlorides, and finally evaporate the chloride solution to give a moist chloride residue. Dissolve this residue in 10 ml of 0.6 N hydrochloric acid, by warming if necessary, and set aside.

Transfer 25 ml of the citrate-bicarbonate solution to a 100-ml separating funnel and add approximately 10 mg of sodium diethyldithiocarbamate. Swirl to dissolve the solid, allow to stand for 15 minutes and then add 25 ml of the pentanol-toluene solvent. Stopper the separating funnel and shake for 1 minute, allow the layers to separate and then run the lower, aqueous layer into the beaker containing the rock solution, previously set aside. Shake the organic layer remaining in the funnel with two successive 10-ml portions of 0.6 N hydrochloric acid and discard these washings.

Now transfer the rock solution to the separating funnel, rinsing the beaker with a few millilitres of the clean pentanol-toluene solvent mixture. Allow to stand for 15 minutes, shake for 2 minutes, allow the layers to separate and then discard the lower, aqueous layer. Wash the organic layer with 25 ml of water and discard the washings. Extract the organic layer with two successive 10-ml portions of 0.6 N hydrochloric acid, shaking for 2 minutes and 1 minute respectively, and transfer the acid extracts to a 50-ml volumetric flask. Rinse the separating funnel with 0.6 N hydrochloric acid, add the washings to the solution in the flask and dilute to volume with 0.6 N hydrochloric acid. Mix well.

Using a safety pipette, transfer 25 ml of the ammonia-cyanide-sulphite reagent to a clean 100-ml separating funnel, add 10 ml of the dithizone solution, stopper and shake for 1 minute. Allow the layers to separate and discard the lower, organic layer. Wash the ammoniacal layer by shaking with 10 ml of carbon tetrachloride and discard the washings. Transfer to the separating funnel an aliquot of the rock solution from the volumetric flask, containing not more than 10 μg of lead, and extract the lead into organic solution by shaking for 2 minutes with 10 ml of carbon tetrachloride. Run off the lower, organic layer, filter through a dry paper into a spectrophotometer cell and measure the optical density against carbon tetrachloride at a wavelength of 520 nm. Determine also the optical density of a reagent blank solution similarly prepared but omitting the sample material.

Calibration. To effect the calibration it is necessary to take aliquots of the standard lead solution containing 2-10 μg lead through the whole procedure, including the extraction with diethyldithiocarbamate, and finally measuring the optical density of each of the dithizone extracts. Plot the relation of the optical densities of the extracts measured against carbon tetrachloride to lead concentration.

Notes: 1. The small residue that is usually obtained can often be neglected. Where an appreciable residue is obtained it should be collected, fused with sodium carbonate, extracted with water and filtered. Discard the filtrate. Dissolve the residue in a little 0.6 N hydrochloric acid and add to the main rock solution.

Determination of Lead by Atomic Absorption Spectroscopy

Atomic absorption techniques for lead have improved considerably since it was earlier reported that they were insufficiently sensitive for application to normal silicate rocks. There is massive interference from molecular absorption of a wide CaOH band. Hence the background measurements have always to be made at the 217 nm wavelength.

A procedure using solvent extraction of lead into a trioctylphosphine oxide-methylisobutyl ketone mixture followed by direct aspiration into an air-acetylene flame is described by Sanzolone and Chao.[6] A relatively simple and straightforward method based upon atomic absorption after an ion-exchange separation is described by Korkisch and Gross,[9] and forms the basis of one of the procedures described in detail below.

The use of graphite furnaces for rock samples that have been decomposed by normal chemical methods have been described by Cruz and Van Loon,[11] and Campbell and Ottaway.[12] Langmyhr _et al_[13] determined lead and other heavy metals in silicate rocks by direct atomisation from the solid state. Small samples (1-20 mg) of the solid silicate material mixed with graphite in the ratio 3:1, were weighed into a graphite furnace,[14] dried at 130° and then atomised at a temperature of 1750° for 60 seconds. An atomic absorption spectrometer set at a wavelength of 283.3 nm was used to measure the lead. The detection limit for lead was given as 10^{-9}g, which in a 1 mg sample corresponds to 1 ppm and in a 50 mg sample corresponds to 50 ppb.

Method

Reagents: Dowex 1 x 8 ion exchange material, soak 4 g of the 100-200 mesh chloride form of the resin in a few ml of 2 M hydrochloric acid and transfer to a small glass column to form a bed approximately 1 cm in diameter and 10 cm in length. Allow 50 ml of 2 M hydrobromic acid to pass through the column to convert the chloride to the bromide form.

Hydrobromic acid, 2 M, dilute 235 ml of concentrated (47% acid, SG about 1.5) to 1 litre with water.

Lead nitrate standard solution, dissolve 33.0 g of lead nitrate $Pb(NO_3)_2$ in 0.5% v/v nitric acid and dilute to 1 litre with 0.5% nitric acid. This solution contains 20.36 mg Pb per ml. The exact lead content may be determined by EDTA titration if required. Working solutions containing 0.2-200 ppm Pb can be prepared from the stock solution by dilution with 0.5% nitric acid as required.

Procedure. Accurately weigh approximately 1 g of the finely powdered rock material into a platinum dish, moisten with water and add 5 ml of concentrated nitric acid and 10 ml of concentrated hydrofluoric acid. Transfer the dish to a sand bath or hot plate and evaporate to dryness. Add 10 ml of concentrated nitric acid and again evaporate to dryness. Take up the residue in 1 ml of water and 2 ml of concentrated perchloric acid, and digest the residue in 20 ml of 2 M hydrobromic acid, allowing to stand for about an hour.

Using 10 ml of 2 M hydrobromic acid as a rinse, transfer the solution to a small beaker. Add 2 g of potassium bromide and let stand for several hours or overnight to allow the potassium perchlorate to crystallise. Filter off the insoluble material, rinsing with a little 2 M hydrobromic acid and dilute the filtrate to about 50 ml with this acid. Transfer the solution to the ion-exchange column and allow to percolate through. Wash the bed with 50 ml of 2 M hydrobromic acid and discard the eluate. Remove the lead from the resin bed by elution with 50 ml of 6 M hydrochloric acid. Transfer this solution to a small beaker and evaporate to dryness on a steam bath, dissolve the residue in 5 ml of 0.5% nitric acid and, after about 30 minutes, transfer to a 10-ml volumetric flask and dilute to volume with 0.5% nitric acid. If the solution contains any insoluble particulate material, filter through a dry paper before aspirating into an air-acetylene flame of an atomic absorption spectrometer, set to a wavelength of 283.3 nm. A lead hollow cathode lamp should be fitted and the instrument set according to the manufacturer's instructions.

A reagent blank solution should be prepared and similarly aspirated together with lead standard solutions covering the range 0.2 to 100 ppm Pb. For greatest sensitivity set to a wavelength of 217 and make a background correction.

Determination by Sublimation Procedures

The sublimation procedure of Marshall and Hess[8] can be used for those rocks containing 1 ppm of lead or less. Approximately 20 g of the silicate rock material is used for each measurement. This sample weight is transferred to a tall-form carbon crucible made from high purity graphite that has previously been cleaned by soaking in concentrated hydrochloric acid, rinsing, drying and baking at 1300-1400° for an hour. This temperature is achieved using an induction furnace and is measured with an optical pyrometer.

The apparatus required is shown in Fig. 22 . It consists of a water-cooled quartz glass tube as the furnace chamber, with the graphite crucible containing the sample material supported on quartz rings, themselves on a graphite pedestal.

A slow stream of nitrogen is used to sweep the gaseous products including volatilised lead out of the hot zone, through a sintered quartz-glass disc and to a plug of quartz-glass wool. The lead is condensed on the walls of the chamber as well as in the disc and quartz-glass wool plug. Marshall and Hess heated their samples for

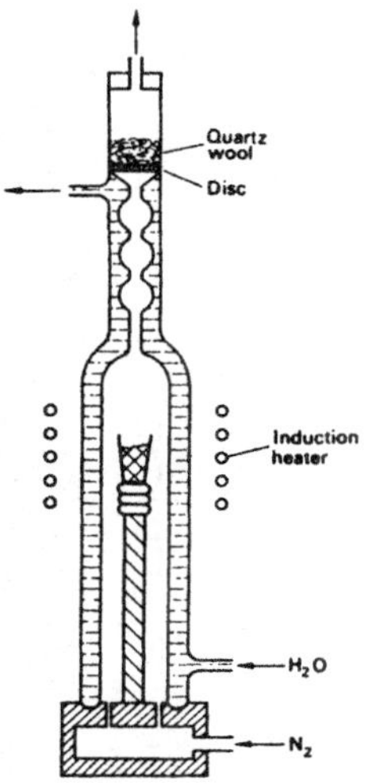

FIG 22 Apparatus for the recovery of lead by distillation.

1 hour, but this is probably longer than is necessary. Some silicates tend to froth excessively, and the rate of increase of temperature must be carefully controlled to keep the melt within the graphite crucible.

After cooling and dismantling of the apparatus, the lead from the sample together with other metals that are also sublimed onto the walls of the apparatus are recovered by washing with concentrated nitric acid, followed by rinsing with water. The lead in this solution can then be determined photometrically with dithizone after a preliminary separation with diethyldithiocarbamate as described above.

The original paper(8) should be consulted for details of the procedure.

References

1. KORKISCH J and FEIK F., Analyt. Chem. (1964) 36, 1793
2. SANDELL E B., Colorimetric Determination of Traces of Metals, Interscience, New York, 3rd ed., 1959, p.572
3. THAMPSON C E and NAKAGAWA H M., US Geol. Surv. Bull. 1084-F, (1960)
4. MAYNES A D and McBRYDE W A E., Analyt. Chem. (1957) 29, 1259
5. BASKOVA Z A., Zhur. Anal. Khim. (1959) 14, 75
6. SANZOLONE R F and CHAO T T., Anal. Chim. Acta (1976) 86, 163
7. IORDANOV N and KOCHEVA L., Bulgar. Akad. Nauk. Izv. Khim. Inst. (1956) 4, 327
8. MARSHALL R R and HESS D C., Analyt. Chem. (1960) 32, 960
9. KORKISCH J and GROSS H., Talanta (1974) 21, 1025
10. GAGE J C., Analyst (1955) 80, 789 and (1957) 82, 453
11. CRUZ R B and VAN LOON J C., Anal. Chim. Acta (1974) 72, 231
12. CAMPBELL W C and OTTAWAY J M., Talanta (1975) 22, 729
13. LANGMYHR F J, STUBERGH J R, THOMASSEN Y, HANSSEN J E and DOLEZAL J., Anal. Chim. Acta (1974) 71, 35
14. LANGMYHR F J and THOMASSEN Y., Z. Anal. Chem. (1973) 264, 122

CHAPTER 27

Magnesium

Magnesium is usually reported as the oxide MgO, and as such can constitute as much as 51 per cent of certain silicate rocks - varieties of dunite for example. 30 to 40 per cent MgO is fairly common, being encountered in certain ultrabasic rocks such as picrites and peridotites, which are among the first to crystallise. Successive rocks contain less as the magma becomes depleted in magnesium, late stage granites for example, contain only trace amounts.

Carbonate minerals include magnesite $MgCO_3$, dolomite $CaMg(CO_3)_2$ and ankerite in which iron and manganese have substituted to some extent for magnesium in dolomite.

Gravimetric determination as pyrophosphate has long formed part of the classical procedure for the main fraction in silicate and carbonate rock analysis. In this scheme it is applied after the separation of almost all other constituents of the sample material including silicon, iron, aluminium, titanium, vanadium, chromium, calcium and part of the manganese. The remaining part of the manganese is precipitated with magnesium as the ammonium phosphate $(Mg,Mn)(NH_4)PO_4$, which is then ignited to pyrophosphate $(Mg,Mn)_2P_2O_7$ as the weighing form and a correction applied for the manganese present. This method is described in detail in chapter 3.

Experience with the pyrophosphate method has shown that with rocks containing very little magnesium, the recoveries by the classical method are both poor and erratic. With some rocks it is impossible to recover any magnesium at all until the excess of ammonium salts is destroyed, and even then the recoveries can be incomplete. For these rocks the gravimetric procedure based upon precipitation with 8-hydroxyquinoline is preferred. This is not suited to the determination of large amounts and in the case of rocks rich in magnesium, the solution should be diluted to a suitable volume and an aliquot taken for the precipitation. As with the pyrophosphate method, it is necessary to remove most other constituents of silicate rocks by ammonia precipitation and preferable also to remove the bulk of ammonium salts before precipitating the magnesium. This latter step is usually accomplished by evaporating the rock solution and heating the residue with concentrated nitric acid, although this is not necessary where only a portion of the solution is taken for the determination of magnesium.

The precipitation of magnesium as the complex with 8-hydroxyquinoline can also form the basis of a titrimetric determination of magnesium. As with the gravimetric method aluminium, iron and other elements of the ammonia group must be absent, whilst any manganese that is not collected in the ammonia precipitate will be collected with the magnesium as the organic complex with 8-hydroxyquinoline. This oxide precipitate is collected, dissolved in warm dilute hydrochloric acid, and a known excess of a standard potassium bromide-potassium bromate solution added to brominate the organic reagent. The excess of oxidising agent is then determined by adding potassium iodide and titrating the liberated iodine with standard sodium thiosulphate solution.

Magnesium and calcium both form stable complexes with EDTA at pH 10, whereas at pH 7.6 only that of calcium is stable. It is therefore theoretically possible, by choosing a pH just above 7.6 and a suitable indicator, to titrate calcium in the presence of magnesium, whereas by using a pH greater than 10.0 calcium and magnesium will be titrated together. In practice, however, calcium is often titrated at pH of 12 or more, where magnesium is precipitated as hydroxide and then does not interfere. The titrimetric determination of magnesium consists therefore of two separate titrations, that of calcium alone and that of calcium plus magnesium, giving the magnesium content by difference. The accuracy obtainable is however poor and this procedure cannot therefore be recommended. An alternative and somewhat better procedure is to precipitate calcium as oxalate and determine magnesium by EDTA titration in the filtrate after removal of ammonium salts.

Many other metals present in the solution will be similarly titrated and some, such as iron, react irreversibly with the recommended indicators. Iron, aluminium and manganese can be complexed by adding triethanolamine but the presence of more than traces of these elements, even when masked in this way, can give rise to colour changes in the indicator, making end-point detection difficult. Even when titrating pure magnesium and calcium solutions the exact end-points are subjective, and for best results it is necessary to complete each titration under the same conditions of dilution, indicator concentration, lighting and viewing.

Indicators that have been suggested for titrimetric determination of magnesium plus calcium include eriochrome black T (solochrome black T,C.I.14645),[1] eriochrome blue black B (solochrome black 6B, C.I.14640),[2] calmagite,[3] metalphthalein with naphthol green B screening,[4] and 1-Dicarboxymethylaminomethyl-2-hydroxy-3-naphthoic acid (DHNA).[5] Where magnesium is present in only very small amounts (as, for example, in many limestones), acid alizarin black S.N.(C.I.21725)[6] and methyl thymol blue[7] can be used to titrate calcium present in solution. In the more general case where calcium and magnesium are both present in quantity, calcein,[8] calcon (eriochrome blue black R, C.I. 15705)[9] or HSN[10] ("Patton and Reeder's

Indicator") can all be used for the titration of calcium. The separation of calcium and magnesium from other elements and from each other by elution from a cation exchange resin[12] simplifies the choice of indicator.

Very few photometric reagents have found general application for the determination of magnesium in silicate or carbonate rocks, partly because of the general lack of good photometric reagents for magnesium and partly because other elements tend to interfere excessively. Two reagents that have been regularly applied to this determination are titan yellow and magon.

A great deal more is known concerning the application of titan yellow, C.I. 19540 (known also as titan yellow 2GS, Clayton yellow, thiazole yellow, acridingelb 5G, azidingelb 5G and brilliant yellow). This compound is a water-soluble triazole dye which forms a reddish coloured lake with freshly formed colloidal magnesium hydroxide. The colloid is protected from precipitation by the addition of other colloids such as agar, starch, gum arabic, and the more recently suggested polyvinyl alcohol, polyacrylate and glycerol. In colloidal solution, the lake formed between magnesium hydroxide and titan yellow has maximum light absorption at a wavelength of 530 nm, although if polyvinyl alcohol has been added, a somewhat higher wavelength of 540 nm is used to prevent interference from a polyvinyl-alcohol complex which has maximum absorption at 490 nm.

Titan yellow is produced by coupling dehydrothio-p-toluidine sulphonic acid with its diazonium salt, and commercial samples have been shown[13,14] to differ considerably in their reactivity towards magnesium. The complexity of the organic product is well known, and inorganic salts (especially sodium chloride) are known to have formed the major part of some batches of the reagent. King and Pruden[14] have suggested fractionation of the acetone-soluble material on a Sephadex G-10 (a dextran gel) column, but more recently[15] a new synthesis was suggested for a titan yellow product that is much superior to that previously available in its reactivity towards magnesium.

Elements that interfere with the determination of magnesium with titan yellow include aluminium and other elements of the ammonia group and also calcium and phosphorus. These interferences may be prevented by the addition of suitable masking agents, but this has been found to affect the stability of the magnesium lake and to impair the reproducibility of the results. The optical density of the colloid is also affected by the ammonium salt concentration, and any scheme for removing the elements of the ammonia group, together with calcium by precipitation with ammonia and ammonium oxalate, must include also a subsequent stage in which the added ammonium salts are removed or destroyed. An alternative procedure for the removal of the elements of the ammonia group described by Evans[16] is based upon their precipitation with sodium succinate at a pH of 6. The quantity of succinate has little effect upon the determination, except to increase slightly the reagent blank value. Calcium is not removed, but interference from it is avoided by adding sucrose.

Magon and magon sulphate (known also as xylidyl blue) form soluble pink coloured complexes with magnesium.[17,18] These reagents have been used in procedures for magnesium in which iron and aluminium were removed by precipitation with ammonia, as described by Abbey and Maxwell,[19] or by precipitation with sodium hydroxide as described by Yoshida et al[20] and an aliquot of the filtrate was evaporated with concentrated hydrochloric and nitric acids to decompose ammonium salts. Triethanolamine was added to complex any residual aluminium and the pink colour was developed in a borax-buffered solution.

The presence of calcium has a slight effect upon the magnesium determination, although interference from this source can be avoided by adding an excess, equivalent to 40 per cent CaO, to give a constant calcium effect.

Magon reagent is blue in colour, and has an appreciable absorption at the wavelength used to measure the absorption of the magnesium complex. A certain instability of the magnesium complex has also been noted.[17] Two or three standards are carried through the procedure with each batch of samples and the results calculated by interpolation.

Flame photometric methods have not achieved any degree of popularity for the determination of magnesium, probably because other elements occurring in silicate rocks seriously interfere with the magnesium emission. Aluminium, silicon, phosphate and sulphate in particular have been implicated, although some improvement can be obtained by adding an excess of calcium or strontium as a releasing agent, and also by working in an aqueous acetone medium.[21]

A similar although very much less severe interference has been noted in the determination of calcium and magnesium by atomic absorption spectroscopy, again principally from the elements silicon, aluminium, phosphate and sulphate. The interference from phosphate and sulphate is considerably reduced by using a high temperature flame, and provided that the sample material is decomposed with a mixture of perchloric and hydrofluoric acid, silicon is removed and the only serious interference is from aluminium.

The addition of calcium has been shown to lessen the depressing effect of aluminium on the magnesium absorption, but very large amounts of calcium are required for the complete release of magnesium and the determination of calcium is then no longer possible. Lanthanum has also been recommended for this purpose. Aluminium oxide is reported as forming a mixed compound with calcium and magnesium which is not dissociated in an air-acetylene flame. The hotter flame obtained with a nitrous oxide burner, is however, sufficient for this[22] and the interference from aluminium can thus be avoided. With this flame many of the difficulties experienced earlier with the determination of both calcium and magnesium by this technique have disappeared.

Gravimetric Determination with 8-Hydroxyquinoline

Method

Reagents: 8-Hydroxyquinoline solution, dissolve 2.5 g of the reagent in 100 ml of 2 N acetic acid.

Procedure. Combine the filtrates and washings from the precipitation of calcium oxalate, make just acid with hydrochloric acid and evaporate to near dryness on a steam bath. Allow to cool, cover with a watch glass and add 50 ml of concentrated nitric acid. Transfer the beaker to a steam bath and gradually raise the temperature until a vigorous reaction sets in, with evolution of brown nitrous fumes. When the reaction has subsided, rinse and remove the cover, wash down the sides of the beaker and again evaporate to dryness. If any appreciable ammonium salt residue remains, repeat the evaporation with a further quantity of concentrated nitric acid, and then evaporate to dryness. Moisten the residue with a little concentrated hydrochloric acid and evaporate to dryness once more, to expel the remaining traces of nitric acid.

Dissolve the residue in a little hot water containing 1 ml of concentrated hydrochloric acid and filter if necessary into a clean 400-ml beaker. Dilute to a volume of about 100 ml with water, heat to boiling and add concentrated ammonia solution until a small excess is present. If any precipitate forms at this stage it should be collected, washed with a little dilute ammonia and discarded (Note 1). Allow the combined filtrate and washings to cool to a temperature of 65-70°, then add 10 ml of 8-hydroxyquinoline solution (Note 2) and 8 ml of concentrated ammonia solution. Stir the solution, transfer to a steam bath for 5 minutes, stir again and then set aside for 30 minutes.

Collect the precipitate on a weighed, sintered glass or porcelain crucible and wash well with warm 0.5 N ammonia solution. Transfer to an electric oven set at a temperature of 140°, and dry to a constant weight. The precipitate of $Mg(C_9H_6ON)_2$

contains 12.91 per cent magnesium oxide, MgO (Notes 3 and 4).

Notes: 1. Traces of alumina that have not been collected in the main ammonia precipitate are usually noted at this stage.

2. This quantity of reagent is sufficient to precipitate about 30 mg of magnesium oxide, corresponding to 3 per cent MgO from a 1 g sample portion. For rocks of higher magnesia content, the amount of reagent should be increased proportionately, or the rock solution diluted to volume and an aliquot taken for the precipitation.

3. If a large excess of reagent has been added, the magnesium oxinate precipitate will be contaminated with reagent. This can be removed by dissolving the precipitate in dilute acid and reprecipitating as described above.

4. Any manganese present in the solution will be recovered with the magnesium. It can be determined photometrically in the precipitate after destruction of the organic matter with sulphuric and nitric acids.

Titrimetric Determination of Calcium and Magnesium

Reagents:

Triethanolamine solution, equal volumes of triethanolamine reagent and water.

pH 10 buffer solution, dissolve 67.5 g of ammonium chloride in water, add 570 ml of ammonia solution (s.g.0.880) and dilute to 1 litre with water.

1-Dicarboxymethylaminomethyl-2-hydroxy-3-naphthoic acid reagent (DHNA), mix together 0.05 g of the solid material with 10 g of sodium chloride.

Diaminocyclohexanetetraacetic acid solution (CyDTA), 0.01 M dissolve 3.3 g of the solid reagent in about 200 ml of water with small additions of M sodium hydroxide solution as necessary. Add acetic acid to bring the pH to 10 and dilute to 1 litre with water. Standardise by titration with the standard magnesium solution using the procedure given below.

Ethylenebis(oxyethylenenitrilo) tetraacetate solution (EGTA), 0.02 M, dissolve 3.81 g of the free acid in 25 ml of 1 M sodium hydroxide solution and dilute to 1 litre with water. Standardise by titration with aliquots of standard calcium chloride solution using the procedure given below.

Potassium hydroxide solution, 1 M, dissolve 14.0 g of the solid material in 250 ml of water.

Calcein reagent, grind together 0.1 g of the solid reagent with 10 g of solid potassium nitrate.

Standard magnesium solution, clean a length of magnesium ribbon and dissolve 0.603 g in about 100 ml of water to which 10 ml of concentrated perchloric acid have been added. Dilute to 1 litre with water. This solution contains 1 mg MgO per ml.

Standard calcium solution, 0.02 M, dissolve 2.002 g of pure, dried calcium carbonate in the minimum quantity of very dilute hydrochloric acid and dilute to 1 litre with water. This solution contains 1.12 mg CaO per ml.

Procdure for magnesium plus calcium. Accurately weigh approximately 1 g of the finely powdered silicate rock material into a platinum or PTFE basin, moisten with water and add 1 ml of concentrated nitric acid, 5 ml of concentrated perchloric acid and 20 ml of concentrated hydrofluoric acid and allow the basin to stand overnight. Complete the decomposition of the rock material by evaporating to dryness. Add 2 ml of concentrated perchloric acid and evaporate almost to dryness. Repeat this evaporation almost to dryness twice more with intervening additions of small quantities of perchloric acid. Finally add 5 ml of concentrated perchloric acid, 50 ml of water and heat on a water bath until all solid material has dissolved, allow to cool and dilute to volume with water in a 200-ml volumetric flask.

Transfer by pipette 10 ml of the rock solution to a titration vessel and add 80 ml of water, 5 ml of the triethanolamine solution, 10 ml of the prepared buffer solution and about 30 mg of the DHNA indicator reagent. Using a fluorimetric titrator, titrate the total calcium plus magnesium with CyDTA solution under filtered ultraviolet illumination to the end point given by the disappearance of the blue-green fluorescence (Notes 1 and 2).

Procedure for calcium. Transfer by pipette 20 ml of the rock solution (Note 3) to a titration vessel, add 4 ml of triethanolamine solution (Note 4) and 2 ml of 1 M potassium hydroxide solution (Note 5). Add about 30 mg of the prepared calcein reagent and titrate slowly with the 0.02 M EGTA solution to the disappearance of the green fluorescence (Note 6).

Notes: 1. The precise end point is difficult to distinguish visually under normal lighting conditions due to a slight residual fluorescence, hence the use of a fluorimetric titrator.

2. Barium and strontium are partially titrated in the presence of calcium and magnesium, but do not themselves form fluorescent complexes with DHNA.

3. The solution should contain not more than 40 mg total magnesium plus calcium, with not more than 50 mg of iron plus aluminium.

4. This quantity can be increased up to 8 ml if necessary to complex the amounts of iron, aluminium, etc present in the sample solution.

5. ie one-tenth of the original sample volume.

6. When much magnesium is present, there is a tendency to obtain slightly high results, possibly by as much as 5 per cent relative. Where magnesium is present in quantity and more accurate results are required than a back titration procedure, as described by Pribil and Vesely,[11] should be used.

Determination of Calcium and Magnesium by Atomic Absorption Spectroscopy

The procedure given below is based upon that described by Esson,[23] in which a mixture of perchloric and hydrofluoric acids is used to decompose the silicate sample material, and lanthanum chloride solution is added to serve as releasing agent.

Method

Reagents:

Lanthanum chloride solution, dissolve 58.6 g of lanthanum oxide by heating with 1 litre of 1.2 N hydrochloric acid. Store in a polyethylene bottle.

Standard magnesium stock solution, dissolve 0.151 g of pure fresh magnesium ribbon in dilute perchloric acid, transfer to a 1 litre volumetric flask and dilute to volume with water. This solution contains 250 μg MgO per ml.

Standard magnesium working solution, pipette 10 ml of the stock solution into a 250-ml volumetric flask and dilute to volume with water. This solution contains 10 μg MgO per ml.

Standard calcium stock solution, dissolve 0.446 g of pure calcium carbonate in dilute perchloric acid, transfer to a 1-litre volumetric flask and dilute to volume with water. This solution contains 250 μg CaO per ml.

Standard calcium working solution, pipette 10 ml of the stock solution into a 250-ml volumetric flask and dilute to volume with water. This solution contains 10 μg CaO per ml.

Procedure. Accurately weigh 0.1 g of the finely powdered silicate rock material into a small platinum crucible, moisten with water and add 5 ml of concentrated hydrofluoric acid and 4 ml of concentrated perchloric acid. Transfer the crucible to a hot plate and decompose the sample material by repeated evaporation with perchloric acid as described above. Dilute the rock solution containing 1 ml of concentrated perchloric acid to volume with water in a 100-ml volumetric flask.

Pipette 10 ml (Note 1) of the rock solution into a clean 100-ml volumetric flask, add 20 ml of the lanthanum solution, dilute to volume with water and mix well. With the instrument operating in accordance with the manufacturer's instructions, spray the solution into the air-acetylene flame (Note 2) of an atomic absorption spectrometer fitted with a calcium/magnesium lamp or in succession with separate calcium and magnesium lamps. Spray also a reagent blank solution, similarly prepared but omitting the sample material. The spectrometer should be set at a wavelength of 285.2 nm for measurement of the magnesium absorption and at 422.7 nm for the calcium absorption.

Calibration. Prepare a set of standard solutions containing 0-200 μg CaO and 0-200 μg MgO each with 20 ml of the lanthanum solution in separate 100-ml volumetric flasks. Spray in turn into an air-acetylene flame of the spectrometer as for the sample and reagent blank solutions, and plot the absorption against the calcium and magnesium concentration. The calibration graphs may be slightly convex to the concentration axis.

Notes: 1. In the procedure described[23] a volume of 1 to 10 ml was suggested depending upon the calcium and magnesium content of the rock material. Aliquots of as small as 1 ml cannot be recommended.

2. As noted earlier, lanthanum is added as a release agent. This is not required if a nitrous oxide-acetylene flame is substituted for the air-acetylene employed in the method as described. At the higher temperature obtained with a nitrous oxide-acetylene flame an ionisation suppressant (eg potassium) is required.

References

1. BIEDERMANN W and SCHWARZENBACH G., Chimia (1948) 2, 56
2. SCHNEIDER F and EMMERICH A., Zucker Beih. (1951) 1, 53
3. LINDSTROM F and DIEHL H., Analyt. Chem. (1960) 32, 1123
4. TUCKER B M., J. Austr . Inst. Agri. Sci. (1955) 21, 100
5. CLEMENTS R L, READ J I and SERGEANT G A., Analyst (1971) 96, 656
6. BELCHER R, CLOSE R A and WEST T S., Talanta (1958) 1, 238
7. KORBL J and PRIBIL R., Chem. and Ind. (1957) p.233
8. DIEHL H and ELLINGBOE J., Analyt. Chem. (1956) 28, 882
9. HILDEBRAND G P and REILLEY C N., Analyt. Chem. (1957) 29, 258
10. PATTON J and REEDER W., Analyt. Chem. (1956) 28, 1026
11. PRIBIL R and VESELY V., Chemist-Analyst (1966) 55, 82
12. ABDULLAH M I and RILEY J P., Anal. Chim. Acta (1965) 33, 391
13. MIKKELSEN D S and TOTH S J., J. Amer. Soc. Agron. (1947) 39, 165
14. KING H G C and PRUDEN G., Analyst (1967) 92, 83
15. KING H G C, PRUDEN G and JAMES N F., Analyst (1967) 92, 695
16. EVANS W H., Analyst (1968) 93, 306
17. MANN C K and YOE J H., Analyt. Chem. (1956) 28, 202
18. MANN C K and YOE J H., Anal. Chim. Acta (1957) 16, 155
19. ABBEY S and MAXWELL J A., Anal. Chim. Acta (1962) 27, 233
20. YOSHIDA S, YOSHIDA M and IWASAKI I., Japan Analyst (1974) 23, 1232
21. DINNIN J I., U S Geol. Surv. Prof. Paper 424-D, p.391, 1961
22. WALSH J N and HOWIE R A., Inst. Min. Metall. Trans. B (1967) 76, 119
23. ESSON J, Unicam Instruments, Atomic Absorption Methods No. Ca-3 and Mg-4.

CHAPTER 28

Manganese

Although there are a number of volumetric methods for determining manganese in manganese ores, these are unlikely to be suitable for the determination in silicate rocks. The method most commonly employed is spectrophotometric determination based upon the oxidation of manganese (II) to permanganate with either potassium periodate or ammonium persulphate in the presence of silver ions as catalyst. Attempts to use silver (II) oxide for this purpose were unsuccessful, giving low results for a number of standard samples. It seems likely that some reduction of the permanganate had occurred after the destruction of excess silver (II) oxide. Sodium perxenate has been suggested[1] for the oxidation of manganese to permanganate but in view of the cost of the reagent, is unlikely to be widely used in the analysis of silicate rocks.

Oxidation with potassium periodate proceeds fairly rapidly in nitric or sulphuric acid solution at or near the boiling point and, provided that more than a trace of manganese is present, complete oxidation can be obtained in about 1 hour. According to Nydahl[2] under these conditions the oxidation is incomplete if only trace amounts of manganese are present, and for this reason he prefers to use ammonium persulphate which gives a much more rapid oxidation. This occurs smoothly in nitric-phosphoric acid solution provided that a catalytic amount of silver ion is present. Just as Nydahl obtained complete oxidation of manganese with persulphate and not periodate, so Langmyhr[3] obtained complete oxidation only with periodate. In the experience of one of the authors both reagents gave complete oxidation, but greater fading occurred when persulphate solutions were subjected to prolonged boiling.

Solutions of permanganate obey the Beer-Lambert Law and slight variations in the reagent concentrations do not affect the optical density values of the solution. The absorption spectrum consists of a series of maxima, in the range 500 to 575 nm and a wavelength of 525 nm is recommended for the determination.

In the determination of manganese, the yellow colour of the ferric ion is removed by adding phosphoric acid. This can introduce difficulties if the sample material contains a great deal of titanium, when titanium phosphate may be precipitated, but this can then be avoided by increasing the sulphuric acid concentration. Of the elements remaining in the rock solution, only chromium has an appreciable absorption at 525 nm. Interference from chromium can be avoided by making the photometric measurement at a wavelength of 575 nm,[4] but at this wavelength, the absorption curve falls steeply, and the wavelength settings must be made and reproduced very accurately. Since chromate solutions are not reduced by sodium nitrite, chromium interference can be avoided by measuring the absorbance relative to a solution produced by adding a little sodium nitrite to a separate portion of the permanganate solution.

All permanganate solutions should be diluted to volume with very dilute nitric acid that has previously been boiled with either potassium periodate or ammonium persulphate and allowed to cool. If this is not done, and distilled water is used, some fading of the permanganate is likely to be noted. Water from polyethylene wash bottles should be avoided, as this has been implicated as one cause of fading.[5]

An alternative method for determining manganese in silicate rocks is that based upon the reaction of manganese (II) with formaldoxime in alkaline solution.[6] The sensitivity of the determination is said to be approximately five times that based upon permanganate formation. The most serious interference appears to be that from iron which also forms a highly coloured complex with the reagent. Methods for overcoming this interference have been described by Riley and Williams,[7] who extracted iron and aluminium as oxinates with chloroform, and by Abdulla[8] who added ascorbic acid, hydroxyammonium chloride and EDTA.

The orange red complex with formaldoxime has a maximum absorption at 490 nm and the Beer-Lambert Law is obeyed over the concentrations range 0-10 μg MnO/ml. The procedure given below is based upon that described by Abdulla.[8]

Sulphuric acid solutions are commonly used for the photometric determination of manganese. These are readily obtained by evaporation of the sample material with hydrofluoric and sulphuric acid. Any accessory minerals (such as garnets and spinels) that remain after this initial attack may contain a large proportion of the total manganese present in the sample material, and provision must be made to recover the manganese in this residual fraction. With many rocks complete decomposition will be obtained by evaporating the sample portion to dryness with hydrofluoric and sulphuric acids and fusing the dry residue with a little potassium pyrosulphate.

Most carbonate rocks can be decomposed by warming the finely ground sample material with dilute nitric acid. A suitable aliquot of the solution can then be taken for a photometric determination by one of the methods given for silicate rocks. When this method is applied to carbonatite rocks, the carbonate mineral fraction is decomposed together with certain of the accessory minerals (some sulphides for example) leaving most of the oxide and silicate minerals which may contain an appreciable proportion, if not the major part of the total manganese content of the sample. For these materials it has been found possible to obtain a complete decomposition by fusing a small sample weight with sodium peroxide in a zirconium crucible. After extraction with water, the material is dissolved in dilute nitric acid, diluted to volume and a suitable aliquot taken for photometric determination of manganese following oxidation with potassium periodate. For carbonatite rocks that contain appreciable amounts of silica it may be necessary to evaporate the aliquot with hydrofluoric and nitric acids before the oxidation of the manganese.

A mixture containing equal proportions of sodium carborate and borax has also been recommended[9] for the analysis of silicate materials rich in manganese. The melt is dissolved in diluted nitric acid, with rapid stirring to avoid polymerisation of silica.

Atomic absorption techniques have been widely applied to the determination of manganese in silicate rocks, using fuel-lean or stoicheiometric air-acetylene flames and a wavelength of 279.5 nm. However Sanzolone and Chao[10] have reported matrix effects with fuel-lean (oxidising) air-acetylene flames from silicon, aluminium, iron, calcium and magnesium, and recommend a reducing nitrous oxide-acetylene flame as a means of minimising or removing these. A stoicheiometric nitrous oxide flame has also been recommended by Price and Whiteside,[11] in a general method for the analysis of silicate material.

Flame emission methods have also been suggested for determining manganese in silicate material[12,13] but do not appear to have been extensively employed, possibly because of direct interference from the potassium line at 404 nm with the manganese emission at 403 nm. Phosphate and sulphate ions tend to decrease, whereas chloride and perchlorate ions enhance the manganese flame emission.

Spectrophotometric Determination of Manganese as Permanganate

In the two spectrophotometric permanganate procedures described in detail below, sulphuric and hydrofluoric acids are used to decompose the sample material and, after decomposition of any remaining accessory minerals, the rock solution is diluted to volume for the determination of manganese. This sample solution can be used also for the determination of total iron, titanium and phosphorus if required.

OXIDATION WITH PERIODATE

Method

Reagents: Potassium periodate.

Sodium nitrite.

Nitric acid solution, boil 1 litre of 0.2 N nitric acid with approximately 0.1 g of potassium periodate, allow to cool and store in an all-glass wash-bottle.

Standard manganese stock solution, accurately weigh 0.155 g of pure manganese into a small beaker, dissolve in 50 ml of 0.5 N sulphuric acid, transfer to a 1-litre volumetric flask and dilute to volume with water. This solution contains 200 μg MnO per ml.

Standard manganese working solution, transfer 25 ml of the stock solution to a 100-ml volumetric flask and dilute to volume with water. This solution contains 50 μg MnO per ml and should be used for the calibration of the 1-cm spectrophotometer cells. If 4-cm cells are to be used, the working solution can be prepared by diluting 5 ml of the stock solution to 100 ml, giving 10 μg per ml.

Procedure. Accurately weigh approximately 0.5 g (Note 1) of the finely powdered silicate rock material into a platinum crucible or dish, moisten with a little water and add 1 ml of concentrated nitric acid, 5 ml of 20 N sulphuric acid and 10 ml of hydrofluoric acid. Transfer the dish to a hot plate and evaporate to fumes of sulphuric acid. Allow the dish to cool, rinse down the sides of the dish with a few ml of water, add 5 ml of concentrated hydrofluoric acid and again evaporate, this time to complete dryness. Allow the dish to cool.

Add a small quantity of potassium pyrosulphate to the residue and fuse gently to give a completely fluid melt. Allow to cool, add a little water to the dish and warm to detach the cake from the dish. Rinse the melt into a 100-ml beaker, add 5 ml of concentrated nitric acid, 2.5 ml of syrupy phosphoric acid and approximately 0.2 g of potassium periodate and dilute to a volume of approximately 45 ml. Add a few granules of fused alumina (or other "anti-bump" device, such as a Gernez boiling tube), cover the beaker with a clock glass and gently boil until the purple permanganate colour forms and then no longer deepens in intensity. If the colour does not develop after boiling for about 30 minutes, add a further 0.2 g of potassium periodate and continue boiling for a further period of 30 minutes.

Allow the solution to cool, transfer to a 50-ml volumetric flask and dilute to volume with the dilute nitric acid, previously boiled with a little potassium periodate. Fill two matched spectrophotometer cells with the coloured sample solution and add to one cell only a small crystal of sodium nitrite. Stir gently with a thin glass rod to complete the decomposition of the permanganate ion. Measure the optical density of the coloured solution, relative to the solution to which the sodium nitrite has been added, using the spectrophotometer set at a wavelength of 525 nm, and hence determine the manganese content of the sample material by reference to the calibration graph or by using a calibration factor (Note 2).

Calibration. Transfer aliquots of 5 to 25 ml of the standard manganese solution containing 0.25-1.25 mg MnO (Note 3) to separate 100-ml beakers, dilute each aliquot to about 40 ml, add concentrated nitric acid, syrupy phosphoric acid and potassium periodate and continue as described above for the sample solution. Measure the optical density of each solution and plot these values against concentration of manganese.

Notes: 1. This quantity of material is recommended for acid rocks such as granites and rhyolites and other similar materials low in manganese. A 0.1 g sample weight is adequate for intermediate and basic rocks.

2. If the measured optical density of the sample solution is beyond the range covered by the 1-cm cell calibration graph, transfer 5 ml of the permanganate solution to a 100-ml beaker, add further amounts of concentrated nitric acid, syrupy phosphoric acid and potassium periodate, dilute to about 45 ml and repeat the oxidation as described.

3. These aliquots are suggested for the calibration of the 1-cm matched spectrophotometer cells. When using 4-cm cells, use the more dilute working standard when these aliquots will contain 50-250 μg MnO.

OXIDATION WITH PERSULPHATE

Method

Reagents: Acid reagent solution, dissolve 36.5 g of mercuric sulphate in a mixture of 200 ml of concentrated nitric acid with 100 ml of water. Add 100 ml of syrupy phosphoric acid and 0.017 g of silver nitrate. When cold, dilute to 500 ml with water.

Dilute acid reagent solution, add about 0.5 g of ammonium persulphate to 25 ml of the acid reagent solution and dilute to 1 litre with water. Boil this solution for 5 minutes,

allow to cool and store in a glass wash-bottle.
Ammonium persulphate, use only fresh, good-quality reagent.

Procedure. Decompose a 0.5 g sample portion of the finely powdered rock material by evaporation with nitric, hydrofluoric and sulphuric acids, and fuse the residue with potassium pyrosulphate as described above. Extract the melt with water, dilute to volume in a volumetric flask and transfer an aliquot containing not more than 1.25 mg MnO to a 100-ml beaker. Dilute to about 40 ml with water. Add 3 ml of the acid reagent solution and about 0.5 g of ammonium persulphate. Using a few granules of fused alumina or some other "anti-bump" device, bring the solution to the boil and boil rapidly for 2 minutes but not longer. Cool the solution rapidly and dilute to volume in a 50-ml volumetric flask with the dilute acid reagent solution.

Prepare also a reference solution by diluting a second aliquot of the rock solution, together with 3 ml of the acid reagent solution to volume in a second 50-ml volumetric flask with the dilute acid reagent solution.

Fill a spectrophotometer cell with the coloured permanganate solution, and measure the optical density relative to the reference solution in a separate cell. Prepare also the calibration solutions as described above, but using the acid reagent and ammonium persulphate as oxidant.

Spectrophotometric Determination with Formaldoxime

The procedure given here is based upon that described by Abdulla.[8] In the original paper details are also given of an automatic procedure using an Auto Analyser.

Procedure. Decompose a 0.5 g sample of the finely powdered silicate rock material by evaporation with hydrofluoric and perchloric acids in the usual way, and dilute the solution of metallic perchlorates to 500 ml with water. Transfer by pipette, a 20-ml aliquot of the solution to a 50-ml volumetric flask and add in succession 3 ml of freshly prepared 0.4 M ascorbic acid solution, 3 ml of a 0.4 M formaldoxime solution and 4 ml of a buffer solution containing 70 g of ammonium chloride and 600 ml of ammonia (s.g. 0.880) per litre. Allow to stand for 2 minutes.

Add 3 ml of a 0.1 M EDTA solution and 4 ml of 2.2 M hydroxyammonium chloride solution. Dilute to volume with water, stopper the flask and mix well. Allow to stand for 5-10 minutes then measure the optical density in 1-cm cells at a wavelength of 490 nm, relative to water. Prepare also a reagent blank and a series of calibration solutions containing up to 0.5 mg MnO in 50 ml.

Determination of Manganese by Atomic Absorption Spectroscopy

As noted earlier, the determination of manganese in silicate and other rocks is almost invariably combined with the determination of other constituents, principally aluminium, iron, titanium, calcium and magnesium. A variety of general methods have been suggested involving hydrofluoric decomposition with or without sulphuric, nitric or perchloric acids. Pressure vessels are now commonly employed. The addition of boric acid is recommended for the dual pruposes of complexing certain of the elements present (eg silicon) and of dissolving precipitated fluorides. An alkali metal salt (usually potassium, or caesium if potassium is to be determined by this technique) is added as ionisation buffer, and the determination is completed by measuring the atomic absorption in a nitrous oxide-acetylene flame with the spectrometer set at the appropriate wavelength - 279.5 nm for manganese.

References

1. BANE R W., Analyst (1965) 90, 756
2. NYDAHL F., Anal. Chim. Acta (1949) 3, 144
3. LANGMYHR F J and GRAFF P R., Norges Geol. Undersokelse No. 230, 1965, p.17
4. SANDELL E B., Colorimetric Determination of Traces of Metals, Interscience, 1950, 2nd ed, p.433
5. RILEY J P., Anal. Chim. Acta (1958) 19, 421
6. SIDERIS C P., Ind. Eng. Chem, Anal. Ed. (1937) 9, 445
7. RILEY J P and WILLIAMS H P., Mikrochim. Acta (1959) 6, 804
8. ABDULLA M I., Anal. Chim. Acta (1968) 40, 526
9. PIRYUTKO M M and MIRONOVITCH V Ya., Zav. Lab. (1975) 41, 395
10. SANZOLONE R F and CHAO T T., Talanta (1978) 25, 287
11. PRICE W J and WHITESIDE P J., Analyst (1977) 102, 664
12. DIPPEL W A and BRICKER C E., Analyt. Chem. (1955) 27, 1484
13. ROY N., Analyt. Chem. (1956) 28, 34

CHAPTER 29

Mercury

The toxic nature of mercury and many of its compounds has long been known, and the chronic effects of prolonged and repeated exposure to low concentrations of mercury vapour are well documented. Recent interest in the hazards posed by the release of mercury to the environment arises from its accumulation in the marine environment in sediments, bio-conversion to alkyl mercury, concentration in fish and hence to chronic poisoning in man where such fish form a significant proportion of his diet. The incidence of this form of mercury poisoning - the so-called Minamata disease - is associated with the discharge of mercury into ocean or lacustine areas in the effluent from chloralkali plants. This interest has led in turn to an interest in the mercury content of sedimentary rocks and unconsolidated sediments, where the element may occur as the metal, or as chloride, alkyl derivatives, sulphide, or combination of these forms.

No special difficulties are encountered in the sampling and preparation for analysis of consolidated sediments such as shales, clays, and mudstones, other than the need to ensure that mercury is not lost by elevated temperatures in the drying of the powdered material. Samples of mud, silt and similar loose material are usually obtained in the form of grab samples that can be homogenised after drying. If core samples are available and since the main interest may be in the most recent part of the sediment, this layer should be removed, dried and analysed separately from the remainder of the core. Despite a report to the contrary, mercury does not appear to be lost in any significant quantity from samples stored at room temperature.[1]

Elemental mercury is appreciably volatile, as are the alkyl compounds. Elevated temperatures will therefore induce mercury loss from the sample material. Maintenance of a temperature of 80° for 24 hours has been recommended for drying sample material, at which no loss of mercury was observed,[2] although Iskandar et al[3] noted a significant loss of organomercury compounds from a lake sediment on drying at 60°.

Determination of Mercury in Silicate Rocks

The normal techniques of optical spectroscopy and spectrophotometry are not sufficiently sensitive to determine mercury in silicate rocks. Neutron activation analysis can be used,[4,5] but wherever chemical processing is undertaken care must be exercised to prevent or restrict loss of mercury by volatilisation.

For rocks associated with mineral deposits containing high concentrations of mercury, Popea and Jemaneanu[6] have proposed a method based upon dithizone extraction into carbon tetrachloride solution. 1-5 g of the sample are heated under reflux with sulphuric and nitric acids, and the mercury dithizonate extracted at pH 4-5 from acetic acid solution containing both EDTA and potassium thiocyanate. This method is useful for recovering 2-20 μg mercury. A combination of dithizone extraction with direct atomic absorption spectroscopy using the organic extract was described by Pyrih and Bisque;[7] the detection limit was given as 0.05 ppm Hg in the rock sample.

Most procedures for the analysis of sediments report total mercury, rather than identified compound. The separate determination of methyl mercury and dimethyl mercury based on steam distillation and atomic absorption was reported by Floyd and Sommers.[18]

Procedures commonly used for the decomposition and recovery of mercury from silicate material, use an acid digestion as the initial stage. A wide variety of conditions have been proposed, including the use of acid mixtures that may contain hydrochloric, nitric, sulphuric, hydrofluoric and perchloric acids. Potassium permanganate, persulphate and dichromate have all been recommended as additions to promote the oxidation of organic matter and the decomposition of mercuric sulphide. There is a possibility that manganese dioxide, formed by the decomposition of potassium permanganate, may retain trace amounts of mercury, giving rise to low and erratic results,[1] although this can be avoided by hydroxyammonium reduction of the manganese dioxide.[9]

An alternative procedure for the recovery of mercury is based upon thermal decomposition of the sample material with distillation of the liberated elemental mercury. This technique, described in detail by Nicholson,[10] and used in combination with a gold trap and vapour absorption, provides a means of rapid and convenient determination of mercury in a wide variety of silicate materials, including those high in organic content.

Vapour Absorption for the Determination of Mercury

This method of determining mercury, known also as flameless atomic absorption and as cold vapour absorption, was used initially[11,12] to determine mercury in rocks down to about 0.05 ppm. However, its usage has been extended[13,14] to rocks containing no more than a few ppb (1 in 10^9). Measurement is based upon the difference in absorption between the vapour from the rock material and the same vapour from which the mercury has been stripped. Measurement is made at a wavelength of 253.7 nm, rather than at the more sensitive 184.9 nm, which suffers from interference from oxygen.

An alternative procedure is to use a purification stage involving the trapping of the evolved mercury in a low temperature (eg liquid nitrogen) trap as described by Aston and Riley,[15] or a metallic gold trap which removes the mercury as an amalgam, as described by Warren et al.[16] The procedures given below are based upon those described by Head and Nicholson[17] and Omang and Paus.[14] See also Ure[18] for a review of this technique.

METHOD USING ACID DECOMPOSITION

In order to ensure that all the mercury present in the rock sample is released, a decomposition procedure based upon the use of hydrofluoric acid is included. A sealed PTFE vessel is used for this, and nitric acid is added to ensure that oxidising conditions are maintained, as mercury is readily lost from reducing solutions. Boric acid is used to complex fluoride ion and stannous chloride to provide the reduced condition at the appropriate moment.

The solution is aerated to remove the mercury which is collected on gold wire. Once collection is complete, the gold wire is heated in a furnace to remove the mercury and the absorption of the vapour measured at 253.7 nm using a standard atomic absorption spectrophotometer with a quartz-window gas cell.

Apparatus. For the sample decomposition a PTFE-lined bomb with an internal volume of about 110 ml is used.

The additional apparatus required is shown in Fig. 23 . It consists of a sample bubbler A, fitted with a magnesium perchlorate drying tube B, a gold wire collector C, furnace D, and quartz-window gas cell E. The gas flow system incorporates a mercury trap F, in the form of a coil of silver wire, a flow meter and a furnace by-pass system. The silica tube furnace, described in detail by Nicholson and Smith,[19] is wire-wound and gives a temperature of 850°. The gas cell is fitted to a standard atomic absorption spectrophotometer in the usual way.

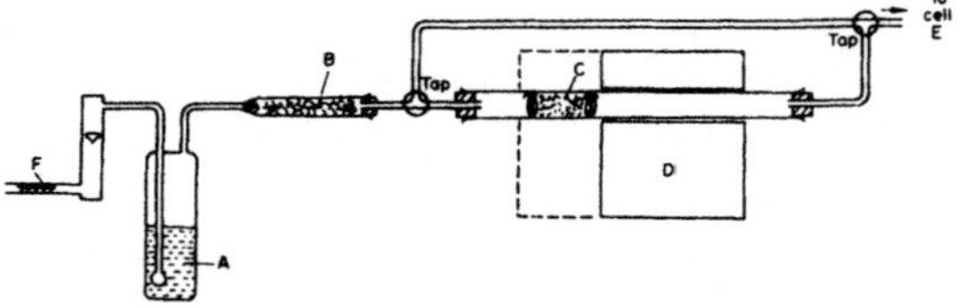

FIG. 23 Apparatus for determining mercury following acid digestion.

The gold wire collector consists of about 10 g of gold wire of 0.5 mm dia. cut into lengths of about 2 mm packed between gold spirals into the quartz tube furnace.

<u>Reagents:</u> Boric acid, saturated aqueous solution.

Stannous chloride solution, dissolve 50 g of stannous chloride in 250 ml of 4 N sulphuric acid.

Standard mercury stock solution, dissolve 0.054 g of mercuric oxide (HgO) in 250 ml of N sulphuric acid. This solution contains 200 μg Hg per ml.

Standard mercury working solution, dilute 10 ml of the stock solution to 500 ml with N sulphuric acid. This solution contains 4 μg Hg per ml. As required, prepare from this a further dilution with N sulphuric acid to give 0.02 μg Hg per ml.

<u>Procedure.</u> Accurately weigh approximately 0.2 g of the finely powdered rock material into the PTFE-lined decomposition vessel, moisten with water and add 5 ml of concentrated hydrofluoric acid and 0.5 ml of concentrated nitric acid. Seal the bomb, transfer to an electric oven and heat at a temperature of 120° for 10 minutes. Allow to cool to room temperature, open the bomb and add 50 ml of saturated boric acid and heat to dissolve any precipitated fluorides.

Set up the apparatus as shown in Fig. 23, with the 250-ml bubbler containing 50 ml of water. Adjust the air flow through the apparatus to a rate of about 2.5 litres per minute and heat the furnace to a temperature of 850°. Push the silica tube with the gold wire collector into the furnace for 2 or 3 minutes and then divert the air flow over the collector. Any absorbed mercury will be removed and appear as a response on the spectrophotometer recorder. Repeat this operation two or three times, when no further mercury response should be obtained. When this point has been reached, move the silica tube so that the gold wire collector is outside the hot zone of the furnace and allow it to cool. Divert the air stream to bypass the furnace.

Replace the bubbler containing water with a fresh bubbler containing the sample solution to which has been added 2 ml of 20 N sulphuric acid and 2 ml of the stannous chloride solution immediately before the replacement. At the same time, divert the air flow back over the gold wire collector and through the furnace, allowing the evolved mercury to be collected on the gold wire.

After about 2 minutes, divert the air flow to bypass the furnace once again and push the silica tube so that the collector is once again in the hot zone of the furnace, heating for 1 minute. Again divert the air flow and sweep the evolved mercury into the gas cell and record the mercury absorption. For calibration use aliquots of the lowest concentration of working standard solution containing up to 120 ng mercury. The calibration is linear up to at least this quantity.

Notes: 1. The silver coil is necessary to remove the small quantity of mercury that is usually found in the air supply.

2. The flow meter, which can be of the Rotameter type, should be calibrated for air in the range 0.5 to 2.5 litres per minute.

METHOD USING THERMAL DECOMPOSITION

Mercury is readily released from most silicate rocks by heating to a temperature of about 650°. Organo-mercury compounds are also evolved, and can be readily oxidised to metallic mercury. Under these conditions mercuric sulphide may be only partially decomposed, and the addition of calcium oxide to the sample is recommended as an aid to decomposition.

Apparatus. This is shown in Fig. 24. It consists of an absorber A, containing silica gel and a gold coil to remove water vapour and any mercury vapour from the air current, a flow meter, two combustion tubes arranged as shown complete with both furnaces and cooling coils. The combustion tube, T_1 used for the sample material contains a silver coil to promote the oxidation of organic compounds, silica gel and alumina. The tube T_2 contains magnesium perchlorate and the gold wire collector. Air can be drawn through the tubes T_1 and T_2 in series in the mercury evolution and collection stage, and through the tube T_2 only in the subsequent release stage. The air flow is completed through a gas cell in the path of an atomic absorption spectrometer.

Reagents: Calcium oxide, powdered, previously ignited.

Silica gel, about 6-22 mesh, non-indicating, dried briefly at 400° before use.

Alumina, chromatographic grade 100-200 mesh, also dried briefly at 400° before use.

Magnesium perchlorate, 14-22 mesh.

Mercury standards, prepared as described above.

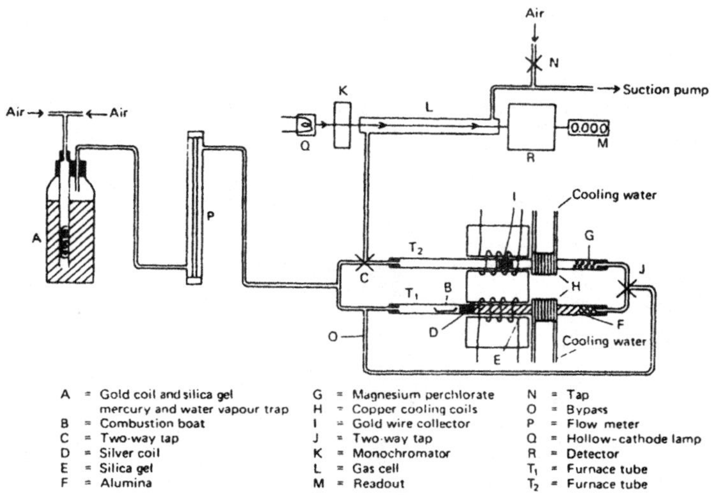

FIG 24 **Apparatus for determining mercury by thermal decomposition (NB monochromator would be better placed between gas cell and detector).**

Procedure. Set up the apparatus as shown, and set the temperature of the furnace T_1 to 650° and T_2 to 750°. Adjust the air flow rate to about 2.5 litres min^{-1}, leave for several minutes until the instrument response indicates that all the mercury has been removed from the apparatus. Slide that part of the tube T_2 containing the gold collector from the furnace and allow it to cool.

Accurately weigh approximately 0.1 g of the powdered sample material into a previously ignited nickel boat, mix with an equal amount of calcium oxide and place into tube T_1. Reconnect the gas flow system. With the air flow passing through the two tubes in series, heat the sample to 650° and maintain at this temperature for about 2 minutes. Mercury evolved from the sample will collect on the gold wire.

Divert the gas flow through T_2 only, heat the gold wire to a temperature of 750°, sweeping the released mercury through the gas cell. Note the instrument response. Once the response has been obtained, the gold wire can be cooled once more, ready for the next determination.

The instrument can be calibrated with known concentrations of mercury vapour or by **using prepared rock standards. The original paper**[10] **should be consulted for further details.**

References

1. CRAIG P J and MORTON S F., Nature (1976) 261, 125
2. SKEI J M., Mar. Poll. Bull. (1978) 9, 191
3. ISKANDAR I K., SYERS J K, JACOBS L W, KEENEY D R and GILMOUR J T., Analyst (1972) 97, 388
4. EHMANN W D and LOVERING J F., Geochim. Cosmochim. Acta (1967) 31, 357
5. KASPAR J and KRAL R., Sbornik Vysoke Skoly Chem-Technol v Praze, Oddil Fak. Anorg. a Org. Techol. (1958) 281-288
6. POPEA F and JEMANEANU M., Acad. R.P.R, Stud. Cercet. Chim. (1960) 8, 607
7. PYRIH R Z and BISQUE R E., Econ. Geol. (1969) 64, 825
8. FLOYD M and SOMMERS L E., Anal. Lett. (1975) 8, 525
9. AGEMIAN H and CHAN A S Y., Anal. Chim. Acta (1975) 75, 297
10. NICHOLSON R A., Analyst (1977) 102, 399
11. VAUGN W W and McCARTHY J M., U S Geol. Surv. Prof. Paper 501-D (1964) p.123
12. JAMES C H and WEBB JS., Bull. Inst. Min. Metall. (1964) 691, 633
13. HATCH W R and OTT W L., Analyt. Chem. (1968) 40, 2085
14. OMANG S H and PAUS P E., Anal. Chim. Acta (1971) 56, 393
15. ASTON S R and RILEY J P., Anal. Chim. Acta (1972) 59, 349
16. WARREN H V, DELAVAULT R E and BARAKSO J., Econ. Geol. (1966) 61, 1010
17. HEAD P C and NICHOLSON R A., Analyst (1973) 98, 53
18. URE A M., Anal. Chim. Acta (1975) 76, 1
19. NICHOLSON R A and SMITH J D., Lab. Practice (1972) 21, 638

CHAPTER 30

Molybdenum and Tungsten

Only a small number of reagents have been suggested for the spectrophotometric determination of molybdenum and tungsten, and of these only thiocyanate and dithiol have been widely used for the analysis of silicate rocks. Both molybdenum and tungsten react with alkali thiocyanate to give intense yellow-to-orange colours. The reactions occur in acid solution in the presence of strong reducing agents such as stannous chloride. The coloured species can be measured directly in the aqueous solution, or after extraction into a water-immiscible solvent such as isopentanol, isopropyl ether, mixtures of isopentanol and carbon tetrachloride. Titanium, vanadium and chromium also form coloured products and can interfere with the determination of both molybdenum and tungsten.

A procedure using this reaction for the determination of molybdenum in silicate rocks was described by Sandell.[(1)] The rock material was decomposed by fusion with alkali carbonate, and the melt extracted with water. Chromium, vanadium and molybdenum were all determined in aliquots of the filtrate. Difficulties encountered include the precipitation of silica, fading of the organic extracts, both leading to low recoveries of molybdenum and the temperature dependence of the colour-forming reaction, leading to erratic values.

For tungsten, Sandell[(2)] has described a more detailed procedure based upon a decomposition of the silicate rock material with hydrofluoric and sulphuric acids. Tungsten was separated from iron and titanium by precipitation with aqueous alkali and from molybdenum by precipitation of the latter as sulphide with antimony as carrier. Chan and Riley[(3)] have however found that at low tungsten levels, some of the tungsten is co-precipitated as sulphide with the molybdenum and antimony. A lower limit of 0.5 ppm (using a 1-g sample weight) has been suggested for this procedure, which is not as sensitive as that for molybdenum, and barely adequate for many basic rocks.

Toluene-3,4-dithiol forms green-coloured complexes with both molybdenum and tungsten soluble in organic solvents to give green solutions that can be used for photometric measurement. Tin, bismuth, copper and other metals form coloured complexes with the reagent, but these are insoluble in most organic solvents. The reagent itself is unstable, and solutions deteriorate slowly on standing.

The conditions necessary for the quantitative formation of the complexes of molybdenum and tungsten have been the subject of considerable investigation, and the reports available are to some extent contradictory. It is, however, clear that the molybdenum complex is formed under conditions of high acidity and that reducing agents are not required. For tungsten, two quite different sets of conditions have been proposed, based respectively upon reaction in hot, strongly acid solution in the presence of a reducing agent, and upon reaction in hot, weakly acid solution without additional reducing agent. The maximum absorption occurs at wavelengths of 630 nm (tungsten) and 680 nm (molybdenum). Chloroform, carbon tetrachloride, light petroleum, isopentyl acetate and n-butyl acetate have all been used as solvents for the molybdenum and tungsten complexes.

The formation of the tungsten complex is completely suppressed by the addition of citric acid,[4] in contrast to the molybdenum complex which is not suppressed and can be extracted into a suitable organic solvent. This enables molybdenum to be separated quantitatively from tungsten, which is not extracted, but can be determined with dithiol after the destruction of the citirc acid remaining in solution.

Neither molybdenum nor tungsten is a particularly easy element to determine by atomic absorption spectroscopy. Sensitivities are poor, the high temperature nitrous oxide-acetylene flame is clearly needed, and precautions are usually necessary to prevent suppression of absorption by such elements as calcium, strontium, manganese and iron. Molybdenum absorption is commonly measured at 313.3 nm, and tungsten at 255.1 nm, although the line at 400.9 of much reduced sensitivity is said to give a better signal-to-noise ratio. The method for molybdenum given below is based upon that described by Hutchison.[5] It can be used also for tungsten, which is similarly extracted as complex with benzoin α-oxime. However, the sensitivity of tungsten is very poor, and even if large (10 g) sample portions are taken, the method is insufficiently sensitive for many silicate rocks.

Molybdenum and tungsten may also be extracted as the thiocyanate complex[6,7,8] and with quaternary long-chain aliphatic amines ("Aliquat 336").[9] However, the problems associated with poor sensitivity of molybdenum and very poor sensitivity of tungsten remain.

Spectrophotometric Determination of Tungsten and Molybdenum in Silicate Rocks

Earlier procedures for molybdenum and tungsten in silicate rocks were based upon decomposition of the sample by fusion with alkali carbonate, followed by an aqueous extraction of the melt. However, erratic recoveries of molybdenum and tungsten are due in part to the failure to recover these two elements quantitatively in the alkaline filtrate. For this reason an acid decomposition procedure based upon evaporation with hydrofluoric acid is now preferred.

Stepanova and Yakumina,[10] who have also advocated this acid decomposition, suggested that for some silicate rocks it is possible to extract the molybdenum and tungsten complexes with dithiol directly from the acid sulphate solution remaining after the removal of silica. For more complex rocks, however, a separation stage is recommended. This involves the extraction of the molybdenum and tungsten complexes with benzoin α-oxime as previously described by Jeffery.[11] Alternative schemes for the separation of molybdenum and tungsten were described by Chan and Riley,[3.12] involving co-precipitation with manganese dioxide followed by ion-exchange separation and by Kawabuchi and Kuroda[13] involving anion exchange separation from acid sulphate solution containing hydrogen peroxide.

The procedure described in detail below is based upon published and unpublished work.

Method

Reagents:

Benzoin α-oxime solution, dissolve 2 g of the reagent in 100 ml of ethanol.

Citric acid solution, dissolve 25 g of the reagent in water and dilute to 100 ml.

Toluene-3,4-dithiol solution, dissolve 1 g of the reagent and 5 g of sodium hydroxide in 500 ml of water. When solution is complete, add 5 ml of thioglycollic acid. Store in a refrigerator and discard excess after 14 days. This reagent can also be prepared from 1.4 g of the zinc complex or from 1.5 g of the diacetyl derivative by dissolution in dilute sodium hydroxide solution.

Standard molybdenum and tungsten stock solutions, dissolve 0.150 g of pure molybdic oxide and 0.126 g of pure tungstic oxide separately in small volumes of aqueous sodium hydroxide and dilute each to a volume of 1 litre with water. These solutions contain 100 μg Mo and W per ml, respectively.

Standard molybdenum and tungsten working solutions, dilute 10 ml of each of the stock solutions to separate 1 litre volumetric flasks and dilute to volume with water. These solutions contain 1 μg Mo and W per ml respectively.

Procedure. Accurately weigh approximately 1 g of the finely ground silicate rock material into a platinum dish, moisten with water and add 1 ml of concentrated nitric acid, 3 ml of 20 N sulphuric acid and 10 ml of concentrated hydrofluoric acid and evaporate to fumes of sulphuric acid. Allow to cool, rinse down the sides of the dish with a little water, add 5 ml of concentrated hydrofluoric acid and again evaporate to fumes of sulphuric acid. Allow to cool, again rinse down the sides of the dish with a little water and evaporate, this time to dryness. Allow to cool and fuse the dry residue with 2 g of potassium pyrosulphate to give a completely fluid melt.

Extract the melt with approximately 100 ml of N hydrochloric acid (Note 1), and transfer the solution to a 250-ml separating funnel. Add 2 ml of the benzoin α-oxime solution and mix by shaking. Add 10 ml of chloroform, stopper the funnel and shake for about 1½ minutes to extract the molybdenum and tungsten into organic solution. Allow the phases to separate, remove the organic layer and repeat the extraction with a 5 ml portion of chloroform. Remove the chloroform layer and repeat the extraction with a further 5-ml portion of chloroform. Discard the aqueous layer.

Combine the chloroform extracts in a platinum crucible, add about 50 mg of sodium carbonate and allow the chloroform to evaporate. Dry the residue, ignite and fuse gently over a burner and allow to cool. Extract the melt with 10 ml of water (Note 2) transfer the solution to a 100-ml conical flask, add 1 ml of citric acid solution and sufficient 20 N sulphuric acid to bring the final acid concentration to 10 N. Add 5 ml of the dithiol solution, mix and allow to stand for 30 minutes. Add 5 ml of carbon tetrachloride, stopper the flask and shake for 1 minute to extract the molybdenum-dithiol complex into the organic solution.

Transfer the contents of the conical flask to a separating funnel and run off the organic layer. Filter if necessary through a small dry paper into a 1-cm cell and measure the optical density using the spectrophotometer set at a wavelength of 680 nm.

After the extraction of the molybdenum, rinse the aqueous solution with 2 ml of carbon tetrachloride, remove and discard the organic layer. Run the aqueous layer into a 100-ml beaker, add 3 ml of concentrated nitric acid and 2 ml of concentrated perchloric acid and evaporate to dryness on a hot plate. A white residue should remain after the sulphuric acid has evaporated. If the residue is not white, add 1 ml of 20 N sulphuric acid, 3 ml of concentrated nitric acid and 2 ml of concentrated perchloric acid and repeat the evaporation. This oxidation should be repeated again if necessary to give a white residue.

Dissolve the residue in 50 ml of water and, using a pH meter, adjust the pH of the solution to a value of 2.0 by the addition of dilute ammonia or dilute sulphuric acid as necessary. Transfer the solution to a 100-ml conical flask and add 3 ml of the dithiol reagent solution. Place the flask on a hot plate and heat to just below boiling for a period of 30 minutes. Allow to cool, add 5 ml of carbon tetrachloride, stopper the flask and shake for 1 minute to extract the tungsten into the organic layer. Allow the layers to separate and filter the green organic solution through a small dry paper into a 2-cm cell. Measure the optical density using the spectrophotometer set at a wavelength of 630 nm.

Calibration. Transfer aliquots of 0-5 ml of the standard molybdenum solution containing 0-5 μg Mo to a series of 100-ml conical flasks and dilute each solution to 10 ml with water. Add 1 ml of citric acid solution to each followed by sufficient 20 N sulphuric acid to bring the final acid concentration to 10 N. Add 5ml of the dithiol reagent solution, mix by swirling and allow to stand for 30 minutes. Extract the molybdenum complex into 5 ml of carbon tetrachloride and measure the optical density at a wavelength of 680 nm as described above. Plot the relation of optical density to molybdenum concentration.

Similarly transfer aliquots of 0-5 ml of the standard tungsten solution containing 0-5 μg W to a series of 100-ml beakers, dilute each solution to about 50 ml with water and adjust the pH to a value of 2 by adding dilute sulphuric acid. Rinse each solution into a 100-ml conical flask with a little water and add 3 ml of the dithiol reagent to each. Transfer the flasks to a hot plate and heat to near boiling for a period of 30 minutes. Allow to cool, extract the tungsten complex into 5 ml of carbon tetrachloride and measure the optical density at a wavelength of 630 nm as described above. Plot the relation of optical density to tungsten concentration.

Notes: 1. With most silicate rocks little or no residue will be obtained at this stage. If a residue is observed it should be collected, washed with water and ignited and fused in a platinum crucible with a little anhydrous sodium carbonate. Extract the melt with water, acidify with a little hydrochloric acid and add to the main rock solution.

2. Many silicate rocks contain less than 5 ppm of molybdenum and tungsten. With these samples the whole of the rock solution should be taken for the formation and extraction of the complexes with dithiol. Rocks containing larger amounts of molybdenum or tungsten should be diluted to volume and an aliquot taken for the subsequent stages of the analysis.

Determination of Molybdenum in Silicate Rocks by Atomic Absorption Spectroscopy

In view of the poor sensitivity of molybdenum, as large a sample as possible should be taken of those rock materials containing only very small amounts of molybdenum. As described below a 1 g sample portion is used; this may not be sufficient for many silicate rocks. The full sensitivity of the spectrometer will be required for these materials.

Method

Reagents:

Benzoin α-oxime solution, dissolve 2 g of reagent in 100 ml of ethanol. Store in a refrigerator and renew every few days.

Ammonium chloride-perchloric acid mixture, dissolve 5 g of ammonium chloride in water, add 10 ml of 60 per cent perchloric acid and dilute to 1 litre.

Standard molybdenum stock and working solutions, prepare as described above.

Procedure. Accurately weigh approximately 1 g of the finely powdered rock material into a platinum or PTFE dish, moisten with water and add 10 ml of hydrofluoric acid. Allow to stand for an hour and then evaporate to dryness. Add a further 5 ml of hydrofluoric acid together with 6 ml of concentrated nitric acid and 4 ml of perchloric acid and evaporate to fumes of perchloric acid. Allow to cool, rinse down the sides of the dish with a little water and again evaporate, this time to dryness. Allow to cool.

Moisten the dry residue with a few drops of concentrated ammonia and digest on a steam bath or hot plate until all fumes of ammonia are removed. Add 2.5 ml of perchloric acid together with 10 ml of water and a few drops of 100-volume hydrogen peroxide (Note 1). Warm gently until the solid material has dissolved, evaporate to fumes of perchloric acid and dilute with 20 ml of water. Any residue remaining can be removed by centrifugation and the clear solution transferred to a 100-ml separating funnel and diluted to approximately 50 ml with water.

Add 2 ml of the benzoin α-oxime solution, shake and allow to stand for 5 minutes. Add 5 ml of chloroform and shake for 1 minute to extract the molybdenum complex into the chloroform layer. Run this layer into a 25-ml beaker and repeat the extractions with two further 5-ml portions of chloroform. Combine the extracts and allow the chloroform to evaporate by standing the beaker in a warm place under a fume hood.

Add to the cool residue 5 ml of a 3 + 1 mixture of concentrated nitric and perchloric acids, cover the beaker with a watchglass and, after any initial vigorous action has subsided, evaporate the contents to dryness. Add a few drops of ammonia to the beaker and again evaporate to dryness.

Add by pipette 5 ml of the ammonium chloride-perchloric acid solution (Note 2) and warm gently for a few minutes. Aspirate the solution into a nitrous oxide-acetylene flame of an atomic absorption spectrometer fitted with a molybdenum hollow cathode lamp and measure the absorption at a wavelength of 313.3 nm (Note 3).

Measure also the absorption of standard amounts of molybdenum in ammonium chloride-perchloric acid solution, covering the concentration range of the sample material.

Notes: 1. This addition of hydrogen peroxide is required for samples high in manganese where molybdenum will be co-precipitated with MnO_2 in the acid decomposition. It may be omitted for most silicate rocks.

2. For samples high in molybdenum, this solution may be diluted to a larger volume with the ammonium chloride-perchloric acid solution.

3. Tungsten is similarly precipitated by benzoin α-oxime and will be collected with molybdenum. If present in high concentrations it may be determined by using a tungsten hollow cathode lamp and measuring the absorption at a wavelength of 255.1 nm.

References

1. SANDELL E B., Ind. Eng. Chem, Anal. Ed. (1936) 8, 336
2. SANDELL E B., Ind. Eng. Chem, Anal. Ed. (1946) 18, 163
3. CHAN K M and RILEY J P., Anal. Chim. Acta (1967) 39, 103
4. JEFFERY P G., Analyst (1957) 82, 558
5. HUTCHISON D., Analyst (1972) 97, 118
6. KIM C H, OWENS C M and SMYTHE L E., Talanta (1974) 21, 445
7. SUTCLIFFE P., Analyst (1976) 101, 949
8. COGGER N., Anal. Chim. Acta (1976) 84, 143
9. KIM C H , ALEXANDER P W and SMYTHE L E., Talanta (1976) 23, 229 and 573
10. STEPANOVA N A and YAKUMINA G A., Zhur. Anal. Khim. (1962) 17, 858
11. JEFFERY P G., Analyst (1956) 81, 104
12. CHAN K M and RILEY J P., Anal. Chim. Acta (1966) 36, 220
13. KAWABUCHI K and KURODA R., Talanta (1970) 17, 67

CHAPTER 31

Nickel

The earliest methods for the determination of nickel in silicate rocks were those based upon precipitation as sulphide from ammoniacal solution. Such methods were suitable only for those very few rocks containing relatively large amounts of nickel, and were soon displaced following the discovery of the formation of the red-coloured complex of nickel with dimethylglyoxime. This reaction was used by Harwood and Theobald[1] for the gravimetric determination of nickel, and their methods provided for the first time, an accurate procedure for nickel in most basic rocks. In view of its importance, this gravimetric method is described in detail below.

The small amounts of nickel in granite and other acidic rocks were difficult, and in some cases impossible to determine by this gravimetric procedure, and for these colorimetric methods were developed, again based upon the reagent dimethylglyoxime. They utilised oxidising agents such as persulphate and bromine to give a red colour with nickel and the reagent,that could be used for spectrophotometric measurement. A number of elements, particularly iron, interfere with the determination, and a prior separation of the nickel is required. This can be done by an ion-exchange separation or by an extraction of the nickel complex into an organic solvent. Such procedures are long, tedious, require a skilled analyst and are unsuited to batch operation. For these reasons, spectrophotometric methods for nickel have, in many laboratories, now been replaced by methods based upon atomic absorption spectroscopy.

The spectrum obtained from a nickel hollow-cathode lamp shows many nickel lines suitable for analysis. The more important in order of sensitivity being at 232.0, 341.5, 352.4, 351.5 and 362.5 nm. These can be used for determining nickel when present in quantity, as for example in nickeliferous and certain basic rocks. The most sensitive line at 232.0 nm is subject to interference from non-atomic species and a background correction should be made. At the nickel concentrations encountered the calibration graph has considerable curvature. Measurements at 341.5 or 352.5 nm are recommended to eliminate this problem.

Care must be taken when using multi-element hollow cathode lamps containing iron, that the wavelength chosen for nickel measurement is not co-incident with, or near to emission from iron. In particular it should be noted that Fe 352.42 nm will interfere appreciably with nickel measurement at 352.45 nm. For this reason such multi-element lamps are not recommended.

There is little need for high temperature flames for the atomic absorption measurement of nickel. Fuel-lean flames have been recommended as helping to reduce the interference that can occur from iron and chromium. For further discussion of interferences in the determination of nickel by atomic absorption spectroscopy, see Sundberg.[2] Flameless (carbon furnace) techniques can also be used.[3] The method given below is based upon that described by Warren and Carter.[4]

Gravimetric Determination with Dimethylglyoxime

This procedure is rather time consuming, but may be preferred where only a small number of samples are to be examined, or where results obtained by other methods are to be checked. It cannot be considered as a suitable method for those rocks which contain less than about 0.2% nickel, although acceptable results can be obtained at levels of not less than 0.02%, using a 2 g sample weight.

The rock material is decomposed and silica removed by evaporation with hydrofluoric and sulphuric acids in the usual way. The nickel present in the sulphate solution is then precipitated with dimethylglyoxime in the presence of citric acid which minimises the co-precipitation of iron and aluminium hydroxides. The organic precipitate is destroyed by evaporation with nitric acid, and the nickel again precipitated with dimethylglyoxime, this time from a small volume, but again in the presence of citric acid. The red nickel complex is collected on a sintered glass crucible dried and weighed.

The reagent, dimethylglyoxime, is not very soluble in water, and is often added in alcoholic solution. When added in this way, some of the excess reagent may contaminate the precipitated nickel complex; for this reason an aqueous solution of the sodium salt is preferred.

Method

Reagents:

Citric acid.

Sodium hydroxide solution, dissolve 25 g of the reagent in 250 ml of water.

Methyl red indicator solution.

Dimethylglyoxime solution, dissolve 2 g of the sodium salt in 100 ml of water.

Procedure. Accurately weigh 1-2 g of the finely powdered silicate rock material into a platinum dish and moisten with a little water. Add 10 ml of 20 N sulphuric acid, followed by 1 ml of concentrated nitric acid and 25 ml of concentrated hydrofluoric acid. Transfer the dish to a hot plate or sand bath and evaporate just to fumes of sulphuric acid, stirring as necessary with a platinum rod and taking care that loss by spitting does not occur, particularly in the latter stages of the evaporation. Allow to cool. Add 10 ml of water to the dish and wash down the sides with a further quantity of water. Stir the solution and again evaporate on a hot plate, this time to copious fumes of sulphuric acid. Allow to fume for at least 10 minutes and then allow to cool.

Dilute the cold solution with water and rinse the solution and any residue into a 600-ml beaker using a total of 200-300 ml of water. Heat on a steam bath to bring all soluble material into solution, and filter if necessary (Note 1). Wash the residue with water and combine the filtrate and washings.

Add approximately 3 g of solid citric acid to the solution and stir to dissolve. Add concentrated aqueous sodium hydroxide until the solution is alkaline to methyl red. If iron and aluminium hydroxides are precipitated, make acid with sulphuric acid, add an additional 3 g of citric acid, stir to dissolve and again make the solution alkaline to methyl red. Now add dilute sulphuric acid until the solution is just acid. Heat the solution to a temperature of about 50°, then add 25 ml of dimethylglyoxime solution followed by dilute ammonia drop by drop until a test with methyl red indicator paper indicates that the solution is just alkaline, and add 2 or 3 drops of ammonia in excess. Transfer the beaker to a steam bath for 30 minutes, then allow to cool. Collect the precipitate on a close-textured filter paper and wash thoroughly with cold water (Note 2).

Using a fine jet of water, rinse the residue into a small beaker and dissolve in a little 8 N nitric acid. Moisten the filter paper with 8 N nitric acid and filter the solution back through the paper, finally washing the paper with small quantities of water. Collect the filtrate and washings in a small beaker.

Add 5 drops of 20 N sulphuric acid to the solution and evaporate on a water bath. Transfer the beaker to a hot plate and heat to fumes of sulphuric acid to complete the expulsion of all nitric acid. If any darkening of the solution occurs, clear this with concentrated nitric acid and again evaporate to fumes of sulphuric acid. Dissolve the residue in a little water and filter if necessary into a 250-ml beaker. The volume at this stage should be about 100 ml. To the solution add approximately

0.1 g of solid citric acid, 10 ml of dimethylglyoxime solution and a few drops of methyl red indicator solution. Now add dilute ammonia until the solution is just alkaline (Note 3), followed by 2 to 3 drops in excess. Allow the solution to stand overnight and then collect the red precipitate of the nickel salt on a sintered glass crucible of medium porosity previously dried and weighed. Dry the crucible and precipitate in an electric oven set at a temperature of 120-130°, and weigh. The precipitate contains 20.31 per cent Ni or 25.8 per cent NiO.

Notes: 1. Some silicate rocks are completely attacked by this evaporation and no residue will be observed, with others only a small, pure white residue consisting largely of barium sulphate will be obtained. This is unlikely to contain nickel and can be discarded. Dark-coloured residues may contain an appreciable proportion of the total nickel content of the rock, and should be ignited, fused with a little sodium carbonate (or peroxide if the rock contains chromite), the melt extracted with water, acidified with sulphuric acid and added to the main solution.

2. Small amounts of iron are usually precipitated with nickel at this stage. With a 1 g sample and less than about 0.04 per cent Ni, it is not always possible to distinguish the slight precipitate of the nickel complex and for this reason the second precipitation should be made, before reporting the absence of nickel at this concentration level.

3. The indicator colour change may be obscured in the presence of much nickel. Methyl red test paper should be used if any difficulty is experienced in neutralising the solution.

Spectrophotometric Determination with Dimethylglyoxime

This determination is based upon the formation of a deep red-coloured complex of nickel with dimethylglyoxime in strongly alkaline solution in the presence of an oxidising agent. This reaction was first noted by Feigl,[5] who used bromine as the oxidising agent. Other agents that have been used include persulphate,[6,7] hypochlorite, periodate and iodine.

Two coloured compounds can be formed in solution. Hooreman[8] has suggested that in these two compounds nickel and dimethylglyoxime are present in the ratios 1:2 and 1:4 respectively. The 1:2 complex is unstable and for this reason analytical procedures for nickel are designed to favour the formation of the 1:4 complex. This is done by using an excess of the reagent, by restricting the quantity of ammonia added and by using sodium hydroxide to increase the alkalinity of the solution. In the procedure described below, broadly based upon that of Rader and Grimaldi,[9] potassium persulphate is used as oxidising agent and a prior extraction of the nickel complex into chloroform is used to separate the nickel from other elements such as iron, that interfere. Citric acid is added to the rock solution to prevent the precipitation of iron and aluminium hydroxides, and hydroxyammonium chloride added to prevent the oxidation of manganese. Any traces of copper that are extracted with the nickel are removed by shaking the chloroform extract with dilute aqueous ammonia.

Method

Reagents:

Hydroxyammonium chloride.

Sodium hydroxide solution, dissolve 25 g of the reagent in 250 ml of water, stirring as necessary.

Dimethylglyoxime solution, dissolve 1 g of the reagent in 100 ml of ethyl alcohol. Store in a stoppered bottle.

Sodium citrate solution, dissolve 10 g of the dihydrate in water and dilute to about 100 ml.

Phenolphthalein indicator solution.

Potassium persulphate solution, dissolve 5 g of the pure fresh reagent in water and dilute to 100 ml in a volumetric flask. This solution deteriorates on keeping, and should be prepared as required.

Chloroform.

Standard nickel stock solution, dissolve 0.405 g of nickel chloride hexahydrate in water containing a few ml of concentrated hydrochloric acid, and dilute to 1 litre. This solution contains 100 µg Ni per ml.

Standard nickel working solution, dilute 25 ml of the stock solution to 250 ml with water in a volumetric flask. This solution contains 10 µg Ni per ml.

Procedure. Accurately weigh 1-2 g of the finely powdered rock material into a platinum dish and moisten with a little water. Add 5 ml of concentrated nitric acid, 5 ml of concentrated perchloric acid and 15 ml of concentrated hydrofluoric acid. Transfer the dish to a hot plate or sand bath and evaporate to copious fumes of perchloric acid. Allow to cool, wash down the sides of the dish with a little water, add a further 5 ml of water to the dish and again evaporate on a hot plate or sand bath - this time until the residue remains just moist. Do not allow to dry completely.

Add 1-2 ml of concentrated hydrochloric acid and 20 ml of water and warm, adding a further quantity of water if necessary to complete the dissolution of all soluble material. If any unattacked residue remains, collect on a small filter paper and wash well with cold water (Note 1). Dilute the combined filtrate and washings to 100 ml in a volumetric flask.

Transfer 25 ml of this solution (for rocks containing more than about 200 ppm Ni use a smaller aliquot containing not more than 100 µg Ni and dilute to 25 ml with water) to a 60- or 100-ml separating funnel and add 10 ml of sodium citrate solution (Note 2). Now add concentrated ammonia to bring the pH of the solution to 9 (just pink to phenolphthalein), making the final adjustments with dilute ammonia or hydrochloric acid. Add approximately 0.1 g of hydroxyammonium chloride and 3 ml of dimethylglyoxime solution to the separating funnel, shake well and allow the solution to stand for 5 minutes.

Add 10 ml of chloroform to the solution and shake for 2 minutes. Allow the phases to separate and then transfer the lower, organic layer to a second separating funnel. Repeat the extraction of the aqueous solution twice more with 5-ml aliquots of chloroform and combine the three extracts. Discard the aqueous layer. Now add 10 ml of the dilute ammonia to the chloroform solution, shake for 2 minutes and again separate the organic layer and transfer it to a third separating funnel. Add 3 ml of chloroform to the aqueous phase remaining in the second funnel, shake for 1 minute and add the chloroform layer to the main extract in the third funnel. Discard the aqueous phase.

Add 10 ml of 0.5 N hydrochloric acid to the chloroform solution and shake vigorously for 2 minutes. Transfer the lower, organic layer to yet another separating funnel and shake it with a further 10-ml portion of 0.5 N hydrochloric acid. Discard the chloroform layer. Combine the acid extracts and filter through a small close-textured filter paper into a 50-ml volumetric flask. Using no more than 10 ml of water rinse the separating funnels, filter paper and filter funnel, adding the washings to the main solution. Add 2 ml of the sodium citrate solution to the flask, followed by 0.6 ml of concentrated sodium hydroxide solution to give a pH of 12 or just over. Add 10 ml of potassium persulphate solution followed by 3 ml of dimethylglyoxime solution. Dilute to volume, mix well and allow to stand for between 30 and 60 minutes for the colour to develop.

Measure the optical density of the solution in 4 or 5 cm cells, using the spectrophotometer set at a wavelength of 530 nm. The optical density of a reagent blank extract should also be measured. This extract is prepared in the same way as the sample extract, but from 25 ml of water in place of 25 ml of the rock sample solution.

Calculate the nickel content of the sample material by reference to the calibration graph.

Calibration. Transfer aliquots of 0-12 ml of the nickel solution containing 0-120 µg Ni, to separate 50-ml volumetric flasks and add to each 2 ml of sodium citrate, 0.6 ml of sodium hydroxide solution, 10 ml of potassium persulphate solution and 3 ml of dimethylglyoxime solution. Dilute each to volume with water, mix well, allow to stand and then measure the optical densities as described above. Plot the relation of optical density to nickel concentration.

Notes: 1. If the residue contains dark gritty particles of unattacked sample material, further treatment to recover any contained nickel is imperative. The dried and ignited residue should be fused with sodium carbonate, leached with water, acidified with hydrochloric acid and added to the main solution. Where the rock sample contains chromite, it may be necessary to sinter or fuse the unattacked residue with sodium peroxide in a zirconium crucible.

2. An increased quantity of sodium citrate solution may be needed to keep all the iron and aluminium of some basic rocks in solution.

Determination by Atomic Absorption Spectroscopy

The procedure given below, based upon that described by Warren and Carter,[(4)] can be combined with the determination of a number of other elements in silicate rocks including copper vanadium, chromium, cobalt and barium. The rock matrix is decomposed with perchloric and hydrofluoric acid in a pressure decomposition vessel, and boric acid added to dissolve any precipitated fluorides. For the measurement of nickel, an air-acetylene flame is used. For preference, the nickel absorption at 341.5 nm is used although (as described by Warren and Carter) the more sensitive line at 232.0 nm can be used with background correction.

Procedure. Decompose a 2 g portion of the finely ground sample material in a PTFE bomb or by evaporation in the usual way with a mixture of nitric, perchloric and hydrofluoric acids (Note 1). Evaporate to fumes of perchloric acid, but do not allow to go to dryness. Cool and add 10 ml of concentrated hydrochloric acid followed in turn by 15 ml of water and 1.6 g of boric acid. Boil gently for a few minutes, when any precipitated fluorides will dissolve. Cool again, transfer to a 200-ml measuring flask, add by pipette 10 ml of the potassium buffer solution (Note 2), and dilute to volume with water. If the atomic absorption measurements are not to be made immediately then the rock solution should be transferred to a clean, dry polyethylene bottle. A reagent blank solution should be similarly prepared, but omitting the rock sample.

For the determination of nickel (Note 3), set the atomic absorption spectrometer according to the manufacturer's instructions. A nickel hollow cathode lamp should be used, and set at the recommended lamp current. A wavelength of 341.5 nm should be used (but see above) and an air-acetylene flame. Measurements of the rock solution should be interspersed between those of the standards and the blank solution.

A background solution can be prepared by dissolving 3.174 g of aluminium filings, 2.238 g of iron sponge, 0.482 g of magnesium turnings, 3.570 g of calcium carbonate, 3.016 g of sodium chloride and 1.898 g of potassium chloride consecutively in the minimum quantity of diluted (1+1) hydrochloric acid, adding 80 ml of perchloric acid, 4 ml of nitric acid, 200 ml of hydrochloric acid and 32 g of boric acid and diluting to 2 litres.

The standard solutions are prepared by transferring aliquots of a standard solution of nickel, plus 100 ml of the background solution and 10 ml of the potassium buffer solution to a 200 ml measuring flask, and diluting to volume with water (Note 4).

Notes: 1. If a pressure decomposition vessel is not used, provision should be made in the usual way for the recovery of any residual, unattacked material.

2. Dissolve 7.628 g of potassium chloride in water and dilute to a volume of 200 ml. This solution contains 20 g potassium per litre. This addition is recommended, particularly if barium is to be determined in the rock solution.

3. The conditions used for the determination of nickel are not necessarily those required for the determination of other elements for which this rock solution may be used.

4. Stock solutions can be used containing all six elements - copper, vanadium, chromium, nickel, cobalt and barium, providing the materials used are not contaminated by any of the other metals to be determined.

References

1. HARWOOD H F and THEOBALD L S., Analyst (1933) 58, 673
2. SUNDBERG L L., Analyt. Chem. (1973) 45, 1460
3. SCHWEIZER V B., Atom. Absorb. Newsl. (1975) 14, 137
4. WARREN J and CARTER D., Canad. J. Spectr. (1975) 20, 1
5. FEIGL F., Ber. (1924) 57, 758
6. HAAR K and WESTERVELD W., Rec. Trav. Chim. (1948) 67, 71
7. CLAASEN A and BASTINGS L., Rec. Trav. Chim.,(1954) 73, 783
8. HOOREMAN M., Anal. Chim. Acta (1949) 3, 635
9. RADER L F and GRIMALDI F S., U S Geol. Surv. Prof. Paper 391-A, 1961

CHAPTER 32

Niobium and Tantalum

A number of reagents have been suggested for the spectrophotometric determination of niobium, although few have found general application to the analysis of rocks and minerals. The high sensitivities of procedures based upon thiocyanate and 4-(2-pyridylazo)-resorcinol (PAR) have made these two reagents of particular importance in the examination of silicate rocks. Procedures based upon the formation of a peroxy complex, although insufficiently sensitive for application to most silicates, can be used for certain rocks and minerals where some degree of niobium enrichment can be expected.[1]

Niobium forms a yellow-coloured thiocyanate that can be used as the basis of a spectrophotometric determination. Two general procedures have been developed involving extraction into organic solution,[2,3,4,5,6] and formation in a homogeneous acetone-water solution.[7,8]

The yellow complex has an absorption maximum at 385 nm, and the calibration graph is a straight line over the range 0-10 μg Nb/25 ml ethyl acetate. The intensity of the colour is time-dependent, although the change in optical density is small when the solutions are stood for 1 hour. The amount of niobium thiocyanate extracted from the aqueous phase into the organic is temperature-dependent, and Grimaldi[5] has recommended that a standard niobium solution should accompany the sample solution through the colour formation and extraction stages when the room temperature differs by more than 2° from that at which the calibration was undertaken. The "ageing" of reagents may also affect the intensity of the colour, and it is recommended that the stannous chloride and thiocyanate solutions should be freshly prepared.

Even the small amount of platinum that can be removed from platinum apparatus in the course of a hydrofluoric-sulphuric acid attack of a rock sample is sufficient to interfere with the determination. For this reason Grimaldi[5] recommends a decomposition procedure based upon fusion with sodium hydroxide. The acid attack in PTFE dishes described by Esson[6] may be more effective for some minerals.

Interfering elements include tungsten, molybdenum, uranium and a number of other rarer elements that must all be separated from niobium. Titanium and tantalum interfere but to a much smaller extent and can be ignored unless present in quantity.

Esson[6] has reported that when a multi-stage separation and extraction procedure is used, considerable loss of niobium occurs and that this loss varies from sample to sample. Samples containing more iron tended to give better niobium recoveries, presumably because ferric hydroxide acts as a carrier for niobium during precipitation.

4-(2-Pyridylazo)-resorcinol(PAR) forms coloured complexes with many metals, but in the presence of EDTA or CyDTA the reagent is highly selective for niobium.[9,10] The niobium complex, red in colour, has a maximum absorption at 550 nm, whilst the reagent itself has a maximum at 410 nm and only negligible absorption at the higher wavelength. Close control of pH in the range 6-7 is required, for which the addition of ammonium acetate has been recommended.

The time required for colour development, normally 25 minutes, is increased to 40 minutes in the presence of EDTA. After this period the optical density of both the reagent and the niobium solution increase slowly with time.

Interference from uranium, vanadium and phosphate can be avoided by making a prior separation, and from tantalum by masking with tartaric acid. In the examination of silicate rocks, silica can conveniently be removed by evaporation with hydrofluoric and sulphuric acids in PTFE dishes in the usual way. The niobium present is readily separated from excess iron, aluminium and other elements by precipitation from the sulphate solution with a carefully controlled amount of cupferron. Vanadium is not precipitated if sulphurous acid is added to the solution, whilst any copper, uranium, tungsten and molybdenum are removed from the filter pad by washing with dilute aqueous ammonia.

The cupferrate residue is ignited in a silica crucible, fused with a little potassium pyrosulphate, extracted and the coloured complex formed in tartaric-CyDTA solution. This procedure, described in detail in the earlier editions of this book, was adapted from that given by Jenkins[11] for niobium in mild steel.

Alternative procedures using PAR as the basis for spectrophotometric measurement, but with ion-exchange separation have been given by Greenland and Campbell[12] and Mazzucotelli <u>et al</u>.[13] The latter forms the basis of the procedure described in detail below.

A slightly more sensitive spectrophotometric method using sulphochlorophenol S has recently been described by Childress and Greenland.[14] Niobium is extracted from a sulphuric-hydrofluoric acid medium into methyl isobutyl ketone and back extracted into water. The niobium-sulphochlorophenol S complex is extracted into amyl alcohol for photometric measurement.

Atomic absorption spectroscopy has not been used extensively for the determination of niobium or tantalum because of inherent low sensitivity, the considerable matrix effects and the serious interelement effects. It cannot yet be recommended for the determination of either niobium or tantalum in normal silicate rocks, although a procedure has been given by Husler[15] for rocks and other materials containing 0.02% Nb_2O_5.

Spectrophotometric Determination of Niobium

Greenland and Campbell[12] used an anion exchange resin to separate niobium in a fluoride medium. The difficulties of working in this medium are well known, and a somewhat simpler procedure described by Mazzucotelli et al[13] is based upon a strongly acidic cation exchange. Niobium and tantalum are eluted together in hydrochloric acid solution containing hydrogen peroxide.

Vanadium and molybdenum are eluted with niobium and tantalum from the cation exchange column. Interference from these elements is prevented by the addition of EDTA and tartaric acid to the solutions prior to colour formation and spectrophotometric determination. Tantalum interferes, but as the molar absorptivity of the niobium complex is about four times that of the tantalum complex, and the wavelengths of maximum absorption are separated by about 50 nm, the interference is small. It can be tolerated for most rocks, ie those in which the niobium content exceeds that of tantalum.

Method

Apparatus. Ion-exchange column - a borosilicate glass column of 22 mm internal diameter filled with Dowex 50-X8 strongly acidic, 200-400 mesh ion-exchange resin. 30 g of the resin gives a bed depth of about 28 cm. Wash the resin thoroughly with water followed by 30 ml of the acid-hydrogen peroxide solution.

Reagents:

Sodium peroxide.

Acid-hydrogen peroxide solution, 0.6 M hydrochloric acid containing 1.5 ml of 100 vol hydrogen peroxide per litre.

Tartaric acid solution, dissolve 1 g in 100 ml of water.

Zinc sulphate solution, dissolve 3 g of zinc sulphate hepta hydrate in water and dilute to 100 ml.

EDTA solution, dissolve 3.7 g of the disodium salt of EDTA in 500 ml of water.

4-(2-Pyridylazo)resorcinol solution, dissolve 0.3 g of the monosodium salt in water and dilute to 1 litre.

Acetate buffer solution, add 2.25 ml of glacial acetic acid to 40 g of ammonium acetate, dissolve in water and dilute to 500 ml.

Standard niobium stock solution, fuse 71.6 mg of pure dry niobium pentoxide with 2 g of potassium pyrosulphate in a platinum crucible. Dissolve the melt in 10 per cent tartaric acid solution and dilute with tartaric acid solution to 500 ml. This solution contains 100 μg Nb/ml.

Standard niobium working solution, dilute 10 ml of the stock solution to 100 ml with water. Prepare this as required. This solution contains 10 μg Nb/ml in 1 per cent tartaric acid.

Procedure. Accurately weigh approximately 0.3 g of the finely powdered sample material into a nickel crucible and mix with 1.5 g of sodium peroxide. Add a further 1.5 g of sodium peroxide to cover the sample material and fuse over a gas burner in the usual way. Allow to cool, stand the crucible on its side in a beaker and dissolve the melt by cautious addition of the acid hydrogen peroxide. Rinse the crucible and remove it. Pass the clear solution together with the washings through the ion exchange column and wash the resin with 50 ml of the acid-hydrogen peroxide.

Collect the eluate and washings in a small PTFE beaker, add 5 ml of concentrated hydrofluoric acid and evaporate almost but not quite to dryness. Dissolve the moist residue by heating with 5 ml of the tartaric acid solution. Allow to cool, add 1 ml of the zinc sulphate solution and 2 ml of the EDTA solution. Add ammonia to bring the pH of the solution to 6.0 ± 1.

Using the minimum quantity of water, transfer the solution to a 25-ml volumetric flask, add 10 ml of the PAR solution followed by 0.5 ml of the acetate buffer solution. Dilute to volume with water, mix well and allow to stand for one hour before measuring the absorbance in 4-cm cells with the spectrophotometer set at a wavelength of 550 nm. If a recording instrument is available, record from 450 to 600 nm. The reagent blank solution should be prepared as for the sample solution, but omitting the sample material.

The calibration graph should be prepared by taking 0 to 3 ml aliquots of the standard niobium working solution, adding further volumes of tartaric acid and proceeding as described above. Plot the relation of absorbance to niobium concentration.

Determination of Tantalum

The reagents that have been suggested for the photometric determination of tantalum are not as sensitive as thiocyanate or PAR for niobium. Patrovsky[16] has described a photometric procedure for 20 to 400 μg Ta, based upon the colour given with p-dimethylaminophenylfluorone. This method can be applied to certain pegmatites and other rare rocks appreciably enriched in tantalum, but is not applicable to normal silicate rocks. A somewhat more sensitive procedure has been described by Pavlova and Blyum,[17] involving first an extraction of tantalum as a complex with butyl-rhodamine S, and then photometric determination with rhodamine 6J.[18]

Unlike spectrophotometric and emission spectrographic methods, procedures based upon thermal neutron activation analysis are particularly sensitive for this element. This is partly because the cross-section for thermal neutron capture is quite large, and partly because the convenient half-life of the product nuclide allows adequate time for the chemical processing of the irradiated material. Procedures for determining tantalum in silicate rocks by this method have been described by Atkins and Smales[19] and by Morris and Olya.[20]

References

1. PICKUP R., Colon. Geol. Min. Res. (1953) 3, 358
2. ALIMARIN I P and PODVAL'NAYA R L., Zhur. Anal. Khim. (1946) 1, 30
3. WARD F N and MARRANZINO A P., Analyt. Chem. (1955) 27, 1325
4. FAYE G H., Chem. in Canada (1958) 10, 90
5. GRIMALDI F S., Analyt. Chem.(1960) 32, 119
6. ESSON J., Analyst (1965) 20, 488
7. FREUND H and LEVITT A E., Analyt. Chem. (1951) 23, 1813
8. MARZYS A E O., Analyst (1954) 79, 327
9. BELCHER R, RAMAKRISHNA T V and WEST T S., Talanta (1962) 9, 943
10. ELINSON S V and POBEDINA L I., Zhur. Anal. Khim. (1963) 18, 199
11. JENKINS N., Metallurgia (1964) 70, 95
12. GREENLAND L P and CAMPBELL E Y., Anal. Chim. Acta (1970) 49, 109
13. MAZZUCOTELLI A, FRACHE R, DADONE A and BAFFI F., Analyst (1977) 102, 825
14. CHILDRESS A E and GREENLAND L P., Anal. Chim. Acta (1980) 116, 185
15. HUSLER J., Talanta (1972) 19, 863
16. PATROVSKY V., Chim. Listy (1965) 59, 1464
17. PAVLOVA N N and BLYUM I A., Zavod. Lab. (1966) 32, 1196
18. PAVLOVA N N and BLYUM I A., Zavod. Lab. (1962) 28, 1305
19. ATKINS D H F and SMALES A A., Anal. Chim. Acta (1960) 22, 462
20. MORRIS D F C and OLYA A., Talanta (1960) 4, 194

CHAPTER 33

Nitrogen

Appreciable quantities of nitrogen gas are released from many rocks by heating under vacuum. Silicate rocks are not likely to contain metallic nitrides (but see for example, Baur[1]), but as the nitrogen is not so easily released as other gases,[2] it is probably not present as free nitrogen either. Whatever method of extraction is employed, the larger part of the nitrogen detected is in the form of ammonia. It is possible that this is the way in which most of it occurs, although the possibility that mixtures of nitrogen and hydrogen could be converted to ammonia at the time of their release from silicates has been noted.[3]

The ammoniacal-nitrogen content of silicates is very variable; average values of 13.8 ppm for dunites, 48.5 ppm for mafic rocks and 26.7 ppm for granitic rocks were reported by Vinogradov et al.[2] Similar values were indicated by Stevenson[4] and Wlotzka,[5] with somewhat higher values for certain shales.

The ammoniacal-nitrogen appears to be concentrated in the mica fraction, where it may replace potassium - an indication of this was obtained by Stevenson,[4] who noted a correlation between nitrogen and potassium in mica minerals.

Wlotzka[5] examined a number of rocks for nitrate-nitrogen. He reported small amounts (5-20 ppm) in surface sediments, some in saline clays and limestones, but none in average clays, sandstones and magmatic rocks.

Determination of Nitrogen

Some nitrogen may be extracted from silicate rocks by leaching with water, but this is unlikely to give any real indication of the total amount of ammonia present in the specimen. Methods reported for this determination include comminution in vacuum, ignition in vacuum and chemical decomposition with both acids and alkalis. Vinogradov et al[1] reported that fusion of silicate samples in vacuum did not give concordant results, and preferred decomposition with a mixture of ortho- and pyrophosphoric acids in a partially sealed quartz ampoule. A temperature of 240° was used, and after the decomposition was complete, the ammonia liberated with alkali was absorbed in dilute hydrochloric acid and determined with Nessler's reagent.

A similar procedure was employed by Stevenson[6] using sulphuric acid in a completely sealed tube at a temperature of 420° for a minimum period of 90 minutes. A micro-

Kjeldahl distillation apparatus was used to recover the ammonia, which was determined with Nessler's reagent. For those samples containing more than 100 μg of nitrogen, the ammonia distillate was collected in boric acid and titrated with 0.01 N sulphuric acid solution.

Wlotzka[5] noted that with an alkaline fusion in a closed system, a part of the ammonia was converted to nitrate. This was reduced with Devarda's alloy, and the total ammonia determined with Nessler's reagent. Wlotzka also decomposed silicate rocks and minerals with hydrofluoric acid in a closed polythene bottle, added sodium hydroxide and distilled the ammonia from the alkaline solution. By adding Devarda's alloy to the solution after removal of ammonia, it was possible to determine also the nitrogen originally present as nitrate.

A high temperature fusion technique using helium as an inert carrier gas was described by Gibson and Moore.[7] Samples of silicate rocks and meteorites were fused in a graphite crucible using a Leco induction furnace and a temperature of 2400°. Any carbon monoxide produced was converted to carbon dioxide by passage over a copper oxide-rare earth oxide catalyst, and removed from the system, together with water vapour, by passage through an Ascarite-magnesium perchlorate trap. The remaining gases were collected in a molecular sieve trap at liquid nitrogen temperature and subsequently separated chromatographically using molecular sieve 5A. The components were detected by thermal conductivity.

The reagent blank was of the order of 4 to 6 μg N, and nitrogen contents of 27 to 59 ppm (μg N/g) were reported for seven USGS rock standards.

References

1. BAUR W H., Nature (1972) 240, 461
2. VINOGRADOV A P, FLORENSKII K P and VOLYNETS V F., Geokhimiya (1963) (9), 875
3. RAYLEIGH L., Proc. Royal Soc. (1939) A170, 451
4. STEVENSON F J., Geochim. Cosmochim. Acta (1960) 19, 261
5. WLOTZKA F., Geochim. Cosmochim. Acta (1961) 24, 106
6. STEVENSON F J., Analyt. Chem. (1960) 32, 1704
7. GIBSON E K Jr, and MOORE C B., Analyt. Chem. (1970) 42, 461

CHAPTER 34

Phosphorus

When present in the amounts encountered in many silicate rocks, phosphorus can be precipitated from nitric acid solution as the complex phosphomolybdate and weighed as such or as magnesium pyrophosphate. These two procedures have been used for many years but they are tedious and not ideally suited to this phosphorus concentration range. They have now been largely displaced by photometric procedures based upon the yellow colour given by phosphorus with a vanadomolybdate reagent, or the blue colour of reduced phosphomolybdate solutions. The latter procedure is the more sensitive, whilst the former can be adapted for determining phosphorus in phosphate rock containing up to 30 per cent P_2O_5.

The formation of yellow-coloured complex of phosphorus with solutions of a vanadomolybdate reagent was used as early as 1908[1] for the determination of small amounts of phosphoric acid. More recently it has been established[2] that the maximum absorption of vanadomolybdophosphate solutions (some authors prefer "molybdovanadophosphate" or "phosphovanadomolybdate") occurs at a wavelength of 315 nm in the ultraviolet part of the spectrum. Most workers have however preferred to use a wavelength in the range 420-470 nm, where the interference from iron is much reduced or eliminated altogether.

The acid concentration required for the formation of this yellow colour has been given as 0.021-0.071 N hydrochloric,[2] and 0.2-1.6 N nitric acid.[3] Sulphuric and perchloric acids have also been used, particularly with sample materials containing large amounts of iron.

Interference has been reported from silicon which forms an analogous yellow-coloured vanadomolybdate[4] from titanium and zirconium when present in large amounts[5] and also from chromium[6] and arsenic. These latter metals are not likely to be present in phosphate rock in amounts sufficient to interfere with the determination, whilst silicon is removed in the early stages of the procedure described in detail below.

The yellow-coloured molybdate complexes of both silicon and phosphorus can be reduced to blue-coloured solutions. These contain a colloidal complex known as molybdenum blue or heteropoly blue.[7] The intensity of the coloured product depends upon the pH of the solution, the reducing agent used, the molybdate-acid ratio, the amount of molybdate and the presence of other ions.[8,9,10] The procedure can be as much as twenty times as sensitive as that based upon the yellow vanadomolybdophosphate colour. A pH in the range 1.9-6.0 has been recommended.[11]

Many reducing agents have been suggested for the production of the blue species, including ferrous sulphate, zinc dust, stannous chloride, hydrazine, 1-amino-2-naphthol-4-sulphonic acid, hydroquinone and many more.[12] Ascorbic acid is used in the procedure described below[13], this produces a stable molybdenum blue solution and provides a more sensitive procedure than those based upon many of the other recommended reductants. The disadvantage is that the colour is developed slowly and the solutions should preferably be allowed to stand overnight.

Using this reagent, the maximum absorption of molybdenum blue solutions occurs at a wavelength of about 830 nm; measurements of optical density can also be made at 650 nm, although with much reduced sensitivity. In the procedure described, the calibration graph is a straight line, indicating that the Beer-Lambert Law is valid over the concentration range used.

Interference from silica can be avoided by evaporation with hydrofluoric acid and either sulphuric or perchloric acid in the usual way. Arsenic forms a yellow-coloured complex with ammonium molybdate that can similarly be reduced to a molybdenum blue. This element is unlikely to be present in silicate or carbonate rocks in amounts sufficient to interfere, and additional steps to volatilise arsenic are not necessary.

Because of the high sensitivity of the method to traces of phosphorus and silicon, all the glassware used for the determination must be scrupulously clean. Riley[13] has recommended that a separate set of volumetric flasks be set aside for this determination only. These flasks should be allowed to stand for several hours filled with concentrated sulphuric acid and then well washed out with distilled water. After use the flasks should be rinsed out, and then kept filled with distilled water.

An indirect atomic absorption method for the determination of phosphorus in silicate rocks is described by Riddle and Turek.[14] The yellow molybdophosphate complex is extracted into a diethyl ether-pentanol mixture, back extracted into an ammonium chloride-ammonia buffer solution at pH 9.3, and aspirated directly into a nitrous oxide-acetylene flame for the determination of the molybdenum content in the usual way.

Determination of Phosphorus using a Vanadomolybdate Reagent

A number of workers have adapted the reaction of phosphorus with vanadomolybdate solutions to the determination in silicate rocks. Bennett and Pickup[15] noted interference from titanium, and devised a procedure for separating both silica and titanium based upon a fusion of the rock sample with sodium carbonate. An aliquot of the rock solution prepared from 5 g of material, and used by them for the determination of barium, zirconium, chromium, vanadium, chlorine and total sulphur can be taken for the determination of phosphorus. It is essentially the procedure of Bennett and Pickup that is described below.

This procedure is adequate for many silicate rocks, but as the total phosphorus does not always appear in the aqueous leach from a carbonate fusion, low results may sometimes be obtained. This is particularly so for rocks containing appreciable amounts of calcium.

A somewhat simpler method that can be used for rocks containing not more than 1.6 per cent TiO_2 or 0.1 per cent ZrO_2 was described by Baadsgaard and Sandell.[5] Silica was eliminated by evaporation of the rock material with hydrofluoric and nitric acids, and a correction determined for the interference from iron. Shapiro and Brannock,[16] in their scheme for rapid rock analysis, determined phosphorus in an aliquot of their "solution B", prepared by evaporation of the rock material with hydrofluoric and perchloric acids.

Method

Reagents:

Sodium carbonate wash solution, dissolve 10 g of solid reagent in 500 ml of water.

Nitric acid dilute, dilute 22 ml of concentrated acid to 100 ml with water.

Vanadomolybdate reagent solution, dissolve 20 g of crushed ammonium molybdate in water and pour slowly into 140 ml of concentrated nitric acid. Add 1 g of ammonium vanadate, stir until solution is complete and then dilute to one litre with water. This reagent is stable for several months.

Standard phosphate solution, dissolve 0.192 g of dried potassium dihydrogen phosphate in water and dilute to one litre in a volumetric flask. This solution contains 100 μg P_2O_5 per ml.

Procedure. Accurately weigh approximately 0.2 g of the rock powder into a small platinum crucible, add 1.2 g of anhydrous sodium carbonate and 0.1 g of potassium nitrate and fuse over a gas burner for an hour to complete the decomposition of all minerals present. Allow the crucible to cool, and then extract the melt with water. Collect the insoluble residue on a small, hardened, open-textured filter paper and wash well with small quantities of hot sodium carbonate wash solution. Discard the residue. Combine the filtrate and washings and transfer to a platinum basin. Cover the basin, cautiously acidify with concentrated nitric acid, and add approximately 5 ml in excess. Add 15 ml of concentrated hydrofluoric acid and evaporate the solution to dryness. Dissolve the residue in a little water, add 5 ml of concentrated nitric acid and 5 ml of concentrated hydrofluoric acid and repeat the evaporation to dryness. Add 2 ml of concentrated nitric acid and again evaporate to dryness.

Add 2 drops of concentrated nitric acid to the dry residue and dissolve in a little water. At this stage a clear solution should be obtained - any small residue may be removed by filtration or by further evaporations with small quantities of concentrated nitric acid.

Transfer the clear aqueous solution to a 50-ml volumetric flask, add 5 ml of dilute nitric acid and 10 ml of the vanadomolybdate reagent solution and dilute to volume with water. Measure the optical density of the solution in 2-cm cells, using the spectrophotometer set at a wavelength of 470 nm, against a reagent blank solution prepared in the same way as the sample solution but omitting the sample material. Obtain the phosphate content of the sample by reference to the calibration graph.

Calibration. Transfer aliquots of 0-30 ml of the standard phosphate solution, containing 0-3 mg P_2O_5, to separate 50-ml volumetric flasks, add 5 ml of dilute nitric acid and 10 ml of vanadomolybdate reagent solution to each, and dilute to volume with water. Measure the optical densities against the solution containing no added phosphorus, in 2-cm cells as described above, and plot these values against the phosphate concentration to give the calibration graph.

There is no difficulty in applying the vanadomolybdate method to the determination of phosphorus in limestone rocks. A suitable weight of rock material is evaporated to dryness with nitric acid to remove carbon dioxide and oxidise organic matter, and twice to dryness with perchloric and hydrofluoric acids to remove any silica present in the sample. After further evaporation with perchloric acid to decompose any remaining fluorides, the dry residue is moistened with perchloric acid and dissolved in water. This solution can be used directly for the determination of phosphorus by the addition of the vanadomolybdate reagent, as described above.

An alternative procedure of particular value in the analysis of those carbonates where iron content is appreciable, is to separate phosphorus from iron and other elements present by using a cation exchange resin.(17) The phosphate iron is not retained by the resin and can be determined photometrically in the eluate with vanadomolybdate reagent. The sample solution should be just acid and, to prevent precipitation of iron and aluminium phosphates on the column, the phosphate should be eluted with 0.015 N hydrochloric acid.

Calcium, magnesium, iron and aluminium are retained by the cation exchange resin and can be subsequently recovered by elution with 4 N hydrochloric acid.

Determination of Phosphorus as a Molybdenum Blue

In the procedure described below, the sample material is decomposed by evaporation with hydrofluoric acid thereby removing silica, which would otherwise interfere. An alternative procedure described by Bodkin(18) is based upon a fusion with lithium borate followed by evaporation with hydrofluoric acid to remove the silica.

Method

Reagents:

Sulphuric acid, 3 N.

Ammonium molybdate solution, dissolve 5 g of crushed analytical grade reagent in water and dilute to 250 ml with water.

Ascorbic acid solution, dissolve 4.4 g of reagent in water and dilute to 250 ml. This solution deteriorates slowly on standing, and is best prepared freshly as required.

Reducing solution, dilute a mixture of 125 ml of 3 N sulphuric acid, 38 ml of ammonium molybdate solution and 60 ml of ascorbic acid solution to 250 ml with water. This reagent solution is faint-green in colour and should be prepared immediately before use.

Standard phosphate stock solution, dissolve 0.096 g of dried potassium dihydrogen phosphate in water and dilute to 1 litre in a volumetric flask. This solution contains 50 µg P_2O_5 per ml.

Standard phosphate working solution, dilute 20 ml of the stock standard phosphate solution to 250 ml with water. This solution contains 4 µg P_2O_5 per ml. For the calibration of 4-cm spectrophotometer cells, prepare a phosphate solution containing

1 μg P_2O_5 per ml by diluting 5 ml of the stock solution to 250 ml.

Procedure. Accurately weigh 0.1 g of the finely powdered sample into a small platinum crucible and allow to stand overnight with 2 ml of concentrated perchloric acid and 5 ml of concentrated hydrofluoric acid. Evaporate on a water bath until no further fumes of hydrofluoric acid are visible, and then under an infrared heating lamp until most of the perchloric acid is removed. Do not allow the residue to become completely dry. Add 1 ml of concentrated perchloric acid to the moist residue and again evaporate almost but not quite to dryness.

Using a calibrated pipette, add 0.8 ml of concentrated perchloric acid to the residue, followed by about 20 ml of water. Warm gently on a hot plate or water bath until all soluble material has passed into solution, and then rinse into a small beaker. If there is no visible solid residue, transfer the solution to a 100-ml volumetric flask and dilute to volume with water.

If the rock material is not completely decomposed by this procedure and a small residue remains, collect this residue on a small, close-textured filter paper and wash it with water. Combine the filtrate and washing in a 100-ml volumetric flask. Carefully dry and ignite the filter and residue in a small platinum crucible and fuse with a small quantity of anhydrous sodium carbonate. Allow the melt to cool, extract with water, acidify with the minimum quantity of diluted perchloric acid and add to the solution contained in the 100-ml volumetric flask. Dilute to volume with water.

Using a pipette, transfer 25 ml of the rock solution to a 50-ml volumetric flask, add 20 ml of the reducing solution, dilute to volume with water and mix well. Allow the solution to stand overnight and then measure the optical density in 1- or 4-cm cells with water as the reference solution, using the spectrophotometer set at a wavelength of 830 nm.

Measure also the optical density of a reagent blank solution prepared in the same way as the sample solution, but omitting the sample material. Calculate the phosphorus content of the sample by reference to the calibration graph, or by using a calibration factor.

Calibration. Pipette aliquots of 0-25 ml of the working standard phosphate solution containing 0-100 μg P_2O_5 (for calibration of 1-cm cells) or 0-25 μg P_2O_5 (for 4-cm cells) into separate 50-ml volumetric flasks, add 20 ml of reducing solution to each and dilute to volume. Allow the solutions to stand overnight and measure the optical densities in the appropriate cell using the spectrophotometer as described above. Plot the values obtained against the concentration of phosphate expressed as µg P_2O_5 per 50 ml of solution.

Notes: 1. This procedure can be used for the determination of phosphorus in carbonate rocks with only minor modifications. Dilute perchloric and nitric acids should be used for the sample decomposition, and if the residue contains silicate minerals it should be followed by the digestion with hydrofluoric acid and perchloric acids as described above for silicate rocks.

2. If arsenic is present in sufficient amount to interfere, the evaporation of the sample material with hydrofluoric and perchloric acids should be replaced by an evaporation with hydrofluoric, hydrochloric and nitric acids. A few mg of metallic zinc is used to reduce the arsenic before volatilisation. Finally the solution is evaporated almost to dryness with perchloric acid before proceeding as described above.

References

1. MISSON G., Chem-Ztg. (1908) 32, 633
2. MICHELSEN O B., Analyt. Chem. (1957) 29, 60
3. KITSON R E and MELLON M G., Ind. Eng. Chem., Anal. Ed. (1944) 16, 379
4. LEW R B and JAKOB F., Talanta (1963) 10, 322
5. BAADSGAARD H and SANDELL E B., Anal. Chim. Acta (1954) 11, 183
6. QUINLAN K P and DE SESA M A., Analyt. Chem. (1955) 27, 1626
7. BELL R D and DOISY E A., J. Biol. Chem. (1920) 44, 55
8. WILLARD H H and CENTER E J., Ind. Eng. Chem., Anal. Ed. (1941) 13, 81
9. WOODS J T and MELLON M G., Ind. Eng. Chem., Anal. Ed. (1941) 13, 760
10. MURPHY J and RILEY J P., J. Marine Biol. Assoc. U K (1958) 37, 9
11. KITSON R E and MELLON M G., Ind. Eng. Chem., Anal. Ed. (1944) 16, 466
12. MACDONALD A M G., Ind. Chemist (1960) 88 and 134
13. RILEY J P., Anal. Chim. Acta (1958) 19, 413
14. RIDDLE C and TUREK A., Anal. Chim. Acta (1977) 92, 49
15. BENNETT W H and PICKUP R., Colon. Geol. Min. Res. (1952) 3, 171
16. SHAPIRO L and BRANNOCK W W., U S Geol. Surv. Circ. 165, 1952, Bull. 1144-A, 1962
17. SAMUELSON O., Ion-Exchangers in Analytical Chemistry, p.146, Wiley, New York, 1953.
18. BODKIN J B., Analyst (1976) 101, 44

CHAPTER 35

Scandium, Yttrium and the Lanthanide Rare Earths

Determination of Scandium in Silicate Rocks

Most of the determinations of scandium in silicate rocks appear to have been made by emission spectrography and neutron activation analysis.[1-3] A large number of photometric reagents have been suggested for the determination of scandium but none of these is specific and even the most selective requires extensive separation procedures to remove interfering elements. Brudz et al[4] examined fourteen reagents that form coloured complexes with scandium, including arsenazo, thoron, alizarin, quinalizarin, carmine and murexide, but recommended sulphonazo as the reagent that combined high sensitivity with maximum selectivity to scandium.

Sulphonazo is dark-red in colour and readily soluble in water to give a violet-red coloured solution, which changes to pink on the addition of dilute hydrochloric acid. At pH 4.0-5.5 sulphonazo forms a stable violet-blue coloured water-soluble complex with scandium. Both this and the similar sulphonazo complex with yttrium have an absorption close to that of the reagent, and for this reason the measurements of optical density are made at a wavelength away from the absorption maximum. In an acetate buffered solution the absorption of the yttrium complex is similar to that of the reagent and does not interfere with the determination of scandium.

Elements that react with sulphonazo and interfere with the determination of scandium include vanadium, cobalt, gallium, copper, indium, nickel, uranium (VI), aluminium and zinc. Iron, titanium and zirconium interfere by hydrolysis. The separation procedure described by Brudz et al[4] includes precipitation with sodium hydroxide to remove aluminium, extraction with ether to remove iron, and precipitation of the scandium with ammonium tartrate in the presence of added yttrium. The precipitate of yttrium-scandium ammonium tartrate is ignited to the mixed oxides, dissolved in hydrochloric acid and the colour with sulphonazo developed in acetate-urotropine (hexamethylenetetramine) solution at pH 5.

A somewhat similar separation procedure involving precipitation as a mixed yttrium-scandium ammonium tartrate was described by Belopol'ski and Popov,[5] who used xylenol orange to complete the photometric determination. Shimizu[6] used arsenazo to determine scandium in silicate rocks but used a more extensive separation procedure based on both cation and anion exchange. The same separation was advocated for the determination of scandium in silicate rocks[7] using 4-(2-thiazolyl)resorcinol as photometric reagent. A simpler separation procedure by Galkina and Strel'tsova[8] is based upon extraction of scandium into an isobutanol solution of butyric acid in the presence of sulphosalicyclic acid. The determination is completed photometrically using arsenazo III.

Precipitation with ammonia followed by a cupferron extraction is used by Bakhmatova *et al*[9] prior to determination with sulphonitrazo R.

Determination of Yttrium

As with scandium, yttrium is most frequently determined in silicate rocks by techniques of emission spectrography. The sensitivity is however poor, although some improvement can be obtained by using a cation exchange enrichment procedure as described by Edge and Ahrens.[10] Photometric reagents suggested for yttrium include methylthymol blue, alizarin red S, thoron, catechol violet and xylenol orange. These reagents do not have the selectivity really required for this purpose, and further work is necessary before they can be applied to the determination of yttrium in silicate rocks.

Atomic absorption spectroscopy has been used for the determination of yttrium in silicate rocks.[11] The procedure is described in greater detail in the following section dealing with the rare earth elements.

Determination of the Lanthanide Rare Earths

Individual rare earth elements are usually determined by emission spectrography, neutron activation analysis, or X-ray fluorescence spectrography. Where the total rare earth content of rocks and minerals is required, gravimetric methods have usually been used.

The final determination step is a precipitation of the rare earths as mixed oxalates followed by ignition to mixed oxides as the weighing and reporting form. The major part of the cerium present will be in the higher valent state as CeO_2, whilst the weighed residue will include also any thorium present in the rock material, as ThO_2 and yttrium as Y_2O_3. If required, cerium and thorium can be separated chemically from the remaining earths, and determined individually.

Before the oxalate can be precipitated, it is necessary to separate the rare earths from most of the other components present in silicate rocks. Methods used for this include precipitation with sodium hydroxide to remove aluminium and the alkaline earth elements, precipitation with hydrofluoric acid to remove iron, titanium, zirconium and other elements forming soluble fluorides, and chlorination to remove elements that form volatile chlorides including iron, titanium, aluminium and zirconium. A variety of procedures for determining total rare earths have been based upon combinations of these separation procedures. In most of them there is a significant loss of rare earths, amounting to as little as 3 per cent or as much as 25 per cent or more. These losses can be observed and corrected by adding an active isotope of one or more of the rare earths before the first separation. Cerium-144 and yttrium-90 have been suggested. A procedure based upon that given by Varshal and Ryabchikov,[11] is given in detail in the earlier editions of this book. Such procedures are now little used.

The determination of total rare earths may be combined with that of chromium, vanadium, chlorine, barium and zirconium by using an alkali carbonate fusion to decompose the sample material. The rare earths are recovered by hydroxide precipitation with ammonia after the recovery of zirconium by phosphate precipitation. The hydroxide precipitate containing phosphate is then dissolved in hydrofluoric acid and the fluoride residue recovered for subsequent conversion to oxalates and then to oxides.

If the determination of zirconium is not required, the ammonium phosphate solution should not be added, and the hydroxide precipitate obtained by adding ammonia to the sulphuric acid solution. This precipitate can be calcined and chlorinated.[12] If the determination of barium can also be omitted, the residue obtained after extracting the original melt with water can be dissolved in dilute hydrochloric acid, the hydroxides precipitated with ammonia, and then calcined prior to chlorination.

Reagents suggested for the photometric determination of total rare earths include alizarin red-S, aluminon, xylenol orange, arsenazo I and III, PAN (1-(2-pyridylazo)-2-naphthol) and PAR (4-(2-pyridylazo)resorcinol). None of these is specific for the rare earths, and the selectivity of all of them is poor. The best combination of high sensitivity with some degree of selectivity is given by arsenazo III. This reagent forms complexes with a large number of other elements, including thorium, uranium and zirconium at low pH, and iron, yttrium, the rare earths and other elements at higher pH. In the procedure described below a pH of 1.8-2.0 is used for the determination of the rare earths.

The complexes formed by yttrium and the rare earths with arsenazo III, all have similar absorption spectra with maximum absorption occurring at a wavelength of about 660 nm. The arsenazo III complex of scandium has a maximum absorption at a slightly higher wavelength and interferes somewhat with the rare earth determination. A broad-band red filter has been recommended for this determination.(13)

An interesting procedure for the determination of cerium in the presence of other rare earth elements and components of silicate rocks is that based upon the light emission (candoluminescence) of cerium in a calcium oxide - calcium sulphate matrix, with sulphuric acid as a co-activator, when the matrix is inserted into a hydrogen-nitrogen-air flame.(14) The preparation of the matrix is described in an earlier paper.(15)

The sensitivity of the rare earth elements by atomic absorption methods is poor, and the technique cannot be recommended for those materials that are not enriched in these elements. Ionisation of rare earths in hot flames (nitrous oxide-acetylene has usually been recommended) necessitates the addition of a suitable suppressor such as additional sodium or potassium(16) to both rock and standard solutions. Little is known of interferences from other elements and an interference suppressor such as lanthanum(17) has been recommended. Prior separation as hydroxide(17) or as oxalate(18) has been suggested. The atomic absorption procedure described in detail below, including the determination of yttrium, is based upon that described by Sen Gupta.(18)

Spectrophotometric Determination of Rare Earths in Silicate Rocks

Before this photometric procedure can be applied it is necessary to separate the rare earth elements from all other elements that react with arsenazo III. This can be accomplished by precipitation as hydroxides with ammonia, followed by precipitation as oxalates with calcium added as carrier. The procedure here is based upon that given by Goryushina, Savvin and Romanova.(13)

Method

Reagents: Hydrogen peroxide solution, 30 per cent.

Ammonium chloride wash solution, dissolve 10 g of the reagent in 500 ml of water and add 10 ml of aqueous ammonia.

Calcium carrier solution, dissolve 0.50 g of pure calcium carbonate in the minimum amount of dilute hydrochloric acid and dilute to

100 ml with water. This solution contains 2.0 mg calcium per ml.

Oxalic acid.

Oxalic acid wash solution, dissolve 5 g of oxalic acid in 500 ml of water.

Potassium chloride buffer solution, mix together 80 ml of 0.2 N hydrochloric acid, 250 ml of 0.2 M potassium chloride solution and 670 ml of water.

Arsenazo III solution, dissolve 0.10 g of the reagent in 100 ml of 0.01 N hydrochloric acid.

Standard mixed rare earth solution, dissolve an accurately weighed quantity of rare earth carbonates (see below) in 0.1 N hydrochloric acid and dilute to volume to give a solution containing 2.5 μg total rare earth oxides per ml.

Procedure. Accurately weigh approximately 1 g (Note 1) of the finely powdered silicate rock material into a platinum dish, moisten with water and add 15 ml of concentrated hydrofluoric acid, 1 ml of concentrated nitric acid and 5 ml of 20 N sulphuric acid. Transfer the dish to a hot plate and evaporate to fumes of sulphuric acid. Allow to cool, rinse down the sides of the dish with a little water, add 5 ml of concentrated hydrofluoric acid and again evaporate to fumes of sulphuric acid. Allow to cool, again rinse down the sides of the dish with a little water and evaporate, this time to remove most of the excess sulphuric acid and to leave a moist residue of sulphates. Allow the dish to cool, add about 10 ml of water, break up the residue with a glass rod and rinse the contents of the dish into a 250-ml beaker with about 100 ml of water. Add 3 drops of hydrogen peroxide solution and 10 ml of concentrated hydrochloric acid and heat on a hot plate until a clear solution is obtained.

If any residue remains it should be collected, washed with water, dried and ignited in a small platinum crucible and fused with a little anhydrous sodium carbonate. Extract the melt with water, acidify with a little dilute hydrochloric acid and add to the main rock solution.

Heat the solution to near boiling and add concentrated ammonia solution until the precipitation of hydroxides is apparently complete and then add 15 ml in excess. Allow the precipitate to settle, then filter whilst still hot onto a hardended, open-textured filter paper and wash well with the ammonium chloride wash solution. Discard the filtrate and washings.

Rinse the residue back into the original beaker with about 100 ml of water and dissolve by warming with 10 ml of concentrated hydrochloric acid. Filter the warm solution through the paper previously used and wash well with water. Add 10 ml of the calcium carrier solution to the filtrate, and 1.5 g of oxalic acid. Stir, and heat the solution almost to boiling. Now add concentrated ammonia to bring the pH of the solution to a value of 5, as shown by close-range pH indicator papers. Allow to digest on a hot plate for 1 hour and then set the beaker aside overnight.

Collect the oxalate precipitate on a close-textured filter paper and wash with the oxalic acid wash solution. Discard the filtrate and washings. Transfer the paper and residue to a small platinum or silica crucible and dry and ignite in an electric furnace at a temperature of 500° in order to convert oxalates to carbonates. Allow to cool and dissolve the residue in a small volume of hydrochloric acid containing a few drops of hydrogen peroxide. Transfer the solution to a 50-ml beaker and evaporate to dryness on a steam bath. Dissolve the chloride residue in 0.01 N hydrochloric acid and dilute to volume in a 100-ml volumetric flask with 0.01 N hydrochloric acid.

Transfer an aliquot of this solution (note 2) containing 5-30 µg of rare earths to a 50-ml volumetric flask, add 5 ml of the potassium chloride buffer solution and 2 ml of the arsenazo III solution. Dilute to volume with 0.01 N hydrochloric acid and mix well. Measure the optical density of this solution using an absorptiometer fitted with a red filter. The reference solution should be prepared from 0.01 N hydrochloric acid to which 5 ml of buffer solution and 2 ml of arsenazo III solution have been added.

Calibration. The calibration graph can be obtained by plotting the optical densities of rare earth solutions containing 0-30 µg rare earth oxides, to which buffer solution and arsenazo III solution have been added as described above for the reference solution. Ideally a mixture of rare earths from the silicate rock under examination should be used for the calibration, but the use of a mixed rare earth precipitate from other silicate rocks or from many other sources will not introduce significant errors.

Notes: 1. This sample weight is suggested for rock materials containing not more than about 500 ppm total rare earths. It should be reduced to 0.2 g for rock materials containing higher concentrations of rare earths.

2. For basic and other rocks containing considerably less than 100 ppm rare earths, it may be necessary to take the whole of the sample solution for photometric determination.

Determination of Yttrium and Certain Rare Earths by Atomic Absorption Spectroscopy

Yttrium and the individual rare earths are not of equal sensitivity by this technique. This is illustrated below, giving suitable working ranges suggested for a number of these elements, together with suitable wavelengths.

Yttrium	5-1500 ppm	410.24 nm
Neodymium	10-3000 ppm	492.45 nm
Samarium	20-1500 ppm	429.67 nm
Europium	0.2-80 ppm	459.40 nm
Dysprosium	0.4-150 ppm	421.17 nm
Holmium	1-200 ppm	410.38 nm
Erbium	0.5-150 ppm	400.80 nm
Thulium	0.2-80 ppm	371.79 nm
Ytterbium	0.1-15 ppm	398.80 nm

In general, the lower end of the working range for these elements is given by ten times the detection limit. Cerium, gadolinium and thorium are insensitive to both flame emission and atomic absorption spectroscopic techniques and are therefore unlikely to be recorded. Praseodymium, although more sensitive by flame emission is unlikely to be present in amounts sufficient to be recorded or to interfere (Pr 492.459 nm) with neodymium absorption at 492.453 nm. Terbium and lutecium are also unlikely to be recorded.

In the method as described, lanthanum is used as a spectroscopic buffer, and is added to the solution prior to atomic absorption measurement. Since, of the total rare earths, lanthanum may constitute an appreciable proportion, the lanthanum content was determined by flame emission spectroscopy, and a quantity added prior to atomic absorption measurement of the remaining rare earth elements, adjusted to bring the total lanthanum concentration to 1.00%.

It is doubtful if the slight variation in lanthanum concentration in this solution that would arise from adding a fixed amount instead would give rise to significant error in the determinations of rare earths at the level at which they occur in silicate rocks.

As lanthanum is used as a buffer, it is not determined by atomic absorption spectroscopy in this scheme. However as noted above, it can be included by making a flame emission measurement of the stock rare earth solutions, prior to adding the buffer solution.

Because of the low sensitivity of the rare earth elements by this technique, it is necessary to take a large sample weight and concentrate the rare earths in a small volume prior to aspiration and measurement. Fluoride, oxalate (Note 1) and hydroxide precipitations are used to separate and concentrate the rare earths together with yttrium and thorium.

Method

Reagents: Lanthanum buffer solution, dissolve by heating 5.85 g of pure lanthanum oxide La_2O_3 in dilute hydrochloric acid, evaporate to a small volume, add 10 ml of concentrated perchloric acid and evaporate to a moist residue. Cool and dissolve in 25 ml of absolute ethanol. Dilute to 100 ml with ethanol. This solution contains 50 mg La per ml.

Standard rare earth stock solutions, dissolve pure oxides in dilute nitric acid, add a little perchloric acid, evaporate to a moist residue, dissolve in absolute ethanol and dilute to volume with ethanol. The calibration solutions (each containing 1% lanthanum) should cover the working ranges given above.

Methyl oxalate solution, dissolve 40 g of dried, anhydrous oxalic acid in 100 ml methanol. Allow to stand and if necessary filter before using.

Procedure. Accurately weigh a suitably large size sample portion (Note 2) of the finely ground rock material into a large (100 ml or more) platinum dish, moisten with water and add 30 ml of concentrated hydrofluoric acid. Cover, heat on a steam bath with occasional stirring until it can be seen that complete decomposition of the rock matrix has been obtained. Evaporate to dryness, add 40 ml of concentrated hydrofluoric acid, cover, digest on a steam bath and then collect the precipitated fluorides on a close-textured paper in a polyethylene filter funnel and wash with hot water. Discard the filtrate and washings.

Transfer the filter and fluoride precipitate to the original platinum dish, add 25 ml of concentrated nitric acid, cover and warm until the paper has decomposed. Evaporate to dryness. Repeat the evaporation to dryness with a further 25 ml of concentrated nitric acid. Add 5 ml of nitric acid and 10 ml of perchloric acid and with great care evaporate to give a moist residue. Dissolve in a little 10 per cent nitric acid containing about 5 per cent of 30-volume hydrogen peroxide. If any residue remains, collect on a small filter, and wash thoroughly with diluted nitric acid containing a few drops of hydrogen peroxide. Reserve the filtrate.

Ignite the residue if any in a platinum crucible at a temperature not exceeding 600°C, fuse with a little potassium pyrosulphate, dissolve the residue in dilute sulphuric acid and add to the main solution. Add calcium nitrate solution equivalent to about 50 mg CaO, heat on a steam bath and using pH indicator paper adjust the pH to 3.8 by adding ammonia. Add 10 ml of methyl oxalate solution slowly and with stirring. Digest on the steam bath for about half an hour, adjust pH to 2 with sodium hydroxide solution and allow to stand for an hour.

Collect the precipitated oxalates on a close-textured filter paper and wash with 0.1% ammonium oxalate solution. Decompose the mixed oxalates by heating with 25 ml of concentrated nitric acid. Add 5 ml of concentrated perchloric acid, evaporate to a moist residue (Note 1).

Dissolve the residue in a little dilute nitric acid containing hydrogen peroxide, add sufficient ferric chloride or nitrate to provide 6 mg ferric oxide, and 2 g of hydroxylamine hydrochloride. Precipitate iron, and the rare earths by adding ammonia to the warm solution. Adjust the pH to 10, collect the precipitate on an open-textured filter paper and wash with hot 1% ammonium nitrate solution containing a few drops of aqueous ammonia. Dissolve the precipitate and decompose the filter paper by evaporation with nitric acid and perchloric acid as described above. Evaporate the solution to small volume, transfer to a 20-ml beaker and evaporate to a moist residue. Cool in a desiccator and dissolve the moist perchlorates in absolute ethanol. Dilute to volume in a 10-ml volumetric flask with absolute ethanol to give the stock rare earth solution. Determine lanthanum in this solution by flame emission at a wavelength of 550.13 nm.

Prepare a solution for atomic absorption spectroscopy by diluting an aliquot of the alcoholic solution to volume (Note 3) with absolute ethanol after adding sufficient lanthanum buffer solution to give 1% La in the final solution, and determine the absorption in the usual way using a nitrous oxide-acetylene flame and the spectrometer set at the appropriate wavelength.

Notes: 1. Sen Gupta recommends a double precipitation of rare earth oxalates.

2. For most rocks a minimum of 5 g should be taken for the analysis. Sen Gupta[18] suggests a quantity in the range 2 to 10 g.

3. 3-, 5- or 10-ml volumetric flasks can be used, depending on the concentration of rare earths expected.

References

1. KEMP D M and SMALES A A., Anal. Chim. Acta (1960) 23, 410
2. DESAI H B, KRISHNAMOORTHY IYER R and SANKAR DAS M., Talanta (1964) 11, 1249
3. HAMAGUCHI H, WATANABE T, ONUMA N, TOMURA K and KURODA R., Anal. Chim. Acta (1965) 33, 13
4. BRUDZ V G, TITOV V I, OSIKO E P, DRAPKINA D A and SMIROVA K A., Zhur. Anal. Khim. (1962) 17, 568
5. BELOPOL'SKI M P and POPOV N P., Zavod. Lab. (1964) 30, 1441
6. SHIMIZU T., Anal. Chim. Acta (1967) 37, 75
7. SHIMIZU T and MOMO E., Anal. Chim. Acta (1970) 52, 146
8. GALKINA L L and STREL'TSOVA S A., Zhur. Anal. Khim. (1970) 25, 889
9. BAKHMATOVA T K, DEDKOV Yu M and ERSHOVA V A., Zh. Anal. Khim. (1976) 31, 292
10. EDGE R A and AHRENS L H., Anal. Chim. Acta (1962) 26, 355
11. VARSHAL G M and RYABCHIKOV D I., Zhur. Anal. Khim. (1964) 19, 202
12. IORDANOV N and DAIEV K H R., Zhur. Anal. Khim. (1962) 17, 429
13. GORYUSHINA V G, SAVVIN S B and ROMANOVA E V., Zhur. Anal. Khim. (1963) 18, 1340
14. BELCHER R, NASSER T A K, POLO-DIEZ L and TOWNSEND A., Analyst (1977) 102, 291
15. BELCHER R, RANJITKAR K P and TOWNSEND A., Analyst (1975) 100, 415
16. OOGHE W and VERBEEK F., Anal. Chim. Acta (1974) 73, 87
17. VAN LOON J C, GALBRAITH J H and ARDEN H M., Analyst (1971) 96, 47
18. SEN GUPTA J G., Talanta (1976) 23, 343

CHAPTER 36

Selenium and Tellurium

Determination of Selenium

Although acid decomposition procedures have been recommended for the determination of selenium in silicate rocks, it is clear from the work of Schnepfe[1], Stanton and McDonald[2] and others that loss by volatilisation can readily occur, particularly if the acid mixture contains perchloric acid. A combination of hydrofluoric, phosphoric and nitric acids would appear to be satisfactory, although an alkaline sinter with a mixture of zinc and magnesium oxides and sodium carbonate is preferred.[1] Cresshaw and Lakin[3] who also noted the losses of selenium that can occur on acid digestion, suggest that part at least of the difficulty is failure to convert the selenium (present as selenide) into a form in which it is retained and extracted. An oxidising roast was proposed as a means of converting all the selenium to an extractable form (selenite and selenate), of removing organic matter, and of giving a turbidity-free concentrated acid extract.

Some authors have recommended a separation step in the determination of selenium. Chau and Riley[4] for example describe co-precipitation with ferric hydroxide followed by a separation from iron on an ion-exchange column, whilst Severne and Brooks[5] describe co-precipitation with arsenic using hypophosphorous acid.

Volatilisation as the tetrabromide was recommended by Blankley[6], Golembeski[7] and Terada _et al._[8]

Most of the photometric methods described for selenium utilise organic reagents containing two ortho-amino groups, with which selenium reacts to form a piazselenol ring structure. One of the most important is 3,3'diaminobenzidine which forms an intense yellow coloured piazselenol, a reaction reported by Hoste and Gillis.[9] This reaction was used as the basis of a determination of selenium by Cheng,[10] who used EDTA as a masking agent to prevent interference from a number of metals, and extracted the complex into toluene for photometric measurement. These solutions of piazselenol have absorption maxima at wavelengths of 340 and 420 nm - the latter wavelength being preferred as the reagent itself absorbs much less light at 420 nm than at 340 nm. The Beer-Lambert Law is followed over the concentration range 5-25 μg selenium in 10 ml of toluene. Approximately 30 minutes are required for complete colour development, which is achieved at a pH of 2 to 3 in the presence of formic acid. For extraction into toluene, the pH of the solution is adjusted to a value between 6 and 7.

Other ortho-diamines used as photometric reagents include 2,3-diaminonaphthalene[9,11] o-phenylenediamine[12] and 4-substituted o-phenylenediamines.[13] The 4,5-benzopiazselenol formed by 2,3-diaminonaphthalene has also been used for the spectrofluorimetric determination of selenium.[1,3]

Atomic absorption spectrometry has not been widely applied to the determination of selenium in silicate and other rocks, largely because this technique does not have sufficient sensitivity for direct application. The most sensitive selenium absorption occurs at a wavelength of 196.0 nm which is in a region of high background flame absorbance, giving a low signal to noise ratio at the concentrations of selenium usually encountered. The volatility of the element selenium frequently results in short hollow cathode lamp life. Electrode-less discharge lamps can be used to permit more energy at 196.0 nm to reach the detector and hence improve the sensitivity, whilst the cool flames obtained with nitrogen/hydrogen and argon/hydrogen give improved signal to noise ratio over that observed with an air/acetylene flame. Atomic abosrption methods for selenium in rocks have been described by Golembeski[7] and Severne and Brooks.[5]

Photometric Determination using 2,2'-diaminobenzidine

The procedure given below is based upon Chau and Riley[4] and involves co-precipitation of selenium with ferric hydroxide, separation from iron by ion-exchange and finally extraction of the piazselenol into toluene. The amount of selenium recovered by this procedure, usually about 95 per cent, is determined by adding radioactive selenium-75 at the decomposition stage, and comparing the activity of the aqueous solution after completing the separation of selenium with that of a similar aliquot of the active solution.

Method

Apparatus. Ion-exchange column, this consists of a cation exchange resin such as Zeo-Carb 225, 52-100 mesh, in the form of a column 10 cm in length and 1.5 cm in diameter. This resin should be washed with N hydrochloric acid until the eluate no longer gives a reaction for iron, then with water until almost free from acid.

Spectrophotometer cells, for this determination 4-cm cells with a capacity of only a few ml are required ("micro cells").

Reagents: Diaminobenzidine solution, dissolve 50 mg of the hydrochloride form of the reagent in 10 ml of water. Store in a refrigerator and discard after 3 days or earlier if darkening occurs. The reagent is expensive and solutions deteriorate. They should therefore be prepared only in small volumes as required.

Ammonium nitrate wash solution, dissolve 10 g of solid ammonium nitrate in 1 litre of water.

EDTA solution, dissolve 3.72 g of the disodium salt of EDTA in 100 ml of water.

Formic acid solution, 2.5 M.

Iron carrier solution, dissolve 2 g of anhydrous ferric chloride in 100 ml of water containing a few ml of concentrated hydrochloric acid to prevent hydrolysis.

Selenium-75 solution, carrier-free selenite solution, dilute with water as required to give a solution with approximately 10,000 counts per minute per ml.

Standard selenium stock solution, dissolve 0.25 g of pure selenium in 1-2 ml of concentrated nitric acid and dilute to 250 ml with water. This solution contains 1 mg selenium per ml.

Standard selenium working solution, containing 2 μg selenium per ml. Prepare by dilution of the stock solution with water as required.

Procedure. Accurately weigh 1-2 g of the finely ground silicate rock material (Note 1) into a platinum dish, moisten with water and add 2-ml of the selenium-75 solution. At the same time transfer a similar 2-ml aliquot of the active solution to a counting tray, evaporate to dryness and set aside. To the sample material in the platinum dish add 10 ml of concentrated hydrofluoric acid and 10 ml of concentrated nitric acid, cover the dish and set aside overnight. Rinse and remove the cover and evaporate to dryness on a steam bath. Add 10 ml of concentrated hydrofluoric acid and 10 ml of concentrated nitric acid and repeat the evaporation to dryness. Add 5 ml of concentrated nitric acid to the dry residue and again evaporate to dryness. Repeat this evaporation with two further 5 ml portions of nitric acid to decompose fluorides and remove fluorine as hydrogen fluoride.

Moisten the residue with nitric acid, rinse into a 100-ml beaker and evaporate to dryness. Add 25 ml of 4 N hydrochloric acid and boil gently for 5 minutes to convert any selenate to selenite. If any residue remains at this stage collect on a small filter, wash with water and combine the filtrate and washings. Discard the residue (Note 2).

Dilute the solution to about 5 litres with water (Note 3) and add dilute sodium hydroxide solution to bring the pH to a value between 3.5 and 4. Add 60 g of solid sodium chloride, stir to dissolve and then with stirring add 3 ml of the iron carrier solution. Now bring the pH to a value between 4.5 and 5.0 by adding dilute aqueous ammonia, stir, and allow to stand for 2 hours. Add a further 3 ml of the iron carrier solution, check the pH and adjust to 4.5-5.0 if necessary, stir and allow to stand for at least 2 days. Decant or siphon off the supernatant liquid and collect the precipitate on a small paper. Wash the ferric hydroxide precipitate with dilute ammonium nitrate solution and discard the filtrate and washings.

Rinse the precipitate into a small beaker, dissolve in 1-2 ml of concentrated nitric acid (Note 3), and dissolve any remaining traces of residue from the paper. Dilute the solution to give an acid concentration of 0.2 N and pass through the cation exchange column. Elute with 350 ml of 0.2 N nitric acid. Combine the percolate and eluate, add 1 ml of 2 N sodium hydroxide and evaporate almost to dryness. Transfer to a counting tray and complete the evaporation to dryness. Compare the activity with that of the active residue previously set aside, and hence calculate the selenium recovery.

Rinse the residue from the counting tray into a small beaker, add 2 ml of 2.5 M formic acid and 5 ml of EDTA solution and dilute to about 25 ml with water. Adjust the pH of the solution to a value in the range 2-3 by adding dilute nitric acid or aqueous ammonia as necessary. Now add 2 ml of the disminobenzidine reagent, allow to stand for 30 minutes then add aqueous ammonia to bring the pH to 6-7, and rinse the solution into a separating funnel.

Add 5 ml of the toluene and shake for 3 minutes. Discard the lower, aqueous layer and measure the optical density of the toluene extract in 4-cm cells, using the spectrophotometer set at a wavelength of 429 nm, with pure toluene as the reference solution.

Calibration. Transfer aliquots of 0-5 ml of the standard selenium solution containing 0-10 μg selenium to separate beakers, add 2 ml of 2.5 M formic acid to each and continue as described above. Plot the relation of optical density to selenium concentration.

Notes: 1. Marine sediments and other rock samples containing water-soluble chlorine should be washed with water until no further chloride ion can be detected in the washings, and then dried before analysis.

2. If the residue appears to contain silicate minerals, re-treat by evaporation with small quantities of hydrofluoric and nitric acids as described, dissolve the residue in hydrochloric acid and add to the rock solution. If the

residue is largely oxide minerals, fuse with a little anhydrous sodium carbonate, extract with water, acidify with dilute nitric acid and add to the rock solution.

3. The choice of volume appears to have been made to match this method to the similar determination in sea water. A somewhat smaller volume would probably suffice.

Determination of Tellurium

Small quantities of tellurium are difficult to determine and the difficulties are compounded by the somewhat lower (than selenium) levels encountered. Hanson[14] has described a spectrophotometric procedure based upon the reduction of tellurium to a colloidal form by adding stannous chloride solution. Anderson and Peterson[15] have used this for natural materials, but give a lower limit of 0.5 ppm, for which 10 g sample portions are used. This concentration is greatly in excess of the tellurium content of most silicate rocks.

Lovering, Lakin and McCarthy[16] determined down to 0.1 ppm tellurium in jasperoid samples by a method based upon the induced precipitation of elemental gold from a 6 N hydrochloric acid solution containing gold chloride, cupric chloride and hypophosphorous acid.

Concentrations of down to 0.005 ppm (5 ppb) of tellurium were determined by Hubert[17] measuring the catalytic effect on the reduction of gold by hypophosphorous acid. The method is suitable only for those materials from which the tellurium can be released by digestion with bromine and hydrobromic acid. An extraction into methyl isobutyl ketone is used to isolate and concentrate the tellurium.

The difficulties encountered in determining selenium in silicate rocks using atomic absorption spectrometry, are also encountered with tellurium. A procedure which combines the determination of tellurium with that of selenium and which uses a precipitation with arsenic as a concentrating and isolation step is described by Severne and Brooks,[5] but the detection limit is given as no more than 0.1 ppm. A somewhat more sensitive procedure is given by Watterson and Neuerburg,[18] but this relies on the use of large samples (12 g). The authors report the recovery of 5 ppb, with the implication that this quantity of tellurium can be determined. It is doubtful if any reasonable accuracy can be achieved at this level by this procedure, although as indicated by the authors, the method is adequate for geochemical exploration purposes.

A determination of as little as 5 ppb in a 0.25 g portion of silicate rock was reported by Greenland and Campbell.(19) The sample material was decomposed with hydrofluoric, nitric and perchloric acids. Sodium borohydride solution was then added to reduce the tellurium to its hydride. This was transferred in a stream of nitrogen gas to a resistance wound quartz cell kept at an operating temperature of about 1000°, in which the hydride was decomposed, and the absorption due to tellurium measured. The method is described as "very rapid, quite sensitive and reasonably accurate and precise".

Such methods as this, based upon the formation of a volatile covalent hydride are receiving great attention at the present time, particularly those flameless absorption methods that incorporate trapping in some suitable stabilising medium. Future developments are likely to give a greater sensitivity and much improved precision for the determination in both silicate and other rocks.

References

1. SCHNEPFE M M., J. Res. U S Geol. Surv. (1974) 2, 631
2. STANTON R E and McDONALD A J., Analyst (1969) 90, 497
3. CRESSHAW G L and LAKIN H W., J. Res. U S Geol. Surv. (1974) 2, 483
4. CHAU Y K and RILEY J P., Anal. Chim. Acta (1965) 33, 36
5. SEVERNE B C and BROOKS R R., Talanta (1972) 19, 1467
6. BALANKLEY M., Tech Note 131, Determination of Selenium in Glass, Brit. Glass Res. Assoc. (1970)
7. GOLEMBESKI T., Talanta (1975) 22, 547
8. TERADA K, OOBA T and KIBA T., Talanta (1975) 22, 41
9. HOSTE J and GILLIS J., Anal. Chim. Acta (1955) 12, 158
10. CHENG K L., Analyt. Chem. (1956) 28, 1738
11. HOSTE J and GILLIS J., Anal. Chim. Acta (1948) 2, 402
12. PARKER C A and HARVEY L G., Analyst (1962) 87, 558
13. DEMEYERE D and HOSTE J., Anal. Chim. Acta (1962) 27, 288
14. HANSON C K., Analyt. Chem. (1957) 29, 1204
15. ANDERSON W L and PETERSON H E., U S Bur. Mines Rept. Invest. 6201, (1963)
16. LOVERING T G, LAKIN H W and McCARTHY J H., U S Geol. Surv. Prof. Paper 550-B p.B 138, (1966)
17. HUBERT A E., U S Geol. Surv. Prof. Paper 750-B, p.B 188, (1971)
18. WATTERSON J R and NEUERBURG C J., J. Res. U S Geol. Surv. (1975) 3, 191
19. GREENLAND L P and CAMPBELL E Y., Anal. Chim. Acta (1976) 87, 323

CHAPTER 37

Silicon

After oxygen, silicon is the most abundant of all the elements, amounting to almost 28 per cent of the rocks of the lithosphere. Even when present in trace amounts, silicon is almost always reported as silica, SiO_2, which is the form used here. The distribution of silica in silicate rocks was investigated by Richardson and Sneesby[1] who showed that maximum values occurred at 52.5 per cent and 73 per cent silica, corresponding to the two most common igneous rocks, basalt and granite. Details of this distribution pattern have been criticised by several workers, including Ahrens,[2] who attributed errors in the distribution relating to the basaltic mode, to the inclusion of an undue proportion of unusual and rare rocks in the total of analyses used. It is also clear that systematic error in the early gravimetric determinations of silica will have been included in all or almost all of the analyses.

Most silicate rocks contain between 35 and 80 per cent of silica, although the range of values that an analyst can expect to meet is extended by sandstones and quartzites at one end, and by siliceous and argillaceous carbonates at the other.

Quartzites and sandstones may contain up to about 99 per cent silica, 90-95 per cent being more frequently encountered. Whilst the silica content of these rocks is rarely of much interest to the petrologist or geochemist, it is of importance to the economic mineralogists, as these materials are used in large quantities in the glass, refractory, chemical and building industries.

The carbonate rocks are a large and varied group of igneous (carbonatite), sedimentary and metamorphic rocks, with silica contents ranging from less than 1 per cent (as, for example, in certain limestones such as chalk and some forms of marble) to 30 per cent or more in some sediments and carbonatites.

In view of the importance of silica in silicate rocks it is not surprising that a very wide variety of methods have been suggested for its determination. Gravimetric methods are particularly important, and great efforts have been made to simplify, speed up and improve the classical separation of silica, ie dehydration with hydrochloric acid, ignition to SiO_2 and volatilisation with hydrofluoric acid. Other gravimetric methods based upon the precipitation of a molybdosilicate with 8-hydroxyquinoline[3] and with 2,4-dimethylquinoline[4] have not been widely used in rock analysis, although Bennett <u>et al</u>[5] describes application to clays and other silicate materials used in the pottery industry. The accuracy obtainable does not match that of the classical procedure, but in good hands is probably similar to that obtainable by most spectrophotometric methods.

Precipitation and gravimetric determination of silicon as potassium silicofluoride (fluosilicate)[6] does not appear to have been applied to any great extent to silicate rocks.

Volumetric methods for silica include that described in detail in the earlier editions of this book, based upon the fluoride reaction:

$$Si(OH)_4 + 6F^- + 4H^+ = SiF_6^{2-} + 4H_2O$$

which occurs in acid solution to which excess of alkali fluoride has been added. In order to apply this reaction to silicate rocks Khalizova et al[7] decomposed the rock material by fusion with potassium hydroxide and sodium peroxide, dissolved the melt in water, acidified with hydrochloric acid and removed aluminium, iron, titanium, calcium, magnesium and other elements by passing the solution through a column containing a cation exchange resin. The solution is at a sufficient dilution to prevent polymerisation of the silica.

The eluate is neutralised to methyl orange and a known excess of standard hydrochloric acid added. Potassium fluoride and potassium chloride are then added, and the excess hydrochloric acid titrated with standard sodium hydroxide solution. Some experience is required to judge the exact end point, although this could probably be improved by the use of a screened indicator.

A more recent procedure by Sinha et al[8] is based upon the precipitation of potassium silicofluoride (fluosilicate) in mineral acid solution containing an excess of potassium and fluoride ions, washing the precipitate free from acid, hydrolysing it in hot water and titrating the liberated hydrofluoric acid with standard alkali.

Spectrophotometric methods and methods based upon atomic absorption spectroscopy have been widely used for the determination of silica. These are described and examples given in detail below.

There is thus no lack of methods for determining silica in silicate rocks. The choice must to some extent depend upon the skills of the analyst concerned, but also on the accuracy needed, the time and the analytical instrumentation available.

Gravimetric Determination of Silica in Quartzites and Sandstones

In the classical procedure for determining silica, a silica residue, obtained by chemical processing of the rock material, is subjected to evaporation with a mixture of hydrofluoric and sulphuric acids and the loss in weight caused by volatilisation of silicon tetrafluoride taken as a measure of the silica content of the residue. With rocks rich in silica, such as quartzites, the silica residue obtained by the usual chemical processing is likely to contain elements, particularly sodium, in amounts greater than in the quartzite itself. Furthermore, not all the silica will have been collected in the silica residue. For these reasons the preliminary separation stage is best omitted, and the loss in weight that occurs on volatilisation of the silica determined directly from the ground rock material.

This method is applicable only to those samples with high silica content, where the remaining constituents amount to no more than 1 or 2 per cent of the total. This method is, however, often used for sands and sandstones containing much less silica, where appreciable errors can arise. Carbonate minerals are rare components of these rocks, but ferruginous carbonate material has been noted cementing the silica grains in some sandstones. The loss of carbon dioxide that occurs during the initial heating stage does not give rise to error in the silica determination, but any ferrous iron will be oxidised to ferric during the decomposition of the sulphates and hence gives a negative error. A similar error has been observed to occur when the sandstone contains grains of magnetite or ilmenite, from which the ferrous iron is converted to the higher valency state. Another source of error arises from the volatilisation of alkali metals if, as frequently happens, the sandstone rock contains grains of felspar.

Method

Reagents: Hydrofluoric acid, concentrated.
Sulphuric acid, 20 N.

Procedure. Ignite a clean platinum crucible of about 30-ml capacity over the full flame of a Meker burner, allow to cool and weigh empty. Transfer to it approximately

1 g of the finely ground quartzitic material and reweigh to give the weight of the sample (note 1). Again ignite the crucible over the Meker burner, gently at first and then finally over the full flame of the burner for a period of 1 hour. Allow to cool and reweigh the crucible and contents. Repeat the ignition until no further loss in weight occurs. Report the change in weight as the "loss on ignition". This loss is due to the oxidation of ferrous iron and organic matter, and to the volatilisation of water, carbon dioxide and other gases.

Add 4 drops of 20 N sulphuric acid to the residue, followed by 20 ml of concentrated hydrofluoric acid. Transfer the crucible to a hot plate or sand bath and remove the silica and excess hydrofluoric acid by volatilisation in the usual way. Gently heat the crucible on the hot plate to remove the excess sulphuric acid and then ignite over the full flame of the Meker burner for 10 minutes. Allow to cool and reweigh the crucible. Repeat the ignition until no further loss in weight occurs. Determine a reagent blank by volatilisation in the same way, but omitting the sample material. The blank value obtained with analytical grade hydrofluoric acid will probably amount to between 0.3 and 1.0 mg. This should be added to the loss on volatilisation to give the total silica content of the sample.

In a very few samples some silicate mineral grains may remain undecomposed. These can usually be attacked by a second evaporation with smaller amounts of hydrofluoric and sulphuric acid.;

Note: 1. Tsubaki[(9)] describes a modification to this procedure in which 3 g of boric acid are added to a 0.5 g sample portion prior to igniting at 1000°. The determination is then completed as described above.

Gravimetric Determination in Silicate and Carbonate Rocks

The classical procedure for the determination of silica in silicate rocks has earlier been described in detail. It consists of an alkali carbonate fusion, extraction with water, dissolution and de hydration with hydrochloric acid and finally determination of the loss on volatilisation with hydrofluoric acid. Two successive dehydrations are usually employed and any silica remaining in the filtrate from the second is recovered from the ammonia precipitate.

This procedure is very time consuming, partly in itself and partly because the subsequent stages of the rock analysis by this classical method cannot begin until the silica determination is complete. This may take up to 3 days. It is therefore not surprising that considerable efforts have been devoted to finding a better and more rapid way of recovering the silica from the hydrochloric acid solution of the rock melt.

Sulphuric and perchloric acids have been used to effect the dehydration of silica. Sulphuric acid is more effective than hydrochloric acid for this purpose, whilst perchloric acid is more effective still, although care must be taken to avoid explosions when perchloric acid is used. These can arise either through a dramatic oxidation of organic matter present in the solution, or more likely through incomplete removal of perchloric acid from the filter paper used to collect the dehydrated silica.

DETERMINATION IN CARBONATE ROCKS

Perchloric acid has been found to be particularly useful for the determination of major amounts (5 per cent and upwards) of silica in carbonate rocks. The following procedure has been used.

Procedure. Accurately weigh approximately 1 g of the finely powdered carbonate rock material into a platinum crucible, partly cover with a platinum lid and ignite strongly over a Meker burner for 30 minutes (Note 1). Allow to cool, moisten the cake with water and rinse the residue into a platinum dish (Note 2). Add 50 ml of water, 5 ml of concentrated nitric acid and 10 ml of concentrated perchloric acid. Rinse the crucible and lid with a little dilute nitric acid and add to the solution in the platinum dish. Transfer the dish to a steam bath and evaporate until most of the water has been removed. Place the dish on a hot plate or sand bath and continue evaporation to fumes of perchloric acid. Allow to fume gently for 10 minutes, then allow to cool and add 50 ml of water. Warm to allow all soluble salts to pass into solution and then collect the residue on a close-textured filter paper. Wash the residue twice with 0.1 N hydrochloric acid and then several times with hot water. A few mg of silica can be recovered from the filtrate and washings which should be reserved for the determination of iron, calcium, magnesium and other constituents of the limestone.

The residue, or combined residues if a second dehydration has been made, is transferred to a weighed platinum crucible and the silica content determined by volatilisation with hydrofluoric acid as described in the classical method for the analysis of silicate

Notes: 1. If required, the "loss on ignition" can be determined at this stage. Carbonate minerals are very largely converted to oxides during the ignition, which also converts acid-insoluble silicate minerals to acid-soluble calcium silicate.

2. Silica dishes and both silica and borosilicate glass beakers have also been described for this dehydration of silica with perchloric acid.

GRAVIMETRIC DETERMINATION WITHOUT DEHYDRATION

The substitution of perchloric acid for hydrochloric acid in the dehydration of silica saves only a little time. A much bigger saving can be obtained by avoiding the dehydration step altogether. This can be done by adding gelatin to the acid solution obtained after the addition of hydrochloric acid to the aqueous extract of the melt. The gelatin serves to coagulate the silica, enabling it to be collected and determined in the usual way. This technique has long been used in industrial practice for silicate slag analysis, where the emphasis is more on speed than accuracy. Some 2-4 per cent of the silica will remain in solution, the exact amount depending upon the conditions used and the composition of the slag. This small amount of silica is not recoverable except by dehydration in the usual way, when some part of it will be collected.

As an alternative to gelatin, Bennett and Reed[10] have suggested the addition of industrial coagulating agents, such as polymers of ethylene oxide. A number of "Polyox" polymers were tried and all were effective at coagulating the silica and obviating the need for dehydration. This procedure has the advantages of removing all but about 0.5 per cent of the silica from the solution (this trace can be determined photometrically), the introduced polyethyleneoxide does not interfere with the subsequent photometric determination of iron, titanium or aluminium, and very little boron will accompany the silica even when borate has been added to the alkali carbonate used for sample decomposition. The following detailed procedure is essentially that described by Bennett and Reed.[10]

Method

Reagents: Fusion mixture, an equimolecular mixture of potassium and sodium carbonates.

Boric acid.

Polyethylene oxide solution, dissolve 0.25 g of a polyethylene oxide coagulant in 100 ml of water. (Material supplied by BDH as polyethylene oxide polymer is suitable).

Procedure. Accurately weigh approximately 1 g of the finely powdered silicate rock material into a platinum dish, add 3 g of fusion mixture and 0.4 g of boric acid (Note 1). Mix using a small spatula, and fuse gently over a burner until the vigorous stage of the decomposition is complete, then transfer to an electric furnace at a temperature of 1200° for 10 minutes. Allow to cool, add to the dish 10 ml of water and 15 ml of concentrated hydrochloric acid. Cover, and warm on a steam bath until the melt has completely disintegrated and no further evolution of carbon dioxide can be detected. Remove the lid and rinse any liquid adhering to it back into the dish with a small amount of water.

Replace the dish on the steam bath and heat for a period of 30 minutes, then stir into the solution, which now contains gelatinous silica, a little macerated filter paper, 5 ml of the polyethylene oxide solution and 10 ml of water. Allow to stand for about 5 minutes and then collect the residue on a fine-textured filter paper (note 2). Wash the residue three times with 0.5 N hydrochloric acid and then with water until the filtrate is free from chloride ion. Combine the filtrate and washings and reserve for the determination of silica traces, iron, titanium and aluminium. Transfer the paper and residue to a weighed platinum crucible, dry and ignite, and determine the silica by evaporation with hydrofluoric acid in the usual way.

Notes: 1. This flux is recommended for clays and for high-silica materials such as silicate rocks. For refractories and other high-alumina materials, 2 g of the fusion mixture and 0.4 g of boric acid should be used, together with a longer period for the final stage of the rock fusion. This should be increased to up to 30 minutes.

2. Difficulties may be encountered at the filtration stage if the polyethylene oxide solution is added before the gelling of the silica has taken place. With material containing less than about 30 per cent silica, this gelling will not be observed, but provided that the sample solution is digested on the steam bath as suggested, polymerisation will occur with these samples and difficulties with filtration will be avoided.

Spectrophotometric Determination of Silica

There are a very large number of papers describing the photometric determination of silica, but almost all of these are variations of two basic methods. These are the determination as yellow silicomolybdate and as the molybdenum blue to which this yellow complex can be reduced.

Phosphorus, arsenic and germanium form similar yellow molybdate complexes that can be reduced to molybdenum blues. Fortunately neither arsenic nor germanium is likely to be present in silicate rocks in amounts sufficient to interfere, whilst the yellow phosphomolybdate can be decomposed by the addition of tartaric, citric or oxalic acid. This addition serves also to prevent interference from iron.

Reagents suggested for the reduction of the yellow silicomolybdate to molybdenum blue include stannous chloride, hydroxylamine, hydroquinone, ascorbic acid, ferrous ammonium sulphate and 1-amino-2-naphthol-4-sulphonic acid. Stannous chloride is a very powerful reducing agent and has been favoured because of its extremely rapid action. The blue colours obtained, however, are not as stable as those produced with 1-amino-2-naphthol-4-sulphonic acid, which, in combination with alkali sulphite and bisulphite, has been recommended by a number of authors.

Solutions of the yellow silicomolybdate have maximum absorption at a wavelength of about 350 nm; some authors have recommended higher wavelengths with some loss of sensitivity. This is not of importance in the analysis of silicate rocks, as advantage can seldom be made of the full sensitivity that is available. Molybdenum blue methods are even more sensitive than those based on the yellow complex. Solutions containing molybdenum blue have a maximum absorption at a wavelength of 810 nm, although a somewhat lower wavelength of 650 nm has been recommended(11) as giving more reproducible values of optical density in the presence of iron.

In order to improve the precision of the determination of silica by spectrophotometry, Abbey et al(12) describe the application of a differential technique. The measurement is the small difference in absorbance between a standard and the sample solution, rather than between standard and reagent blank.

The procedure given in detail below is based upon the molybdenum blue procedure given by Shapiro and Brannock(13) for the determination of silica in their compilation of methods for the rapid analysis of silicate rocks.

Method

The combination of good sensitivity to silica with its presence as the major component - in many rocks exceeding the total for the remaining oxides - results in the need to take only a very small sized sample portion for the determination. Care in weighing this small sample must be combined with attention to detail at all the subsequent stages of the analysis if accurate results are to be obtained.

Some care must also be exercised in the choice of silicate material to be used as standard. Previously analysed silicate rock samples have been recommended for this, but any error in the earlier analyses will be repeated in the subsequent photometric determinations. Rock standards such as granite G-1, diabase W-1 or the NBS felspar, No.99, provide a better standard material. Other good standards include suitably prepared quartzite or glass sand (99+ per cent silica) or pure crystal quartz (99.5+ per cent silica), for which accurate silica determinations can be made by evaporation with hydrofluoric acid.

Reagents: Sodium hydroxide solution, dissolve 30 g of sodium hydroxide pellets in water and dilute to 100 ml. Store in a polyethylene bottle.

Ammonium molybdate solution, dissolve 7.5 g of crushed ammonium molybdate in 80 ml of water and add 20 ml of 9 N sulphuric acid. Store in a polyethylene bottle.

Tartaric acid solution, dissolve 25 g of the reagent in water and dilute to 250 ml.

Reducing solution, dissolve 0.7 g of sodium sulphite and 0.15 g of 1-amino-2-naphthol-4-sulphonic acid in 100 ml of water, add 9 g of sodium metabisulphite and stir until solution is complete. Store in a cool, dark cupboard and prepare freshly every few days.

Procedure. Transfer 5-ml aliquots of the sodium hydroxide solution to a series of six nickel crucibles using either a polypropylene pipette or more simply by direct weighing on a laboratory rough balance. Transfer all six crucibles to a hot plate and evaporate the solutions to dryness. Care should be taken to avoid any great loss of reagent, although a small amount of spattering can be ignored. Allow to cool in a desiccator.

Accurately weigh approximately 0.05 g of the finely powdered rock material onto the sodium hydroxide in three of the crucibles, 0.05 g of each of two finely powdered silicate rock standards into two more of the crucibles, and reserve the last crucible for the reagent blank. Cover all six crucibles with nickel lids and transfer to a hot plate set at its highest temperature to melt the sodium hydroxide and allow the fusion process to begin. Take each crucible in turn and heat to dull redness over a gas burner for a period of about 5 minutes, swirling to give a good melt (Note 1). Allow to cool.

Pour a little water into each crucible and warm on a hot plate until complete decomposition of the cake is obtained. Rinse the contents of the crucible into a small polypropylene beaker and add water to give a volume of about 100 ml. Using a rubber-tipped rod remove all particles of residue adhering to the crucible and lid and transfer these to the beaker with a fine jet of water. Set the crucible and lid aside for the next determination (Note 2). Pour the rock solution containing the hydroxide residue into a 600-ml beaker containing 400 ml of water and 10 ml of concentrated hydrochloric acid. Rinse the polypropylene beaker and add the washings to the main rock solution. If a clear solution is not obtained, warm gently for a few minutes until it clears, then cool, transfer to a 1-litre volumetric flask and dilute to volume with water. Mix well. The acid rock solution is fairly stable in glass apparatus, but if the colour development is likely to be delayed, the solution should be transferred to a polyethylene bottle for storage.

Transfer 10 ml of each of the three sample solutions, the two standard solutions and the reagent blank solution to separate 100-ml volumetric flasks, dilute each to 60 ml with water, add by pipette 2 ml of the ammonium molybdate solution, swirl to mix, and set aside for 10 minutes. After exactly 10 minutes add by pipette 4 ml of the tartaric acid solution to each flask, swirling again to mix. Now add 1 ml of the reducing solution, dilute to volume with water and mix well. Set all six solutions aside for at least 30 minutes and preferably for 1 hour. Measure the optical density of each solution relative to water in 1-cm cells, with the spectrophotometer set at a wavelength of 650 nm.

Calibration. As the Beer-Lambert Law is valid, the average optical density of the two standard rock solutions, after deduction of the reagent blank value, can be used to establish the position of the calibration line. Alternatively, the silica content of each sample portion can be calculated directly from the average values:

$$\text{Per Cent } SiO_2 = \frac{S \times A_1 \times w_1}{A_2 \times w_2}$$

where w_1 and w_2 are the weights of sample and standard and A_1 and A_2 the respective average optical densities. S is the silica content in per cent, of the standard.

Notes: 1. On remelting the sodium hydroxide, some of the sample material may float on the surface of the melt and remain unattacked by it. When this occurs, gently swirl the melt to disperse the powder as much as possible, then allow the melt to solidify. On further remelting the material complete fusion should occur. Alternatively, the crucible may be tapped on the bench!

2. The crucible and lid should be rinsed with dilute hydrochloric acid before being used for the next determination.

3. The method described here for the determination of silica in silicate rocks can be applied directly to carbonate rocks. A sample weight of up to 0.2 g can be taken for the analysis, which is otherwise carried out as for silicate rocks. A silicate rock may be used as the reference standard.

Combined Gravimetric and Photometric Determination of Silica

The precision of the photometric determination of silica is controlled to a large extent by the limitations inherent in spectrophotometry. Such determinations cannot by their nature be as precise as those obtained by good gravimetric methods, although of course the possibility of systematic errors exists in both methods. In order to combine the advantages and avoid some of the disadvantages of both methods Jeffery and Wilson[11] have suggested a new procedure based upon a single dehydration with hydrochloric acid, giving a major silica fraction as an insoluble residue, and a minor fraction in the filtrate that can be determined photometrically. As only one dehydration is made, this procedure is more rapid than the classical method, and as all the silica not collected in the residue is determined photometrically, it is also more accurate. Moreover, since the photometric determination is now of a minor constituent (2-8 mg, equivalent to less than 1 per cent of the rock composition), the limiting accuracy of photometric methods is no longer restrictive.

A disadvantage that has been introduced is that in the filtrate from the collection of the major silica fraction, all the remaining elements have been increased in concentration relative to silica. Under these circumstances elements that are not usually present in amounts sufficient to interfere may be concentrated up to and beyond this point. Interference has been noted from titanium and phosphorus, although up to 5 per cent TiO_2 in the rock material does not have a significant effect, and up to 10 per cent P_2O_5 can also be tolerated.

Fluorine in trace amounts does not interfere, although if present in major amounts, some of the silica may be lost by volatilisation during the dehydration stage, as in the classical method for determining silica. Three per cent fluorine in the rock material is likely to give rise to a loss of as much as 0.5 per cent silica.[14]

Method

Reagents:

Ammonium molybdate solution, dissolve 10 g of crushed ammonium molybdate in N aqueous ammonia and dilute to 100 ml with N ammonia.

Oxalic acid solution, dissolve 10 g in water and dilute to 100 ml.

Reducing solution, dissolve 0.15 g of 1-amino-2-naphthol-4-sulphonic acid, 0.7 g of anhydrous sodium sulphite and 9 g of sodium metabisulphite in 100 ml of water. Prepare freshly each month.

Standard silica stock solution, fuse 0.20 g of pure crushed and dried crystal quartz with 1 g of anhydrous sodium carbonate in a platinum crucible. Dissolve in water and dilute to 1 litre. This solution contains 200 µg silica per ml. Dilute with water to give a solution containing 20 µg per ml as required.

Procedure. Full details for the fusion of silicate rock material, dissolution of the melt, dehydration of the silica and determination by volatilisation with hydrofluoric acid are given earlier. This part of the procedure is given below in outline only.

Fuse 1 g of the rock material with 5 g of anhydrous sodium carbonate in a platinum crucible and digest the melt with water. Acidify with concentrated hydrochloric acid, adding approximately 10 ml in excess and evaporate to dryness in a large platinum dish. When completely dry add a further 10 ml of concentrated hydrochloric acid and sufficient water to dissolve all soluble material. Collect the silica residue on a small filter, wash well with hot water and determine the major silica fraction in the usual way. Collect the filtrate and washings from the silica residue in a 200-ml volumetric flask and dilute to volume with water.

Using a pipette, transfer 5 ml of this solution to a 100-ml volumetric flask and add 10 ml of water. Now add 1 ml of ammonium molybdate solution and set aside for 10 minutes to complete the formation of the silicomolybdate complex. After exactly 10 minutes, add 5 ml of the oxalic acid solution, gently swirl the flask to mix the contents, and then add 2 ml of the reducing solution. The addition of the reducing solution should not be delayed. Dilute the solution to volume with water and set aside for at least 30 minutes, but preferably for 1 hour, and then measure the optical density of the solution in 2-cm cells with the spectrophotometer set at a wavelength of 650 nm.

Calibration. Transfer aliquots of the standard silicate solution containing 0-200 µg of silica to separate 100-ml volumetric flasks, dilute to 15 ml with water and add 1 ml of 3 N hydrochloric acid to each. Now add 1 ml of ammonium molybdate solution to each, allow to stand for 10 minutes and then reduce the silicomolybdate to molybdenum blue as described above. Dilute each solution to volume with water and measure the optical density in 2-cm cells with the spectrophotometer set at a wavlength of 650 nm. Plot the relation of optical density to silica concentration. As the calibration graph is a straight line passing through the origin, a single point method can be used for subsequent calibrations.

Determination of Silica by Atomic Absorption Spectroscopy

Although attempts were made, soon after the general introduction of atomic absorption spectrophotometers into rock analysis laboratories, to determine silica by this technique, the results obtained tended to be poor in both accuracy and precision. Some improvement was obtained with the introduction of the higher temperature nitrous oxide burner, and the relative freedom from interference from other elements at the concentrations found in silicate rocks was noted. However, both consistently high[15] and consistently low results[16] were reported - both possibly arising from a failure to appreciate the chemistry of silicon in the rock solutions used.

Satisfactory silica results were reported by Van Loon and Parissis[15] using a decomposition technique similar to that recommended by Ingamells.[17]

Just as Jeffery and Wilson[11] obtained the advantages of gravimetric and spectrophotometric methods by a combination of both, so also may a gravimetric determination of the major part of the silica be combined with a determination of the remaining minor part by atomic absorption spectroscopy.[18] Flameless atomic absorption spectroscopy using a graphite furnace has been shown[19] to give an increased sensitivity to silica.

Procedure. Accurately weigh 0.2 g of the finely powdered silicate rock material into a small platinum crucible and mix with 1 g of pure lithium metaborate. Fuse at a temperature of 1000° for 30 minutes, swirl the contents and quench by placing the crucible upright in a 100-ml beaker containing 25 ml of diluted (1+24) nitric acid. Keeping the crucible in an upright position, add 50 ml of the diluted nitric acid and stir without heating until dissolution is complete. Transfer the solution to a 250-ml volumetric flask and dilute to volume with the diluted nitric acid.

Measure the absorption, spraying the solution into an atomic absorption spectrophotometer using a nitrous oxide-acetylene flame, at a wavelength of 251.6 nm. Measure also the absorption of standard silica solutions similarly prepared from pure silica to give the calibration.

In the analysis of rocks containing silica as the major component, the use of atomic absorption spectroscopy (and to a lesser extent spectrophotometry) for the direct determination of total silica (as distinct from the minor fraction) cannot yet be regarded as ideal. Slight changes in the solutions used and in instrumental conditions may have a disproportionate effect on the results obtained. The interelement effects and other interferences cannot yet be said to be fully understood, and the choice of standard may also introduce an appreciable error.

Clearly a great deal more work is needed before full confidence can exist in such methods. The establishment of the accuracy and precision achieved in day to day practice is seen as the first stage of this requirement.

References

1. RICHARDSON W A and SNEESBY G., Min. Mag. (1922) 19, 303
2. AHRENS L H., Geochim. Cosmochim. Acta (1964) 28, 271
3. BRABSON J A, MATTRAW H C, MAXWELL G E, DARROW A and NEEDHAM H F, Analyt. Chem. (1948) 20, 504
4. MILLER C C and CHALMERS R A, Analyst (1953) 78, 24
5. BENNETT H, EARDLEY R P and HAWLEY W G, Trans. Br. Ceram. Soc. (1956) 57, 1
6. SAJO I. Acta Chim. Acad. Sci. Hung. (1955) 6, 243
7. KHALIZOVA V A, ALEKSEEVA A Ya and SMIRNOVA E P, Zavod. Lab. (1964) 30, 530
8. SINHA B C, SAHA M R and ROY S K, Talanta (1979) 26, 827
9. TSUBAKI I, Bunseki Kagaku (1967) 16, 610
10. BENNETT H and REED R A, Analyst (1967) 92, 466
11. JEFFERY P G and WILSON A D, Analyst (1960) 85, 478
12. ABBEY S, LEE N J and BOUVIER J L, Geol. Surv. Canada, Paper 74-19 (1974)
13. SHAPIRO L and BRANNOCK W W, U S Geol. Surv. Circ. 165, 1952, Bull. 1036-C, 1956 and Bull. 1144-A, 1962
14. HARMER W C E, Mineralog. Mag. (1973) 39, 112
15. VAN LOON J C and PARISSIS C M, Anal. Lett. (1968) 1, 519
16. KATZ A, Amer. Mineral. (1968) 53, 283
17. INGAMELLS C O, Analyt. Chem. (1966) 38, 1228
18. WALSH J N, Analyst (1977) 102, 51
19. LO D B and CHRISTIAN G D, Canad. J. Spectrosc. (1977) 22, 45

CHAPTER 38

Silver, Gold and the Platinum Metals

Determination of Silver in Silicate Rocks

At the concentration levels encountered in silicate rocks, silver cannot be determined directly by normal gravimetric or titrimetric methods. Gravimetric methods have however long been employed in conjunction with a prior concentration step, such as a fire assay procedure which enables large sample weights to be taken for the analysis. Emission spectrographic and spectrophotometric methods have also been combined with fire assay procedures or an extraction step to increase the sensitivity.[1] Such methods, although widely used, have now largely been superseded by methods based upon atomic absorption spectroscopy which are relatively straightforward and highly sensitive.

An air-acetylene flame is commonly used and the absorption measured at 328.1 nm. The interferences are few, the most serious being from aluminium. For this reason, and also to give increased sensitivity, most analysts have reported a prior concentration step which can be a fire assay pot fusion,[2,3] or an extraction with organic reagent. Reagents suggested include triisooctyl thiophosphate,[4] triphenylphosphine,[5] diphenylthiourea,[6] dithizone[7] and diethyldithiocarbamate.[8]

Procedures have also been described for the atomic absorptiometric determination following direct volatilisation from silicate rocks.[9,10]

The method given below is based upon that described by Chao et al.[4] In order to ensure recovery of any silver that is present in silicate structures, the acid digestion stage has been replaced by an evaporation with hydrofluoric acid.

Method

Reagents: Triisooctyl thiophosphate solution, transfer 70 ml of methyl isobutyl ketone to a measuring flask and dilute to 100 ml with triisooctyl thiophosphate.

Standard silver stock solution, dissolve 0.157 g of pure dry silver nitrate in a little water containing a few drops of nitric acid and dilute to 1 litre with 0.1 M nitric acid. This solution contains 100 µg silver per ml. Store in a cupboard or dark bottle.

Standard silver working solution, prepare as required by dilution of suitable aliquots of the stock solution with 6 M nitric acid.

Procedure. Accurately weigh approximately 1 g of the finely powdered sample material into a platinum dish, moisten with water and add 5 ml of 20 N sulphuric acid, 2 ml of concentrated nitric acid and 10 ml of concentrated hydrofluoric acid. Evaporate to fumes of sulphuric acid. Allow to cool, wash down and dilute with a little water, add 2 ml of concentrated nitric acid and again evaporate, this time to dryness.

Moisten the residue with concentrated nitric acid, and using a total of 4 ml of acid, transfer to an extraction vessel. This may be a tube, conical or other flask with a capacity of about 25 ml, fitted with a stopper or other close-fitting closure. Rinse the dish with 6 ml of water, adding the washings to the flask. Add by pipette 3 ml of the triisooctyl thiophosphate solution, close the tube or flask and shake for one minute to extract the silver. Allow the phases to separate.

Aspirate the organic extract into the air-acetylene flame of an atomic absorption spectrometer fitted with a silver hollow cathode lamp, recording the absorbance at a wavelength of 328.1 nm.

Record also the absorption of extracts similarly prepared from aliquots of the standard working solution containing 0.05 to 20 µg Ag. Each aliquot should be diluted to 10 ml with 6 M nitric acid prior to extraction with triisooctyl thiophosphate as described above.

Note: 1. As described, the method can be used for rock samples containing 50 ppb to 20 ppm Ag. A somewhat larger sample weight can be taken for rocks of lower silver content. For rocks of higher silver content, the nitric acid solution of the rock material can be diluted to volume with 6 M nitric acid, and a 10 ml aliquot taken for the extraction.

Determination of Gold in Silicate Rocks

The economic importance of gold, and its winning from silicate ores in which it is present in only small amounts, has provided the underlying need for accurate methods of analysis at the ppm and ppb level. Traditionally fire assay procedures[11] have been used, often completed gravimetrically ie by direct weighing of the metallic gold after collection in a lead button, scorification and cupellation to remove lead, and parting to remove silver. Some improvement in sensitivity can be obtained by completing the determination by emission spectrochemical or spectrophotometric methods, reviewed by Chow <u>et al.</u>[12] Fire assay techniques have however been shown by Mihalka and Resmann[13] to be subject to considerable error, some at least will remain in procedures using fire assay pot fusion as a collecting and concentrating mechanism for determination by atomic absorption.[2,3,13]

Fluorimetric methods for gold include those based upon rhodamine B[14] and butyl rhodamine S.[15] These methods have not been widely adopted, possibly because of the relative ease with which gold can be determined by atomic absorption spectroscopy. The most sensitive line is at 242.8 nm, and few interferences have been reported when using an air-acetylene flame, particularly following extraction into organic solution. There is an extensive literature on this application, including the use of a graphite furnace.[16,17,18,19] Concentration procedures described include extraction with methyisobutyl ketone,[20,21] dibutyl sulphide[22] and isoamyl alcohol,[23] as well as precipitation using selenium[24] and tellurium[17] as collectors.

The method described below is based upon that of Rubeska et al[22] as modified by Parkes and Murray-Smith,[25] which uses an acid attack of the finely ground sample material. This procedure is commonly used for the recovery of gold and other precious metals from silicate material and appears to be largely justified by the observed nature of the distribution of gold in silicate rocks and minerals, summarised by Allmann and Crockett,[26] provided that sulphides and certain other accessory minerals are first decomposed. It must be remembered however that very small quantities of gold will remain in the discarded silicate fraction. A modified procedure is described, also by Rubeska et al[22] involving evaporation with hydrofluoric acid for the decomposition of silicate materials.

Method

Reagents:

Sodium chloride.

Dibutyl sulphide solution, dissolve 14.6 g of the reagent in 500 ml of diisobutyl ketone. Store the reagent in a closely stoppered bottle in the dark to retard oxidation.

Standard gold stock solution, dissolve 0.1 g of gold in a mixture of 8 ml of concentrated hydrochloric acid and 2 ml of concentrated nitric acid. Evaporate on a water bath almost to dryness, dissolve in 2 M hydrochloric acid, transfer to a 100-ml volumetric flask and dilute to volume with 2 M acid. This solution contains 1 mg gold per ml.

Standard gold working solution, prepare as required by diluting an aliquot of the stock solution with 2 M hydrochloric acid to give a working standard containing 1 μg gold per ml.

<u>Procedure.</u> Accurately weigh approximately 25 g of the finely powdered sample material into a silica or porcelain dish and mix intimately with 10 g of ammonium nitrate (Note 1). Transfer the dish to a cold muffle furnace and raise the temperature over a period of an hour or so to 650°, maintaining this for 30 minutes. Remove the dish, allow to cool and break up the sinter with a flattened glass rod. Transfer the material to a 600-ml beaker.

Add 3 g of sodium chloride and 60 ml of concentrated hydrochloric acid. Allow to stand for 20 minutes and then add 20 ml of concentrated nitric acid. Boil the material, with stirring if necessary to avoid bumping, for 30 minutes, and evaporate almost to dryness on a steam bath (Note 2).

Dissolve soluble salts by heating with 50 ml of 2 M hydrochloric acid, allow the residue to settle and pour the supernatant liquid into a 100-ml centrifuge tube. Wash the residue with further quantities of hot 2 M hydrochloric acid, adding the washings to the main solution and centrifuge to remove undissolved material (Note 3).

Transfer the clear rock solution to a 250-ml separating funnel, add by pipette 5 ml of the dibutyl sulphide solution and shake for 5 minutes. Allow the phases to separate. Aspirate the organic phase into an air-acetylene flame of the atomic absorption spectrometer fitted with a gold hollow cathode lamp and measure the absorption at a wavelength of 242.8 nm (Note 4).

Measure also the absorption of a reagent blank solution prepared in the usual way, and of standards prepared by extraction from 2 M hydrochloric acid of aliquots of the standard gold working solution containing 1-5 μg gold.

<u>Procedure involving hydrofluoric acid decomposition.</u> Proceed as described in the first paragraph of the method given above, but transfer the broken sinter to a PTFE dish. Moisten with water and add 30 ml of concentrated hydrofluoric acid followed by 20 ml of concentrated perchloric acid. Evaporate first to fumes of perchloric acid and then to dryness. Allow to cool, moisten with water and add 10 ml of concentrated hydrofluoric acid and 5 ml of perchloric acid and evaporate to dryness

once more. Dissolve the residue remaining by heating with 20 ml of concentrated hydrochloric acid and 8 ml of concentrated nitric acid. Evaporate almost but not quite to dryness, dissolve the residue in 2 M hydrochloric acid and complete the determination as described above.

Notes: 1. Ideally finely powdered ammonium nitrate should be used, but in view of the explosion hazard, coarsely crystalline material should on no account be ground in a mortar - use the finest powdered material that is available. The controlled decomposition of ammonium nitrate provides nitric oxide which promotes oxidation and keeps the sample material in a porous state.

2. Parkes and Murray-Smith[25] report reduction to a lower valency state and hence loss on extraction, when solutions are evaporated to dryness.

3. Filter papers have been reported[21] to reduce gold in solution, for this reason centrifugation is preferred to filtration.

4. In addition to gold, palladium can also be determined in the dibutyl sulphide extract. The atomic absorption spectrometer should be fitted with a palladium lamp and the absorption measured at a wavelength of 247.6 nm.

Determination of the Platinum Metals in Silicate Rocks

As with silver and gold, the traditional methods of determining the six metals of this group are based upon fire assay pot fusion.[11] The final gravimetric step has, in some laboratories, been replaced by emission spectrography or atomic absorption spectroscopy.[26,2,27,28]

Acid soluble palladium in the parts per billion range was determined by Grimaldi and Schnepfe[29] using a 10-g sample, extraction with aqua regia, co-precipitation with added platinum and tellurium and spectrophotometric determination with para-nitrosodimethylaniline. Other colour-forming reagents include mepazine hydrochloride for osmium,[30] 1-phenyl-2-tetrazoline-5-thiolone for platinum and palladium[31] and rhodamine 6G for platinum.[32] Procedures for the determination of acid-soluble ruthenium and osmium[33,34] based upon the volatilisation of RuO_4 and OsO_4 from oxidising perchloric acid used spectrophotometric measurement with sulphanilic acid and N,N-dimethylaniline and thiocyanate.

An atomic absorption procedure for platinum and palladium (with gold) is described by Fishkova,[23] based upon acid extraction and for palladium (also with gold) by Rubeska <u>et al</u> using either acid extraction or evaporation with hydrofluoric acid. This latter method is given in detail in the preceding section. A method based on a hydrofluoric acid attack and solution in aqua regia has been described by Simonsen[35] for the determination of platinum in basic rocks. The sensitivity by atomic absorption spectroscopy is improved by using a dithizone extraction, but remains insufficient for most silicate rocks.

The improved sensitivity obtained in atomic absorption by using a graphite furnace is used by Fryer and Kerrich[18] to determine platinum and palladium (with silver and gold) in large samples, but once again the method gives only acid-soluble metals.

Although procedures are available for the determination of the metals of this group in ores, concentrates and metallurgical products, their application to silicate and other rocks with little or no metal enrichment cannot yet be regarded as satisfactory.

References

1. MARKOVA N V, SUMAKOVA N S, YAKUBTSEVA T V and POLTORYKHINA A K., Tr, Tsent. Mauch.-Issled. Gornorazved. Inst. (1969) 82, 244
2. HOCHN R and JACKWERTH E., Erzmetall (1976) 29, 279
3. MOLOUGHNEY P E., Talanta (1977) 24, 135
4. CHAO T T, BALL J W and NAKAGAWA H M., Anal. Chim. Acta (1971) 54, 77
5. FISHKOVA N L and PETRUKHIN O M., Zh. Anal. Khim. (1973) 28, 645
6. VALL G A, USOL'TSEVA M V, YEDELEVICH I G, SERYAKOVA I V and ZOLOTOV Yu A., Zh. Anal. Khim. (1976) 31, 27
7. CHOWDHURY A N, DAS A K and DAS T N., Z. Anal. Chem. (1974) 269, 284
8. TERASHIMA S., Japan Analyst (1976) 25, 279
9. LANGMYHR F J, STUBERGH J R, THOMASSON Y, HANSSEN J E and DOLEZAL J., Anal. Chim. Acta (1974) 71, 35
10. PCHELINTSEVA N F., Zavod. Lab (1977) 43, 693
11. SMITH E A., The Sampling and Assay of the Precious Metals, Griffin, London, 2nd ed. 1947.
12. CHOW A, LEWIS C L, MODDLE D A and BEAMISH F E., Talanta (1965) 12, 277
13. MIHALKA S and RESMANN A., Rivta Chim.(1975) 26, 875
14. FISHKOVA N L, ADOROVA E P and POPOVA N N., Zh. Anal. Khim. (1975) 30, 806
15. MARINENKO J and MAY I., Analyt. Chem. (1968) 40, 1137
16. BLYUM I A, PAVLOVA N N and KALUPINA F P., Zh. Anal. Khim. (1971) 26, 55
17. MARTIN L., An. Quim. (1976) 72, 217
18. FRYER B J and KERRICH R., At. Absorpt. Newsl. (1978) 17, 4
19. TORGOV V G and KHLCHNIKOVA A A., Zh. Anal. Khim. (1977) 32, 960
20. SIGHINOLFI G P and SANTOS A M., Mikrochim. Acta (1976) II, 33

21. HILDON M A and SULLY G R., Anal. Chim. Acta (1971) 54, 245

22. RUBESKA I, KORECKOVA J and WEISS D., At. Absorpt. Newsl. (1977) 16, 1

23. FISHKOVA N L., Zh. Anal. Khim. (1977) 32, 1776

24. BAZHOV A S and SOKOLOVA E A., Zh. Anal. Khim. (1972) 27, 2442

25. PARKES A and MURRAY-SMITH R., At. Absorpt. Newsl. (1979) 18, 57

26. ALLMANN R and CROCKET J H in Handbook of Geochemistry (Editor K H Wedepohl) II-5, Section 79, 1978

27. MOLOUGHNEY P E and FAYE G H., Talanta (1976) 23, 377

28. COOMBES R J, CHOW A and WAGEMAN R., Talanta (1977) 24, 421

29. GRIMALDI F S and SCHNEPFE M M., U S Geol. Surv. Prof. Paper, 575-C, p.C141, 1967

30. GOWDA H S and KESHAVAN B., Curr. Sci. (1977) 46, 443

31. RADUSHEV A V and STATINA L A., Zh. Anal. Khim. (1972) 27, 1344

32. KOTHNY E L., J. Geochem. Expl. (1974) 291

33. SIL'NICHENKO V G and DOLININA Yu V., Methody Khim. Anal. Gorn. Porod Miner. (1973) 30

34. KHVOSTOVA V P and SHLENSKAYA V I., Anal. Tekhnol. Blagorod. Metal (1971) 149

35. SIMONSEN A., Anal. Chim. Acta (1970) 49, 368

CHAPTER 39

Strontium

The earliest most popular chemical method appears to have been that based upon the precipitation together with calcium as oxalate in the filtrate from the removal of iron, aluminium, titanium and other elements by precipitation with ammonia. The mixed oxalates were converted to the anhydrous nitrates, and the strontium nitrate separated from calcium nitrate by dissolution of the latter in concentrated nitric acid. Groves[1] has suggested that the error involved in making the separation is not as great as that arising from the failure to collect the strontium from the rock material in the oxalate precipitate.

An alternative gravimetric procedure is based upon the precipitation of barium and strontium together with calcium in the form of their sulphates, from the rock solution to which dilute sulphuric acid and ethanol have been added. The sulphates are converted to carbonates by alkali carbonate fusion, and the calcium removed by dissolution of the nitrate in concentrated nitric acid as described above. Barium is then precipitated as chromate and the strontium determined in the filtrate by precipitation as sulphate. This procedure is long and tedious. It has been reported by Groves[1] but appears to have been little used.

Spectrophotometric methods have not found wide application to the determination of strontium, possibly because of interference from other alkaline earth elements. Reagents that have been used include o-cresol-phthalein complexone,[2] murexide[3] and chlorophosphonazo III.[4] For the determination of strontium in silicate and other rocks, flame photometric and atomic absorption techniques now appear to have displaced both photometric and gravimetric procedures.

In common with calcium and barium, strontium has a very characteristic flame emission. The strongest emission of strontium is at the resonance line of 460.7 nm. Other strontium lines are at 407.8 nm and 421.6 nm, and bands occur in the orange and red parts of the spectrum. As with other alkaline earth elements, the flame emission depends not only upon the flame conditions, but also upon the nature of the solution being sprayed and the concentration of other elements in it. The calibration curve for strontium, based upon the resonance line, approximates to a straight line. Where accurate results are required the procedure known as the "method of additions" should be used.

Sodium and potassium in high concentrations tend to enhance the strontium emission at 460.7 nm, as does appreciable amounts of calcium. Magnesium, iron, silica, phosphate, sulphate and more particularly aluminium depress the strontium emission.

To overcome most of these interferences Fabrokova and Isaeva[5] have suggested making a prior separation of the calcium, strontium and magnesium from iron, aluminium and phosphorus by a double precipitation of the latter with aqueous ammonia. The combined filtrates from the two precipitations are evaporated to small volume and the emission at 460.7 nm measured. The method is described as giving only approximate results, with a deviation from the mean of up to ±8.5 per cent in parallel determinations.

An air-acetylene flame has been used by a number of workers[6,7,8,9] for the atomic absorptiometric determination of strontium, although the matrix effect may be considerably lessened by using nitrous oxide-acetylene.[10] The addition of calcium or more particularly of lanthanum[11] has been suggested as a means of eliminating the depressant effect on the strontium absorption of aluminium, silicon and phosphate. An alkali metal ionisation suppressor, eg sodium iodide, is also recommended. The highest sensitivity is given by absorption at 460.7 nm; the line at 407.73 nm has also been used for strontium measurement.

A flameless technique for the determination of strontium involving volatilisation from a sulphuric acid medium in a graphite furnace, has been described by Katz[15] and from hydrochloric acid solution by Barredo et al.[16]

Determination by Flame Photometry

In the following procedure, all separations other than that of silica, are avoided and the determination is made by adding a standard solution of strontium to separate aliquots of the rock solution. The depressing effect of aluminium and other elements is masked by adding a considerable excess of calcium to the rock solution as described by Fornaseri and Grandi[13] and more recently by Kirillov and Alkhimenkova[14]. Silica is removed by evaporation with hydrofluoric and perchloric acids.

Method

Reagents: Calcium perchlorate solution, dissolve 25 g of pure calcium carbonate in dilute perchloric acid, avoiding an excess, and dilute to 500 ml with water. This solution contains 20 mg calcium per ml.

Standard strontium stock solution, dissolve 0.168 g of pure strontium carbonate in dilute perchloric acid, also avoiding excess acid, and dilute to 500 ml with water. This solution contains 200 μg Sr per ml.

Standard strontium working solution, prepare the working standard solution by dilution with water as required.

Procedure. Accurately weigh approximately 2 g of the finely powdered silicate rock material into a small platinum basin, moisten with water, add 20 ml of concentrated perchloric acid and 20 ml of concentrated hydrofluoric acid and evaporate to dryness in the usual way. Cool, add 3 ml of perchloric acid and again evaporate to dryness. Cool, add 1 ml of perchloric acid and evaporate again, this time to give a residue of the moist perchlorates. Allow to cool, dissolve the residue in about 10 ml of water, add 10 ml of the calcium perchlorate solution and dilute to volume in a 100-ml volumetric flask (Note 1). Pipette 25 ml of this solution into a clean 50-ml volumetric flask and dilute to volume with water. Measure the flame emission of this solution at a wavelength of 460.7 nm using a flame photometer set according to the maker's instructions. If a recording attachment is available, trace the emission from about 480 nm to 450 nm. From the emission obtained and a strontium calibration graph calculate the approximate strontium content of the dilute solution.

Prepare two new solutions of the rock sample by transferring two 25-ml portions of the rock solution into separate 50-ml volumetric flasks, adding aliquots of the standard strontium solution, and diluting to volume with water. The strontium addition should be chosen to give new concentrations of strontium approximately two and three times that of the original dilute rock solution.

Measure the flame emission of all three dilute solutions of the rock together with that of a reagent blank solution (Note 2), and hence determine the strontium content of the sample material.

Notes: 1. With normal silicate rocks and minerals, the precipitation of potassium perchlorate is not likely to occur at this dilution. For rocks and minerals rich in potassium where this does occur, allow the crystalline precipitate to settle before removing the 25-ml aliquot.

2. Even so-called "pure" grades of calcium carbonate may contain small amounts of strontium, and a blank determination is therefore essential.

Determination of Strontium in Silicate Rocks by Atomic Absorption Spectrometry

Most of the standard methods for the decomposition of silicate rocks can and have been used in the preparation of the rock solution for strontium determination. These include fusion with alkali carbonate and lithium borate as well as evaporation with acid mixtures containing hydrofluoric acid. Suitable procedures have been given by Abbey[10] and by Abbey et al.[12] Interference from aluminium and silicon can be avoided by using fluoride solutions in which they form fluoranions. Precipitation of potassium fluoborate (derived from a borate flux) can be avoided by choosing sodium as the ionisation buffer. The procedure given below is based on Abbey et al.[12]

Method

Reagents: Sodium chloride solution, dissolve 10 g of sodium chloride in water and dilute to 100 ml.

Boric acid solution, dissolve 50 g boric acid crystals in 1 litre of warm water and allow to cool.

Procedure. Ignite a covered graphite crucible in an electric furnace at a temperature of 1000° for 15 to 20 minutes. Allow to cool, and without disturbing graphite dust within the crucible, add 1 g of lithium metaborate and an accurately weighed approximately 0.2 g of the sample material. Cover the crucible, replace in the electric furnace and heat at 950 to 1000° for 15 minutes. Pour the molten contents into a plastic (eg trimethylpentene) jar containing 40 ml of water. After the crucible has cooled examine it for any retained material from the fusion. If any is found, transfer it to the plastic jar. Add a PTFE-coated stirring bar followed by 25 ml of 5 N hydrofluoric acid, cap tightly and stir until the melt has fully disintegrated. Chill the jar in a refrigerator, then open, add 100 ml of boric acid solution, re-cap and stir until a clear solution is obtained. Collect graphite particles on an open-textured filter paper in a plastic funnel, and the filtrate in a 200-ml volumetric flask, rinse the filter several times with water and dilute the combined filtrate and washings to volume, mix well and transfer to a polyethylene bottle (Note 1).

Transfer by pipette 0.5 ml of the sodium chloride solution to a small polyethylene or similar bottle or other vessel of about 15-ml capacity, add by pipette 10 ml of the prepared rock solution, mix and aspirate into the nitrous oxide-acetylene flame of an atomic absorption spectrometer in the usual way. A strontium hollow cathode lamp should be used, and the absorption measured at a wavelength of 460.7 nm (Note 2).

The determination can be completed by measuring also the absorption at 460.7 nm of standard solutions of strontium similarly prepared with sodium as an ionisation buffer. However, such a procedure can only give approximate results as the matrix effects are quite pronounced. International reference rock material may be used, and the standard solutions for the strontium calibration prepared in the same way as the sample solution. Alternatively, aliquots of a standard strontium solution may be added to a rock matrix solution (resembling that obtained from the sample) from a single rock that is either standardised or for which the strontium content is determined.

Notes: 1. This solution can be used also for the determination of the major elements and certain minor elements in the rock matrix using atomic absorption spectroscopy (for aluminium, manganese, chromium, nickel, iron, magnesium, calcium, sodium, potassium, titanium and barium) and spectrophotometry (silicon and phosphorus).

2. In the procedure as described by Abbey et al,[12] the solution used for the determination of strontium, to which sodium has been added as ionisation buffer, is used also for the determination of barium by measurement at 553.6 nm with a barium hollow cathode lamp and a nitrous oxide-acetylene flame.

References

1. GROVES A W., Silicate Analysis, Allen & Unwin, London, 2nd ed., 1951
2. POLLARD F H and MARTIN J V., Analyst (1956) 81, 348
3. RUSSELL D S, CAMPBELL J B and BERMAN S S., Anal. Chim. Acta (1961) 25, 81
4. LUKIN A M, ZELICHENOK S L and CHERNYSHEVA T V., Zhur. Anal. Khim. (1964) 19, 1513
5. FABROKOVA E A and ISAEVA A G., Trudy Inst. Mineral Geokhim, i Kristallokhim. Redk. Elementov. Akad. Nauk SSSR (1963) (18), 175
6. DAVID D J., Analyst (1962) 87, 585
7. BELCHER C B and BROOKS K A., Anal. Chim. Acta (1963) 29, 202
8. DINNIN J I., U S Geol. Surv. Prof. Paper 424-D, 1961
9. BECCALUVA L and VENTURELLI G., Atom. Absorpt. Newsl. (1971) 10, 50
10. ABBEY S., Geol. Surv. Canada Paper 71-50, 1972
11. CARTER D, REGAN J G T and WARREN J., Analyst (1975) 100, 1195
12. ABBEY S, LEE N J and BOUVIER J L., Geol. Surv. Pap. Canad. (1974) No 74-19
13. FORNASERI M and GRANDI L., Geochim. Cosmochim. Acta (1960) 19, 218
14. KIRILLOV A I and ALKHIMENKOVA G I., Zavod. Lab. (1965) 31, 57
15. KATZ A., Chem. Geol. (1975) 16, 15
16. BARREDO F Bea, VASALLO P Arias and DIEZ L Polo., Chem. Geol. (1978) 23, 171

CHAPTER 40

Sulphur

In the great majority of silicate rocks the sulphur content is very small, amounting to no more than a few hundred parts per million. This is in contrast to the meteoric abundance where, in the order of frequency, sulphur is the fifth element by weight, coming before aluminium.[1]

Determination of Free Sulphur

Methods for the determination of free sulphur are based upon extraction from the ground rock material with an organic solvent. In the procedure described by Volkov,[2] a soxhlett-type apparatus is used and the sample refluxed for 16 hours with acetone. The elemental sulphur, plus any soluble organic compounds containing sulphur, are recovered by evaporation of the acetone and are converted to sulphuric acid by oxidation with a solution of bromine in carbon tetrachloride. The sulphate is then reduced with hydriodic acid to hydrogen sulphide, which is distilled from the solution and collected in a solution of cadmium acetate. An excess of iodine is added and the amount of the excess determined by titration with standard sodium thiosulphate solution.

Extraction times of only 10 minutes were suggested by Ozawa[3] using carbon disulphide or benzene as solvent. The organic solutions of sulphur obey the Beer-Lambert Law, and the optical densities can be measured directly at 390 or 395 nm (CS_2 solution) and 330 or 360 nm (benzene solution).

Determination of Sulphide-Sulphur

For most purposes the determination of nitric acid-soluble sulphur can be regarded as a measure of the sulphide content of silicate rocks. This determination will, however, include sulphur from those sulphate minerals that are soluble in nitric acid, and traces of sulphur from those that are not - barite, for example. An alternative approach to the determination of sulphide-sulphur that does not include sulphate-sulphur is to reduce the sulphides to hydrogen sulphide with hydriodic acid.[4] The gases evolved are bubbled through a suspension of cadmium hydroxide, and the cadmium sulphide determined with iodine as described below. Most sulphides are decomposed, but the presence of mercury is necessary[5] to complete the decomposition of pyrite and chalcopyrite.

Method

Reagents: Hydriodic acid, mix equal volumes of concentrated hydrochloric acid and a 50 % (w/v) solution of potassium iodide. Add a few crystals of sodium hypophosphite to reduce any liberated iodine (but avoid an excess) and decant from the precipitated potassium chloride.

Cadmium acetate solution, dissolve 2 g of cadmium acetate in water and dilute to 100 ml.

Iodine, standard 0.1 N solution.

Acetic acid, N.

Sodium thiosulphate standard 0.1 N solution.

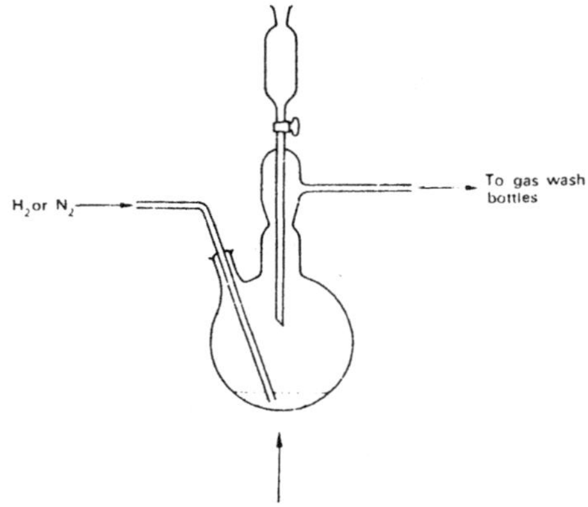

FIG 25 Apparatus for the reduction, distillation and collection of sulphur in sulphide minerals.

Procedure. Accurately weigh 2 g or more of the finely powdered rock material into the reaction flask and add a few mg of mercury. Transfer 25 ml of cadmium acetate solution to two gas wash bottles and add 5 ml of N sodium hydroxide solution to each. Assemble the apparatus as shown in Fig. 25 and displace the air from the flask with nitrogen or hydrogen. Add 10 ml of hydriodic acid to the flask and warm gently to assist the decomposition of the sulphide minerals. Continue warming and passing gas through the apparatus for about 1 hour.

Allow to cool, remove the two gas wash bottles and combine their contents. Add an excess of standard iodine solution and sufficient acetic acid to give a final acid concentration of about 0.5 N and titrate the excess iodine with standard thiosulphate solution.

Determination of Water-soluble Sulphur

The extraction of more than a trace of sulphate ions from silicate rocks into aqueous solution indicates the presence of evaporite minerals such as kieserite $MgSO_4.H_2O$, kainite $KCl.MgSO_4.3H_2O$, etc. The sulphate ion is easily determined gravimetrically by precipitation with barium chloride as described below.

Determination of Acid-soluble Sulphur++

As noted above, the decomposition of metallic sulphides can be effected with nitric acid. Certain sulphates are also dissolved, although barite remains largely unattacked. Sulphate-containing silicates such as lazulite appear to be completely decomposed. Complete dissolution of silicates can usually be obtained only by adding hydrofluoric acid. This procedure has been used by Wilson *et al.*(6) for the determination of total sulphur in silicate rocks. By including perchloric acid, organic matter and resistant sulphides are completely dissolved. Vanadium pentoxide is added to expedite the oxidation, for which a PTFE dish can be used.

The presence of barium gives rise to precipitation of the insoluble barium sulphate, and the recovery of a high sulphur content can be significantly diminished by the occurrence of as little as 0.2 per cent barium oxide in the rock sample. It is unlikely that sulphur will be lost by evaporation on heating the perchloric acid solution in PTFE, particularly in the presence of an excess of calcium salts.

Method

Reagents:

Vanadium pentoxide.

Calcium chloride.

Cupferron solution, dissolve 0.5 g of the reagent in 50 ml of chloroform. Prepare as required and discard if not used.

Barium chloride solution, dissolve 3 g of the dihydrate in 100 ml of water.

++ Acid-soluble sulphur in this context includes that sulphur which is directly soluble in nitric acid together with sulphur present as sulphide which is oxidised to sulphate by the nitric acid.

Procedure. Accurately weigh 1 g of the finely powdered rock material into a PTFE dish, moisten with water and add in succession, 15 ml of concentrated nitric acid, 2 ml of concentrated hydrochloric acid and 10 ml of concentrated hydrofluoric acid. Cover the dish and set aside overnight. Remove the cover and evaporate to dryness on a steam bath. If the rock powder is completely decomposed, add a few ml of water and 10 ml of concentrated nitric acid and again evaporate to dryness. Repeat this last operation once more. If the decomposition is not complete after the first evaporation, add 5 ml of concentrated perchloric acid (60% w/v) and, if organic matter is present, 100 mg of vanadium pentoxide. In addition if the material is low in calcium, add not less than 100 mg of calcium chloride. (There should be present at least two equivalents of calcium for each equivalent of sulphur). Evaporate to dryness, and, if any organic or black mineral particles remain, evaporate with further portions of perchloric acid until the decomposition is complete.

After the final evaporation, add 3 ml of concentrated hydrochloric acid and about 50 ml of water. Digest on the steam bath for 15 minutes with occasional stirring, then cool. Collect the residue on a small close-textured filter paper and wash well with warm water. Discard the residue. Transfer the filtrate to a 250-ml separating funnel. Dilute to a volume of about 100 ml, and extract iron and titanium with successive 50-ml portions of cupferron solution until the extract is no longer brown. Wash the solution twice with 50-ml portions of chloroform, and finally with 50 ml of light petroleum. Run off the aqueous layer into a 400-ml beaker and wash the light petroleum twice with 10-ml portions of water. Dilute the combined aqueous layer and washings to about 200 ml, and filter if necessary.

Heat the solution to boiling and add a slight excess of a hot barium chloride solution. Allow the solution to stand for an hour on a steam bath and then set the beaker aside overnight. Collect the precipitated barium sulphate on a small close-textured filter paper and wash with successive small quantities of cold water. Transfer the paper to a weighed platinum crucible, dry, ignite over the full flame of a Bunsen burner and weigh as $BaSO_4$.

After ignition, the barium sulphate residue should be perfectly white. Blank values are usually insignificant.

Determination of Total Sulphur

All sulphur in the sample material, whether present as free sulphur, organic sulphur, sulphide or sulphate-sulphur, is converted to alkali sulphate by fusion with sodium carbonate containing a little potassium nitrate. The melt is extracted with water and the sulphate in the aqueous filtrate recovered by adding acid and precipitating with barium chloride. High and somewhat variable blanks can sometimes be obtained, particularly if gas burners are used for the fusion - the use of an electric furnace is recommended. Barium does not interfere. This procedure is given in detail below.

Combustion techniques have been described for total sulphur, in which the sample material is heated with or without an oxidising flux such as vanadium pentoxide, in a stream of oxygen[7,8] or nitrogen.[9] The oxides of sulphur are collected and determined by a titrimetric, photometric or colorimetric method. High temperatures are required and the recoveries of sulphur are reported as not always quantitative.[10] A combustion method has also been described for the determination of total sulphur in limestones.[11]

In an alternative approach covering the range 5 to 2000 ppm of sulphur, sodium chlorate is used to oxidise all other forms of sulphur to sulphate, which is then reduced to hydrogen sulphide by refluxing with sodium iodide, red phosphorus, hypophosphorous acid, phosphoric and propionic acids. The liberated hydrogen sulphide is absorbed in aqueous potassium hydroxide and titrated with 2-(hydroxymercuri)benzoic acid. This procedure, given by Murphy and Sergeant,[12] is also described in detail below. Low results may be obtained in the presence of barite.

GRAVIMETRIC DETERMINATION

Method

Procedure. Accurately weigh approximately 2 g of the finely powdered rock material into a large platinum crucible and mix with 10 g of anhydrous sodium carbonate and 0.25 g of potassium nitrate (Note 1). Transfer the crucible to a small electric furnace, cover with a platinum lid, and slowly raise the temperature to about 1000°. Maintain at this temperature for about 30 minutes and then allow to cool. Extract the melt with hot water and rinse the solution and residue into a 250-ml beaker, add a few drops of ethanol to reduce any manganate formed in the fusion, and dilute to about 150 ml with water. Cover the beaker and digest upon a steam bath, using a glass rod to reduce the insoluble material to a fine state of subdivision.

Filter the solution through an open-textured filter paper and wash the residue with hot sodium carbonate solution. Discard the residue (Note 2). Collect the filtrate and washings in a 800-ml beaker and dilute to a volume of approximately 500 ml.

Add a few drops of methyl red indicator solution and concentrated hydrochloric acid until the neutral point is reached, followed by 10 ml in excess. Boil the solution to expel carbon dioxide and precipitate the sulphur from the boiling solution with a slight excess of barium chloride solution as described in the previous section. Collect, ignite and weigh the precipitate. If the rock contains an appreciable amount of sulphur, it may be necessary to purify the barium sulphate precipitate (see determination of barium).

Notes: 1. The amount of potassium nitrate should be increased up to 1 g for rocks containing much ferrous iron or organic matter.

2. This residue may be used for the determination of barium, zirconium, etc if these are required.

TITRIMETRIC DETERMINATION

Method

Apparatus. This is shown in Fig. 26 . The third neck of the flask is used to introduce the solid sample material. When the apparatus is in use, a weight or preferably retaining springs must be used to keep the stopper in position.

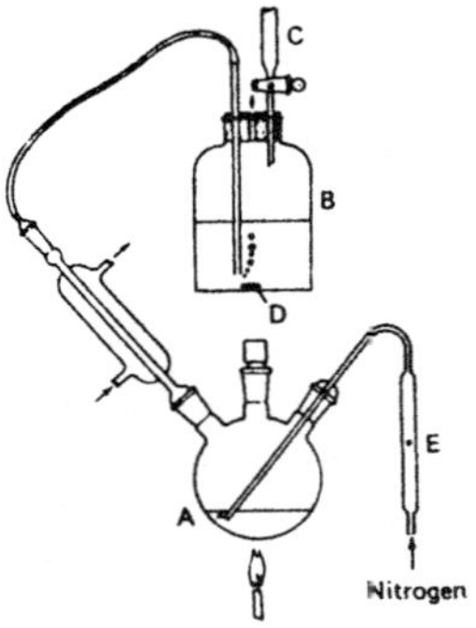

FIG. 26 Reduction and titration apparatus: A, 100-ml distillation flask; B, absorption vessel (150 ml); C, 50-ml burette; D, magnetic stirrer follower; and E, gas flow meter (5 to 190 ml min^{-1} air)

Reagents: Potassium hydroxide solution, 0.2 N.

Dithizone indicator, mix together 0.1 g of dithizone and 10 g of potassium nitrate.

2-(Hydroxymercuri)benzoic acid solution, dissolve 0.2 g of the solid reagent in 1 litre of 0.2 N potassium hydroxide solution. One ml is equivalent to approximately 10 μg sulphur.

Standard sulphate solution, dissolve 1.0872 g of dry potassium sulphate in water and dilute to 100 ml. This solution contains 2 mg S per ml. Prepare also a diluted solution containing 200 μg S per ml by diluting 5 ml of the stock solution to 50 ml with propionic acid.

Procedure. Accurately weigh 0.2 g of the finely powdered rock material into a 50-ml PTFE beaker, add 0.2 g of solid sodium chlorate and 5 ml of concentrated hydrochloric acid. Cover the beaker, allow to stand overnight and then heat on a steam bath. After 30 minutes rinse and remove the cover, wash down the sides of the beaker with a little water and evaporate to dryness under a heating lamp. Break up the residue with a spatula or glass rod and transfer to a desiccator until ready to proceed with the next stage.

Add 1 g of sodium iodide, 0.2 g of red phosphorus, 2 ml of hypophosphorous acid, 8 ml of orthophosphoric acid and 10 ml of propionic acid to the distillation flask, and pass nitrogen through the apparatus at a rate of about 60 ml per minute. Heat the flask over a small flame to reflux the contents for 30 minutes and, with the supply of nitrogen continuing, allow to cool for 15 minutes before proceeding. The apparatus is now ready for the determination. A total of about 12 determinations can be made before it is necessary to replace the reagents.

Add 50 ml of potassium hydroxide solution and about 10 mg of the dithizone indicator to the absorption vessel and add the 2-(hydroxymercuri)benzoic acid solution drop by drop from the burette until the indicator colour changes from yellow to pink.

With the aid of a small, wide-necked funnel and a small stiff brush transfer the sample residue from the PTFE beaker to the distillation flask, and stopper without delay. Replace the burner and reflux the contents of the flask for a period of 30 minutes. Allow the flask to cool for 10 minutes then titrate the contents of the absorption vessel with the 2-(hydroxymercuri)benzoic acid reagent to regain the pink colour. The reagent blank is determined by carrying out the full procedure without the sample.

Finally, calculate the sulphur content of the sample after standardising the 2-(hydroxymercuri)benzoic acid reagent by titration against 1 or 2 ml of the diluted standard sulphate solution added directly to the distillation flask and recovered in the absorption vessel as described. The titration of the standard sulphate solution should be corrected for the small reagent blank obtained from pure propionic acid.

References

1. GOLDSCHMIDT V M., Geochemistry, Oxford, 1954, p.524
2. VOLKOV I I., Zhur. Anal. Khim. (1959) 14, 592
3. OZAWA T., J. Chem. Soc. Japan. Pure Chem. Sect. (1966) 87, 587
4. MURTHY A R V, NARAYAN V A and RAO M R A., Analyst (1956) 81, 37
5. MURTHY A R V and SHARADA K., Analyst (1960) 85, 299
6. WILSON A D, SERGEANT G A and LIONNEL L J., Analyst (1963) 88, 138
7. SEN GUPTA J G., Analyt. Chem. (1963) 35, 1971
8. LANGE J and BRUMSACK H-J., Zeit. Anal. Chem. (1977) 286, 361
9. ARIKAWA Y, OZAWA T and IWASAKI I., Japan Analyst (1975) 24, 497
10. SEN GUPTA J G., Anal. Chim. Acta (1970) 49, 519
11. RUNDLE L M., Analyst (1974) 99, 163
12. MURPHY J M and SERGEANT G A., Analyst (1974) 99, 515

CHAPTER 41

Thallium

There is no lack of photometric reagents for thallium, although none of them is sufficiently selective to be used without a prior separation. Dithizone,[1] brilliant green,[2] crystal violet,[3] methyl violet[4] and rhodamine B[5] have all been suggested for particular applications. The procedure given in detail below has been adapted from that described by Voskresenskaya[2,6] using brilliant green. It is first necessary to separate thallium from interfering elements, particularly antimony, tin, mercury, cadmium, chromium and tungsten. For this a solvent extraction of thallic bromide into diethyl ether is described.

The reaction between the anion $TlBr_4^-$ and the brilliant green cation gives an organic-soluble green coloured complex with maximum absorption at a wavelength of 627 nm. The Beer-Lambert Law is obeyed in the range 0.1-5 µg Tl per ml of extract.

Somewhat higher sensitivities to thallium have been claimed for methods based upon the quenching of the fluorescence of cochineal red and uranyl sulphate,[7] or on the fluorescence given by thallium with rhodamine B,[8,9,10] or of the thallous chloro complex.[11,12] In all cases a prior separation is required, making use of ion-exchange, solvent extraction or co-precipitation techniques.

Atomic absorption spectroscopic methods based on the aspiration of aqueous acid solution of the rock material are not of sufficient sensitivity for determining the thallium content of most silicate rocks. Langmyhr et al[13] have described a direct method based upon atomisation from the solid state, but it is clear that the sensitivity required for many silicate rocks was not reached. Fratta[14] and Hannaker and Hughes[15] both used solvent extraction procedures to concentrate thallium, the latter completing the determination using a flame technique, the former a flameless one, achieving a detection limit of 3 ppb (3×10^{-9}). A somewhat similar detection limit of 1 ppb was claimed by Heinrich[16] using fractional distillation combined with flameless atomic absorption spectroscopy for thallium in 33 international standard reference rocks. Sighinolfi[17] also used a flameless technique, but made use of an acid attack of the silicate material and followed this by a solvent extraction stage to effect a prior concentration.

Spectrophotometric Determination with Brilliant Green

Silicate rock material is decomposed by evaporation with hydrofluoric acid or a mixture of hydrofluoric and sulphuric acids in the usual way. The dry residue is converted to bromides by evaporation with hydrobromic acid. Fusion with alkali carbonates must be avoided as this leads to volatilisation of any thallium present.[18]

Method

Reagents: Hydrobromic acid, concentrated and N.

Hydrobromic acid-bromine reagents, both concentrated and N hydrobromic acid saturated with bromine.

Potassium bromide.

Diethyl ether reagent, saturated with N hydrobromic acid.

Brilliant green solution, 10 mg of reagent dissolved in 100 ml of water.

Amyl acetate.

Standard thallium stock solution, dissolve 0.062 g of thallous sulphate in water and dilute to 500 ml to give a solution containing 100 μg Tl/ml.

Standard thallium working solution, dilute 5 ml of the stock solution to 500 ml with water to give a working standard containing 1 μg Tl/ml.

Procedure. Accurately weigh approximately 1 g of the finely powdered rock material into a platinum basin or crucible, and evaporate to dryness with 5 ml of concentrated hydrofluoric acid. Cool, add a further 5 ml of hydrofluoric acid together with 2 ml of 20 N sulphuric acid, and repeat the evaporation, first to fumes of sulphuric acid then to complete dryness. Add a few ml of water to the residue, warm gently and break up the residue. Rinse the residue with water into a 100-ml beaker, add 5 ml of concentrated hydrobromic acid and evaporate to dryness. Moisten the dry residue with concentrated hydrobromic acid saturated with bromine and again evaporate to dryness. Repeat this evaporation with hydrobromic acid and bromine, but this time do not allow the residue to dry out. Dissolve this residue in 25 ml of N hydrobromic acid saturated with bromine, and transfer to a 100-ml separating funnel.

Add to the solution 25 ml of diethyl ether saturated with N hydrobromic acid, and shake for 1 minute to extract the thallium. Allow the layers to separate and remove the ether extract. Add a further 25 ml of ether saturated with N hydrobromic acid and repeat the extraction. Discard the aqueous layer. Combine the ether extracts, wash with 2-3 ml of N hydrobromic acid and transfer the ether layer to a 100-ml beaker. Remove the ether by evaporation on a water bath.

Add 2 ml of concentrated hydrochloric acid to the small residue followed by 2 ml of bromine water, and evaporate to dryness. Repeat this sequence of additions and evaporations twice more, finally dissolving the dry residue in 3 ml of N hydrochloric acid. Now add 2ml of bromine water and heat on a hot plate to remove all free bromine. Cool the solution, transfer to a 100-ml separating funnel, dilute to 25 ml with water and add 1 ml of brilliant green solution. Shake for 1 minute, then add exactly 5 ml of amyl acetate and shake again to extract the green thallium complex. Allow to stand for 20 to 30 minutes then separate the organic layer and measure the optical density in 1-cm cells using the spectrophotometer set at a wavelength of 627 nm. Measure also the optical density of a reagent blank solution prepared in the same way as the sample solution but omitting the rock powder.

Calibration. Transfer aliquots of 1-5 ml of the standard thallium solution containing 1-5 μg Tl, to separate 100-ml beakers. Add to each 2 ml of concentrated hydrochloric acid and 2 ml of bromine water. Evaporate each solution to dryness on a hot plate, and then repeat the sequence of additions and evaporations twice more. Add 3 ml of N hydrochloric acid, 10 mg of potassium bromide and 2 ml of bromine water. Heat the solution to remove free bromine, and continue as described above for the sample solution. Plot the relation of optical density to thallium concentration.

Fluorimetric Determination in Silicate Rocks

The complex of thallium (III) with rhodamine B exhibits a violet fluorescence in benzene solution that has been used for the fluorimetric determination of thallium. Other elements behave similarly including gold, gallium and antimony (V). A procedure for the determination of thallium in silicate rocks based upon the use of this reaction was described in detail by Matthews and Riley.[9] The rock material was decomposed by evaporation with nitric and hydrofluoric acids, thallium oxidised to the trivalent state with bromine and separated by ion exchange chromatography.

Method

Reagents:
Diethyl ether, redistilled.
Sulphur dioxide, saturated aqueous solution.
Bromine water, saturated.
Rhodamine B solution, dissolve 0.1 g of the solid reagent in 100 ml of 3.5 M hydrochloric acid.

Ion exchange column, remove the fine material from 50-100 mesh Deacidite FF or similar resin by decantation in water. Digest twice with 2 M hydrochloric acid, wash well with water and transfer to a column 6 mm diameter with a bed depth of about 75 mm.

Standard thallium stock solution, dissolve 0.065 g of thallous nitrate in 500 ml of water. This solution contains 100 μg Tl per ml.

Standard thallium working solution, prepare from the stock solution by dilution as required, to give 0.5 μg Tl per ml and to be 2.7 M in hydrochloric acid. (25%v/v).

Procedure. Accurately weigh approximately 1 g of the finely powdered silicate rock material (Note 1) into a 50-ml PTFE beaker and add 10 ml of concentrated hydrofluoric acid and 5 ml of concentrated nitric acid. Cover the beaker, warm on a steam bath and allow to stand overnight. Rinse and remove the cover and evaporate the contents of the beaker to dryness on the water bath. Moisten the dry residue with a little concentrated nitric acid and evaporate to dryness. Repeat the evaporation to dryness with a little nitric acid. Repeat the evaporation twice more but with 6.5 M hydrochloric acid to remove the nitrate ion.

Add to the dry residue 7.7 ml of 6.5 M hydrochloric acid and 20-25 ml of water. Warm until all soluble salts have dissolved, transfer to a 800-ml beaker, dilute to 500 ml with water and then add 5 ml of bromine water. Allow this solution to percolate through the ion exchange column at a rate of about 3 ml per minute. Wash the column with 20 ml of water followed by 350 ml of 0.5 M nitric acid and 250 ml of 0.5 M hydrochloric acid to elute the interfering elements. Rinse with 25 ml of water and remove the thallium from the column by elution with 35 ml of sulphur dioxide saturate solution. Collect the eluate in a small silica beaker, add 1 ml of 6.5 M hydrochloric acid and evaporate on a water bath to a volume of about 15 ml. Remove the last traces of sulphur dioxide by adding 4 ml of bromine water and warming.

Cool, add a further 3 ml of bromine water, transfer to a 50-ml separating funnel using no more than 5 ml of 0.3 M hydrochloric acid to effect the transfer. Add 15 ml of redistilled ether and shake to extract the thallium. Remove the organic layer and repeat the extraction with a further 15 ml of ether. Combine the extracts and wash with 5 ml of 0.3 M hydrochloric acid. Discard the acid layer and transfer the ether extract to a small beaker containing 5 ml of 2.7 M hydrochloric acid. Allow the ether to evaporate and oxidise the thallium with 1 ml of bromine water. Warm gently to remove the excess bromine (Note 2), and transfer to a 50-ml separating funnel using 5 ml of 2.7 M hydrochloric acid. Add 2 ml of rhodamine B solution and 5 ml of benzene and shake for 2 minutes to extract the complex. Measure the fluorescence of the extract at 600 nm, using excitation at about 546 nm.

Measure also the fluorescence of a reagent blank extract prepared in the same way but omitting the sample material. Prepare also a calibration graph from aliquots of the thallium solution containing 0 to 1.5 µg Tl. These should be transferred to small beakers containing 5 ml of 2.7 M hydrochloric acid and 1 ml of bromine water. Warm gently to remove excess bromine and continue as described above. Plot the relation of optical density to thallium concentration.

Notes: 1. Use a sufficient quantity of the powdered rock material to give 0.05 to 1.5 µg of thallium.

2. Do not overheat at this stage. A temperature of 70° should not be exceeded.

References

1. CLARKE R S Jr, and CUTTITTA F., Anal. Chim. Acta (1958) 19, 555
2. VOSKRESENSKAYA N T., Zavod. Lab. (1958) 24, 395
3. PATROVSKY V., Chem. Listy (1963) 57, 961
4. OSHMAN V A., Trudy Ural'sk. Nauch-Issled i Proekt. Inst. Medn. Prom. (1963) (7), 417
5. MINCZEWSKI J, WEITESKA E and MARCZENKO Z., Chem. Anal. Warsaw (1961) 6, 515
6. VOSKRESENSKAYA N T., Zhur. Anal. Khim. (1956) 11, 623
7. GOTO H., Chem. Abstr. (1941) 35, 1721
8. FEIGL F, GENTIL V and GOLDSTEIN D., Anal. Chim. Acta (1953) 9, 393
9. MATTHEWS A D and RILEY J P., ibid. (1969) 48, 25
10. SCHNEPFE M M., ibid. (1975) 79, 101
11. BOCK R and ZIMMER E., Zeit. Anal. Chem. (1963) 198, 170
12. BOHMER R G and PILLE P., Talanta (1977) 24, 521
13. LANGMYHR F J, STUBERGH J R, THOMASSEN Y, HANSSEN J E and DOLEZAL J., Anal. Chim.Acta (1974) 71, 35
14. FRATTA M., Can. J. Spectrosc. (1974) 19, 33
15. HANNAKER P, HUGHES T C., Anal. Chem. (1977) 49, 1485
16. HEINRICHS H., Z. Anal. Chem. (1979) 294, 345
17. SIGHINOLFI G. PAOLO, At. Absorb. Newsl. (1973) 12, 136
18. KUZNETZOV V I and MYASOEDOVA G V., Zhur. Prikl. Khim. (1956) 29, 1875

CHAPTER 42

Thorium

Beach sand deposits and pegmatites containing appreciable amounts of thorium can be decomposed by fusion with potassium hydroxide, and the thorium recovered by precipitation as the insoluble fluoride. After separation from rare earth and other elements, thorium is precipitated as oxalate, ignited and weighed as oxide. This gravimetric procedure can be applied to normal silicate rocks only by taking a very large sample, and has been replaced by photometric methods.

Only a few reagents are known that form coloured complexes with thorium suitable for photometric determination. Of these the four most commonly used are thoron, and arsenazo I, II and III. These compounds all contain arsenic and have related structures.

Thoron is known by several other names including thorone, thoronol, thorin, APANS and naphtharsen. This material reacts also with a number of other elements, although the selectivity can be improved by the masking action of mesotartaric acid.[1] Arsenazo I and II are more sensitive to thorium than is thoron, but are generally similar in that they form stable complexes with thorium and other elements in weakly acid solution. Arsenazo III is appreciably more sensitive, and in addition can be used in fairly strongly acid solution. This gives increased selectivity to thorium, and only zirconium, hafnium and uranium (IV) interfere. Within limits zirconium and hafnium can be masked with oxalic acid, whilst uranium can be oxidised to uranium(VI), which does not react with arsenazo III.

The reagent arsenazo III is a dark-red powder that dissolves in water to give a rose-red solution. The thorium complex is emerald green in colour, although solutions are usually coloured purple from the presence of excess reagent. There is little absorption of the reagent at 665 nm, the wavelength of maximum absorption of the thorium complex. As with other complexes of arsenazo III, the absorption band of the thorium complex is very narrow, and care must therefore be taken to ensure that all measurements of optical density are made at their maximum value.

The calibration graphs given by Savvin[2] and by Savvin and Bareev[3] show a curious change of slope occurring at a concentration of about 0.4 ppm thorium. This feature has been confirmed by Abbey[4] who noted that in a mixed hydrochloric-perchloric acid solution, this break in the slope of the graph occurred at a concentration of about 0.6 ppm. May and Jenkins[5] obtained linear calibration graphs up to 2 ppm thorium, but suggested using a calibration over the range 0-0.4 ppm.

Ion-exchange column chromatography, solvent extraction and precipitation methods have all been advocated for the separation of thorium. None of these methods gives a complete separation in a single operation, and a number of authors have suggested or recommended combinations of two or more procedures. Anion exchange resins have been used to remove those elements that form anionic complexes in chloride solution from thorium and other elements that do not.[6] Uranium and zirconium - elements that may interfere with the photometric determination of thorium with arsenazo II - can be separated in this way. Rare earths, which also interfere with the determination of thorium, accompany thorium in the eluate.

To avoid this difficulty, Korkisch and Dimitriadis[7] used an anion-exchange resin in the nitrate form, eluting rare earths and other elements with nitric acid before recovery of thorium by elution with 6 M hydrochloric acid. A procedure for determining zirconium, thorium and uranium in silicate rocks, all with arsenazo III following anion-exchange chromatography in sulphate media has been described by Kiriyama and Kuroda.[8] Culkin and Riley[9] use a solvent extraction procedure with tributyl phosphate to recover thorium, zirconium (+ hafnium) and cerium from silicate rocks, and separate these elements on a column of cation exchange resin. Oxalic acid solutions are used to elute zirconium (+ hafnium) and thorium, and hydrochloric acid to elute cerium.

Column chromatography using cellulose pulp with and without alumina[10,11] has been used for the analysis of thorium ores and minerals, but does not appear to have been applied to any extent to the analysis of silicate rocks. Solvent extraction procedures used include extraction with mesityl oxide[12] trioctylphosphine oxide[13] and as noted above extraction with tributyl phosphate to recover thorium with zirconium and cerium.[9]

One of the oldest procedures used to recover thorium is precipitation as oxalate. This was used by Abbey,[4] with calcium as carrier prior to a photometric determination. Rare earths accompany thorium in the oxalate precipitate. Rare earths and thorium can be precipitated together as insoluble fluorides, for which calcium can also be used as carrier.[5] A separation of thorium from the rare earths can be achieved by precipitation of the former as iodate from a nitric acid solution.[14,15]

Extraction with a long-chain amine was used by Pakalns[16] to remove zirconium and large amounts of iron, but with many silicate rocks the titanium remaining with the thorium is likely to interfere with the determination.

The procedures given in detail below are based on those described by May and Jenkins[5] and by Korkisch and Dimitriadis[7] respectively.

Spectrophotometric Determination using a Fluoride Precipitation

The silicate material is decomposed by evaporation with hydrofluoric and nitric acids and the thorium, rare earths and calcium precipitated together as fluorides after the addition of a calcium carrier solution. Precipitations with potassium hydroxide and potassium iodate are used to complete the separation of thorium from other elements before photometric determination with arsenazo III.

Reagents:

Calcium carrier solution, dissolve 5 g of calcium carbonate in a mixture of 100 ml of concentrated nitric acid and 200 ml of water. Boil to expel carbon dioxide, cool and dilute to 500 ml with water.

Hydrofluoric acid wash solution, dilute 5 ml of concentrated hydrofluoric acid with 95 ml of water.

Iron carrier solution, dissolve 0.875 g of ferric nitrate hexahydrate in 100 ml of water containing 1 ml of concentrated nitric acid.

Mercury carrier solution, dissolve 0.158 g of mercuric nitrate monohydrate in 2 ml of 8 N nitric acid and dilute to 100 ml with water.

Oxalic acid solution, dissolve 50 g of the dihydrate in a hot mixture of 500 ml of water and 500 ml of concentrated hydrochloric acid.

Potassium hydroxide concentrated solution, dissolve 50 g of potassium hydroxide in 50 ml of water.

Potassium hydroxide wash solution, dilute 5 ml of the concentrated solution to 1 litre with water.

Potassium iodate solution, dissolve 15 g of potassium iodate in water and dilute to 250 ml.

Potassium iodate wash solution, mix together 30 ml of concentrated nitric acid, 3 ml of 30 per cent hydrogen peroxide and 100 ml of potassium iodate solution. Dilute to 500 ml with water.

8-hydroxyquinoline solution, dissolve 0.5 g of the solid reagent in 100 ml of water containing 1 ml of concentrated nitric acid.

Hydrogen peroxide solution, 3 and 30 per cent solutions.

Arsenazo III solution, grind 50 mg of the reagent with 1 ml of water and dilute with water to 100 ml. Prepare freshly each day.

Standard thorium stock solution, fuse 0.114 g of pure thorium oxide with 2 g of potassium hydroxide. Extract the melt with water, acidify with hydrochloric acid, adding 25 ml in excess. Heat until the solution is complete, cool and then dilute to volume with water in a 1-litre flask. This solution contains 100 µg Th per ml.

Standard thorium working solution, transfer 5 ml of the stock solution to a 500-ml volumetric flask, add 25 ml of concentrated hydrochloric acid and dilute to volume with water. This solution contains 1 µg Th per ml.

Procedure. Accurately weigh approximately 1 g of the finely powdered sample material into a platinum dish, moisten with water, add 10 ml of concentrated nitric acid and 10 ml of concentrated hydrofluoric acid and evaporate to dryness on a steam bath. Allow to cool, add 5 ml of concentrated nitric acid and 5 ml of concentrated hydrofluoric acid and again evaporate to dryness on a steam bath. Repeat the evaporation to dryness four or five times with small quantities of concentrated nitric acid and finally dissolve the nitrate residue by warming with 5 ml of concentrated nitric acid and 30 ml of water.

If any residue remains at this stage, collect on a small filter, wash with hot water, dry, ignite and fuse in a platinum crucible with a small amount of anhydrous sodium carbonate. Extract the melt with water, acidify with nitric acid, heat to complete the dissolution and add to the main rock solution (Note 1).

Transfer the rock solution, or an aliquot of it (Note 2) to a platinum dish, add 5 ml of the calcium carrier solution and evaporate to dryness. Moisten the dry residue with water, add 20 ml of concentrated hydrofluoric acid and 20 ml of water and allow to stand on a steam bath for 2 hours. Allow to cool. Stir into the solution a macerated 9-cm filter paper and collect the paper pulp together with the fluoride precipitate on a small close-textured filter paper previously washed with the hydrofluoric acid wash solution. Wash the residue thoroughly with the wash solution and discard the filtrate and washings.

Transfer the paper and residue to a platinum crucible, dry, ignite and fuse with 2 g of potassium pyrosulphate. Allow to cool and dissolve the melt in about 10 ml of water containing 2 ml of nitric acid. Rinse the solution into a 400-ml beaker and dilute with water to a volume of about 150 ml. Add 1 ml of concentrated potassium hydroxide solution to neutralise the acid present and then 15 ml in excess. Transfer the beaker to a steam bath, digest for about 15 minutes and then collect the precipitate on an open-textured filter paper. Rinse the beaker and wash the precipitate several times with the potassium hydroxide was solution.

Rinse the residue into a 100-ml beaker with a little water and add 1 ml of the 8-hydroxyquinoline solution. Moisten the paper with 1 ml of the dilute hydrogen peroxide solution followed by 2 ml of hot 8 N nitric acid, added slowly drop by drop. When the paper has drained, add 5 ml of hot water and again allow the paper to drain. Repeat this sequence of additions to the paper twice more to ensure complete removal of the residue from the paper.

Using pipettes, add 10 ml of potassium iodate solution to the beaker, followed by 5 ml of the mercury carrier solution, stirring continuously. Allow the beaker to stand in an ice-bath for 45 minutes. Mix in a small quantity of macerated filter paper pulp and collect the pulp and precipitate on a small close-textured filter paper. Rinse the beaker and wash the precipitate eight or ten times with ice-cold potassium iodate wash solution. Discard the filtrate and washings. Dissolve the precipitate from the filter by adding 5 ml of hot 6 N hydrochloric acid and rinsing with 5 ml of hot water. Repeat this addition of hydrochloric acid and water twice more, collecting the solution and washings in a 50-ml beaker. Transfer the beaker to a steam bath and evaporate to dryness. Moisten the residue with 2 ml of concentrated hydrochloric acid and again evaporate to dryness.

Add 1.6 ml of 6 N hydrochloric acid to the beaker, moistening the residue and the walls of the beaker. Add 3 ml of water, cover with a watch glass and warm for 5 minutes on a steam bath. Allow to cool, and rinse the solution into a 25-ml volumetric flask using several small portions of water with a total volume of not more than 8 ml. Add 10 ml of the oxalic acid solution followed by 2 ml of arsenazo III solution. Dilute to volume with water and mix well.

Measure the optical density of the solution in 4-cm cells with the spectrophotometer set at a wavelength of 665 nm, using a reference solution without added thorium, prepared as described below for the calibration graph.

Calibration. Transfer aliquots of 0-10 ml of the standard solution containing 0-10 μg of thorium to separate 25-ml beakers. Evaporate each to dryness and then add 1.6 ml of 6 N hydrochloric acid. Transfer each solution to a separate 25-ml volumetric flask and dilute to volume with water after adding 10 ml of the oxalic acid solution and 2 ml of the arsenazo III solution. Measure the optical density of each relative to the solution containing no added thorium, as described above for the sample solution. Plot the relation of optical density to thorium concentration.

Notes: 1. Zircon and other refractory silicate minerals are best decomposed by fusion with sodium carbonate as described. If the residue is composed largely of ilmenite, rutile or other oxide mineral, it should be fused with potassium pyrosulphate. Extract the melt with water, dissolve by adding dilute nitric acid, add 1 ml of the iron carrier solution and precipitate iron, thorium and other elements with ammonia. Collect the precipitate, dissolve in dilute nitric acid and add to the main rock solution.

2. This procedure is designed for silicate materials containing 0-10 ppm of thorium. Many silicate rocks contain higher concentrations of thorium, and for these a smaller sample portion can be taken. Alternatively the nitric acid solution of the rock material can be diluted to volume, and an aliquot of the solution containing 10 μg or less of thorium taken for the subsequent stages of the analysis.

Spectrophotometric Determination using an Anion-Exchange Separation

An alternative procedure for the separation and determination of thorium which includes also the separation and determination of uranium and zirconium in the same sample portion is described in the chapter dealing with uranium.

In the procedure described below, after decomposition of the silicate material, fluorine is complexed with boric acid and the nitrate solution transferred to a strongly basic anion exchange resin. The resin is washed with diluted nitric acid to remove interferring elements, and the thorium recovered by elution with hydrochloric acid. Either thoron or arsenazo III is used to complete the determination spectrophotometrically.

Reagents: Ion exchange resin, Dowex 1x8, 100-200 mesh, chloride form. Convert with 8 M nitric acid until chloride ions can no longer be detected in the effluent. Use in the form of a short column containing approximately 5 g of resin. After use, regenerate by washing with 8 M nitric acid.

Thoron solution, 0.1% (w/v) in water.

Arsenazo III solution, 0.2% (w/v) in water.

Potassium permanganate solution, 2% (w/v) in water.

Nitric acid, 8 M and M, dilute concentrated nitric acid (16 M) with water. Allow to cool before use.

Procedure. Accurately weigh a sample of from 0.5 to 2.0 g, depending upon the amount of thorium expected in the sample material, of the finely powdered rock material into a platinum or PTFE dish, moisten with water, add 10 ml of concentrated nitric acid and 10 ml of concentrated hydrofluoric acid and evaporate to dryness on a steam bath. Repeat the evaporation with further quantities of nitric acid and hydrofluoric acid as required to remove all silica. Repeat the evaporation with 5 ml portions of nitric acid to remove most of the fluorine. Add 50 ml of concentrated nitric acid and 5 g of boric acid and again evaporate to dryness. Dissolve the residue in 10 ml of M nitric acid, filter off any excess boric acid, washing the residue with a little M nitric acid, and dilute the filtrate with an equal volume of concentrated nitric acid (Notes 1 and 2).

Transfer the solution to the prepared anion-exchange column and elute at a flowrate of about 2-3 ml per minute with 200 ml of 8 M nitric acid. Elute thorium with 200 ml of 6 M hydrochloric acid, and evaporate the eluate to dryness on a steam bath. Add 1 ml of M hydrochloric acid and 1 ml of potassium permanganate to the residue and again evaporate to dryness on a steam bath. Dissolve the residue, which may include precipitated manganese dioxide, in 10 ml of concentrated hydrochloric acid and again evaporate to dryness on a steam bath. Complete the determination using either of the following methods.

Determination with thoron: Dissolve the residue in 5 ml of 0.1 M hydrochloric acid, add 1 ml of thoron solution and a few crystals of ascorbic acid and dilute to volume in a 10-ml volumetric flask with 0.1 M hydrochloric acid. Measure the optical density of this solution at a wavelength of 545 nm using a reagent blank solution as reference and determine the thorium content by reference to a calibration graph prepared from solutions containing 0 to 10 µg/ml.

Determination with arsenazo III: Add a few drops of concentrated formic acid and 10 ml of concentrated hydrochloric acid to the residue and evaporate to dryness on a steam bath to remove traces of nitrate. Dissolve in 5 ml of concentrated hydrochloric acid, add 1 ml of arsenazo III solution and dilute to volume with water in a 10-ml volumetric flask. Measure the optical density of this solution relative to a reagent blank at the wavelength of maximum absorption (Note 3), and hence determine the thorium content by reference to a calibration graph prepared from solutions containing 0 to 1 µg/ml.

Notes: 1. Thorite and other accessory minerals containing appreciable amounts of thorium may not be completely decomposed by this method. If the residue collected on the filter contains gritty particles, any thorium present in it should be recovered by fusion of the ignited residue with a little sodium carbonate, acidification with a little nitric acid and addition to the main rock solution.

2. The solution at this stage should be clear and free from precipitated oxides of iron and titanium. If a hazy solution is obtained, allow to stand overnight and filter before the ion-exchange separation.

3. This has been determined as 665 nm, however in view of the narrowness of the absorption band, the maximum absorption should be determined with the instrument used.

References

1. FLETCHER M H, GRIMALDI F S and JENKINS L B., Analyt. Chem. (1957) 29, 963
2. SAVVIN S B., Dokl. Akad. Nauk. SSR (1959) 127, 1231
3. SAVVIN S B and BAREEV V V., Zavod. Lab. (1960) 26, 412
4. ABBEY S., Anal. Chim. Acta (1964) 30, 176
5. MAY I and JENKINS L B., US Geol. Surv. Prof. Paper 525-D, p 192 (1965)
6. ARNFELT A-L and EDMUNDSSON I., Talanta (1961) 8, 473
7. KORKISCH J and DIMITRIADIS D., Talanta (1973) 20, 1199
8. KIRIYAMA T and KURODA R., Anal. Chim. Acta (1974) 71, 375
9. CULKIN F and RILEY J P., Anal. Chim. Acta (1965) 32, 197
10. WILLIAMS A F., Analyst (1952) 77, 297
11. KEMBER N F., Analyst (1952) 77, 78
12. LEVINE H and GRIMALDI F S., Geochim. Cosmochim. Acta (1958) 14, 93
13. POLLOCK E N., Anal. Chim. Acta (1977) 88, 399
14. GRIMALDI F S, JENKINS L B and FLETCHER M H., Analyt. Chem. (1957) 29, 848
15. FURTOVA E V, SADOVA G F, IVANOVA V N and ZAIKOVSKII F V., Zhur. Anal. Khim. (1964) 19, 94
16. PAKALNS P., Anal. Chim. Acta (1972) 58, 463

CHAPTER 43

Tin

Those rocks that contain tin within the silicate lattice present no problem in sample decomposition. Such samples may be evaporated with hydrofluoric acid and either sulphuric or perchloric acid in the usual way. The residue may be fused with potassium pyrosulphate and the melt obtained dissolved in hydrochloric or sulphuric acid.

Where the rock contains discrete particles of cassiterite, the dissolution is considerably more difficult. This is because of the unreactive nature of cassiterite, particularly towards acids and acidic fluxes, and also from the difficulty in transferring very fine grains of unattacked cassiterite from one vessel to another following the decomposition of the silicate phase with hydrofluoric acid. These grains are easily decomposed by fusion with sodium peroxide, but because of the attack of platinum apparatus that occurs it is not possible to use this procedure directly after the volatilisation of silica with hydrofluoric acid. Attempts to mix the dry residue with sodium peroxide and sinter at 500° have not given complete decomposition of residual cassiterite.

Fusion of the rock material with a mixture of sodium carbonate and sulphur converts all the tin present into a thiosalt, from which it is recovered together with other heavy metals as sulphide. This procedure has been used by Popov(1) for determining tin in rocks and tin ores, but is unsuitable for those silicates containing only very small amounts of tin.

One of the simplest ways of decomposing cassiterite present in rock samples is to ignite with ammonium iodide.(2,3) In its simplest form, the sample material is intimately mixed with ammonium iodide, transferred to a borosilicate test tube and ignited over a gas burner. Attempts to obtain a quantitative recovery of tin from the tonalite T-1 was not successful and a more elaborate version of this decomposition was devised.(4) The apparatus used is shown in Fig. 27 and the procedure used as follows.

Five ml of concentrated hydrofluoric acid and 0.5 ml of concentrated sulphuric acid were added to 0.5 g of the sample material in a small platinum crucible, and the silica removed by evaporation to dryness. The residue was mixed with 0.25 g of ammonium iodide, and the crucible placed within the silica tube as shown in Fig. 27. The tube was slowly raised to dull red heat, and this temperature was maintained for

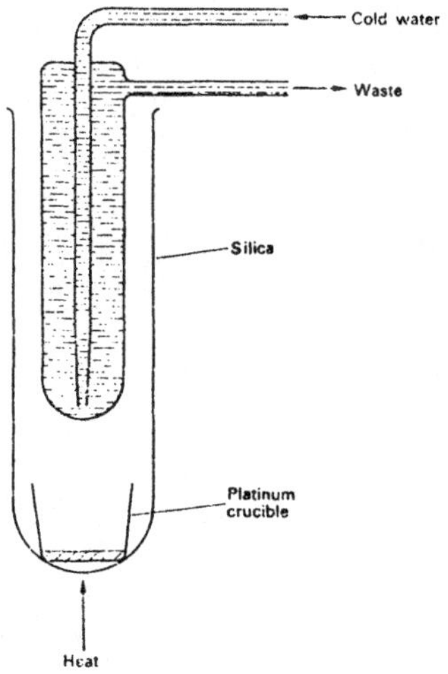

FIG.27 Apparatus for the decomposition of silicate rocks containing cassiterite

5 minutes. After cooling, the residue was extracted by heating with N hydrochloric acid. The sides of the tube and the cold finger were washed down with hot N hydrochloric acid and the washings combined with the acid extract. Free iodine was removed with ascorbic acid and the tin present in the solution determined with salicylideneamino-2-thiophenol.

Although this way of decomposing cassiterite grains in silicate rocks was found to give better recoveries of tin than the simpler form of decomposition, occasionally very low recoveries were noted for which no explanation has yet been suggested. In theory at least, the total tin content may be recovered by this method after first volatilising the silica with hydrofluoric and mineral acids.

Ion-exchange separations have been described by Huffman and Bartel(5) and by De Laeter and Jeffery.(6) The former used an oxalate form of an anion exchange resin and absorbed the tin from a solution containing hydrochloric and oxalic acids. The tin was recovered by elution with dilute sulphuric acid. The latter workers used the chloride form of the resin and absorbed the tin from 6 M hydrochloric acid solution. Nickel and cobalt were eluted with 3 M acid, iron with M acid and tin with 0.1 M hydrochloric acid. A second anion exchange separation was recommended.

A similar anion exchange separation in 6 M medium was described by Smith[7] in a method for determining traces of tin in rocks, sediments and soils similar to that described below; the determination was completed using phenyfluorone.

A highly selective procedure for the recovery of tin from acidic solution was described by Ross and White.[8] This involved the extraction of the tin complex with tris-(2-ethylhexyl)phosphine oxide into cyclohexane solution. The extraction of stannic iodide into nitrobenzene was reported by Gilbert and Sandell[9] and by Tanaka,[10] and into toluene by Newman and Jones.[11]

Stannic bromide is appreciably volatile, and a method for the separation of tin, based upon a distillation from hydrobromic acid solution has been used by Onishi and Sandell[12] for determining small amounts of tin in silicate rocks. Arsenic, antimony and germanium were first removed by distillation from a hydrochloric acid solution. The only element to accompany tin in the bromide distillation that gave rise to interference in the subsequent determination was selenium.

Reagents advocated for the determination of tin include cacotheline, gallein, 8-hydroxyquinoline, haematoxylin, diethyldithiocarbamate, phenylfluorone, dithiol, methyl violet, catechol violet and salicylideneamino-2-thiophenol.[13] In general, both selectivity and sensitivity is poor and there is still a need for a better photometric reagent for tin.

Spectrophotometric Determination with Catechol Violet

In the procedure described below, silica is removed by evaporation with hydrofluoric acid in the usual way. Any cassiterite in the acid-insoluble residue is recovered by fusion with sodium peroxide. Tin present in the solution is concentrated by extraction of stannic bromide into toluene solution, and the determination completed photometrically with catechol violet.

Method

Reagents:

Potassium iodide solution, dissolve 83 g of potassium iodide in water and dilute to 100 ml. Prepare freshly each day.

Potassium iodide wash solution, mix 25 ml of 9 N sulphuric acid with 2.5 ml of the potassium iodide solution.

Toluene.

Sodium hydroxide solution, approximately 5 M and 0.1 M.

Ascorbic acid solution, dissolve 5 g of the reagent in 100 ml of water. Prepare only as required.

Catechol violet solution, dissolve 50 mg of the reagent in 100 ml of water. Prepare freshly each week.

Sodium acetate solution, dissolve 20 g of the trihydrate in water and dilute to 100 ml.

Standard tin stock solution, dissolve 0.100 g of pure tin by evaporation to fumes with 20 ml of concentrated sulphuric acid. Cool, cautiously dilute with water, add 65 ml of concentrated sulphuric acid, cool again and dilute with water to 500 ml in a volumetric flask. This solution contains 200 μg Sn per ml.

Standard tin working solution, dilute 5 ml of the stock tin solution to 500 ml with water as required. This solution contains 2 μg Sn per ml.

Procedure. Accurately weigh approximately 1 g (Note 1) of the finely powdered silicate rock material into a platinum crucible, moisten with water and add 1 ml of 20 N sulphuric acid, 1 ml of concentrated nitric acid and 10 ml of concentrated hydrofluoric acid. Transfer the crucible to a hot plate and evaporate first to fumes of sulphuric acid, then to dryness. Allow the crucible to cool, add 2 g of potassium pyrosulphate and fuse over a burner, taking care to avoid loss by spitting. Allow the crucible to cool, dissolve the melt by warming with dilute sulphuric acid and rinse into a separating funnel to give a solution of about 25 ml in volume in 9 N sulphuric acid (Note 2).

Add 2.5 ml of the potassium iodide solution and mix. Now add 10 ml of toluene, stopper the funnel and shake vigorously for 2-3 minutes. Allow the phases to separate. The toluene layer now contains stannic iodide together with a small quantity of iodine which gives a pink colour to the solution. Separate and discard the aqueous layer. Without shaking rinse the toluene layer with 5 ml of the potassium iodide wash solution and discard the washings.

Add to the toluene solution 5 ml of water and 5 M sodium hydroxide solution drop by drop, with shaking, until the pink colour of the toluene layer is discharged, then add 2 drops in excess. Stopper the funnel, shake for 30 seconds, allow the phases to separate and run the aqueous layer into a small beaker. Rinse the organic layer by shaking with 3 ml of 0.1 M sodium hydroxide solution and add the wash solution to the aqueous solution in the small beaker. Retain the toluene layer in the separating funnel.

Add 2.5 ml of 5 N hydrochloric acid to the solution in the beaker and add ascorbic acid solution drop by drop to decolorise any iodine that may have been liberated. Using a pipette, add 2 ml of the catechol violet solution and mix. Without shaking wash the toluene in the separating funnel with 5 ml of the sodium acetate solution and add the washings to the beaker. Add dilute ammonia drop by drop to the solution in the beaker to bring the pH to a value of 3.8 ± 0.1 (use a pH meter). Transfer the solution to a 25-ml volumetric flask, dilute to volume with water, mix well and allow to stand for 30 minutes. Measure the optical density of the solution in a 4-cm cell (Note 3) using a spectrophotometer set at a wavelength of 552 nm. Prepare also a reagent blank solution in the same way as the sample solution but omitting the sample material (Note 4).

Calibration. (4-cm cells). Transfer aliquots of the standard tin solution containing 0-8 μg tin to a series of 50-ml beakers and dilute to 10 ml with water. Add 1 ml of 5 M sodium hydroxide solution, mix and add 2.5 ml of 5 N hydrochloric acid and by pipette, 2 ml of catechol violet solution. Adjust the pH to 3.8 ± 0.1, and transfer each solution to a separate 25-ml volumetric flask. Add 5 ml of the sodium acetate solution, dilute to volume with water, mix well and set aside for 30 minutes. Measure the optical densities relative to the solution containing no tin, in 4-cm cells at a wavelength of 552 nm, as described above.

Notes: 1. Use a 1-g sample portion for rocks containing up to 30 ppm tin, and smaller weights for samples of higher tin content.

2. This procedure is used only for those rocks that are completely decomposed by evaporation with hydrofluoric and sulphuric acids followed by fusion of the residue with potassium pyrosulphate. If the sample material contains cassiterite, the residue remaining after evaporation to dryness should be digested with dilute sulphuric acid and any remaining residue collected, ignited and fused in a nickel crucible with a small quantity of sodium hydroxide. The melt is extracted with water, acidified and added to the main rock solution. Although some loss may occur, the greater part of the tin present as cassiterite can be recovered in this way.[12] For rocks containing several hundred ppm or more of tin, it may be necessary to fuse the residue with sodium peroxide.

3. Use a 4-cm cell for up to 8 μg of tin, and a 1-cm cell for up to 30 μg.

4. Measure both the rock and the reagent blank solutions against a reference solution prepared without added tin as described in the calibration section.

Spectrophotometric Determination using Salicylideneamino-2-Thiophenol

Salicylideneamino-2-thiophenol is one of the most valuable reagents for tin. It is easily prepared and gives a yellow-coloured complex with tin that can be extracted into any of a number of organic solvents for photometric measurement (xylene is preferred). The maximum optical density occurs at a wavelength of 415 nm and the Beer-Lambert Law is valid up to about 45 µg Sn in 10 ml of xylene. The molar extinction coefficient is about 15,000 which enables the reagent to be used for determining the small amounts of tin that occur in normal silicate rocks. A simple procedure for this, based upon a method used for tin ore analysis, gave erratic recoveries of tin. An initial separation and concentration stage should therefore be used.

Method

Reagents: Salicylideneamino-2-thiophenol solution, dissolve 1 g of ascorbic acid in 100 ml of warm ethanol. Add 0.1 g of the reagent and stir until dissolved. Store in a brown glass bottle away from direct sunlight and prepare freshly each day.

Lactic acid solution, mix 20 ml of lactic acid with 80 ml of water.

Xylene.

Procedure Decompose and recover the tin in a 1-g portion of the rock material as described above by evaporation with hydrofluoric, nitric and sulphuric acids and fusion of the residue with potassium pyrosulphate (or sodium hydroxide if necessary). Extract stannic iodide into toluene solution and back extract into very dilute sodium hydroxide solution.

Dilute to about 20 ml with water and add 2 ml of lactic acid solution. Check the pH and adjust to a value of 2 ± 0.1. Using a pipette add 5 ml of the salicylideneamino-2-thiophenol reagent solution, mix and allow to stand for 20 minutes. Add exactly 10 ml of xylene, shake vigorously for 20 seconds and allow the phases to separate. After 5 minutes, measure the optical density of the organic extract in 4-cm cells, with the spectrophotometer set at a wavelength of 415 nm. Prepare a reagent blank extract and use this as the reference solution. For the calibration use aliquots of the standard solution containing 0-10 µg tin.

Determination of Tin by Atomic Absorption Spectroscopy

The use of atomic absorption spectroscopy for the determination of tin in geological materials, including silicate rocks and sulphide minerals was explored by Moldan et al.[14] Interferences including complex interelement effects were noted, and even with a long path - (45 cm) absorption tube, the sensitivity was scarcely adequate for even average tin contents. An improvement in sensitivity is obtained by using the hotter nitrous oxide-acetylene flame, which can dissociate the highly stable SnO structure, or by using the reducing fuel rich air-hydrogen or argon-hydrogen flame, in which SnH can be formed and dissociated. Under the latter conditions, greatest sensitivity appears to be at 224.6 nm, although a wavelength of 235.5 nm has also been used. This procedure, as described by Guimont et al[15] is of barely adequate sensitivity for application to normal silicate rocks.

References

1. POPOV M A., Byull. Nauch.-Tekh. Inform. Tsentr. Nauch.-Issled. Inst. Olova, Sur'my i Rtuti (1962) (3), 38
2. STANTON R E and McDONALD A J., Trans. Inst. Min. Metall. (1961) 71, 27
3. MARTINET B., Chim. Anal. (1961) 43, 483
4. KERR G O., unpublished work.
5. HUFFMAN C Jr. and BARTEL A J., U S Geol. Surv. Prof. Paper 501-D, 1964
6. DE LAETER J R and JEFFERY P M., J. Geophys. Res. (1965) 70, 2895
7. SMITH J D., Anal. Chim. Acta (1971) 57, 371
8. ROSS W J and WHITE J C., Analyt. Chem. (1961) 33, 421
9. GILBERT D D and SANDELL E B., Microchem. J. (1960) 4, 491
10. TANAKA K., Japan Analyst (1962) 11, 332
11. NEWMAN E J and JONES P D., Analyst (1966) 91, 406
12. ONISHI H and SANDELL E B., Anal. Chim. Acta (1956) 14, 153
13. GREGORY G R E C and JEFFERY P G., Analyst (1967) 97, 293
14. MOLDAN B., RUBESKA I, MIKSOVSKY M and HUKA M., Anal. Chim. Acta (1970) 52, 91
15. GUIMONT J, BOUCHARD A and PICHETTE M., Talanta (1976) 23, 62

CHAPTER 44

Titanium

In the course of chemical analysis, titanium minerals are considerably more resistant to decomposition than the silicate matrix in which they occur, and care must be taken to ensure that all the mineral grains are completely attacked. Most of the accessory minerals can be decomposed by evaporation to fumes with a mixture of sulphuric and hydrofluoric acids. Some grains of perovskite are likely to remain unattacked if this minerals is present in quantity, although a second evaporation is usually sufficient to remove these. Mixtures of hydrofluoric acid with either nitric or perchloric acid are less effective for this decomposition. The most effective decomposition is probably a single evaporation to dryness with sulphuric and hydrofluoric acids followed by a fusion with potassium pyrosulphate. This can be done in a single platinum crucible and serves also to remove fluorine which can otherwise interfere with the determination of titanium.

Gravimetric and occasionally titrimetric methods were formerly used for determining titanium in those rocks containing 2 or 3 per cent or more, but even at the higher level spectrophotometric and atomic absorption methods are now commonly employed. There is no lack of reagents for titanium, although few of these are specific. They are all based upon the formation of coloured complexes with titanium in acid solution. Hydrogen peroxide is probably the most widely used, and although the method is capable of high precision, it is not very sensitive. Two more sensitive reagents are chromotropic acid (1,8-dihydroxynaphthalene-3,6-di-sulphonic acid) and tiron (1,2-dihydroxybenzene-3,5-disulphonic acid). Least subject to interference from other metallic ions is the reaction with diantipyrylmethane.[1] These three reagents are some 20 times as sensitive as hydrogen peroxide.

Several authors, notably Walsh[2] have recorded interference by other elements in the determination of titanium by atomic absorption spectroscopy, and great care must be taken if accurate figures are to be obtained. In particular, the effect of iron, aluminium, manganese, calcium, sodium and potassium in the rock solution must be taken into account. In one such procedure, Van Loon and Parissis[3] add aluminium to the rock solution to bring the total aluminium level to 750-1000 ppm. This addition was shown to stabilise the titanium absorption and to mask the interference from the other constituents. A nitrous oxide-acetylene flame is invariably used in the determination of titanium and most authors have measured the absorption at 364.3 or 365.3 nm.

Spectrophotometric Method using Hydrogen Peroxide

This method is based upon the yellow-coloured complex formed by titanium with hydrogen peroxide in acid solution. The maximum value for the optical density occurs at a wavelength in the range 400-410 nm. There is evidence that the position of the peak depends to some small extent upon the conditions used, but as the absorption band is fairly wide any wavelength within this range can be used for measurement without appreciable loss of sensitivity.

A number of other metals form coloured complexes with hydrogen peroxide including vanadium, uranium, niobium, molybdenum and under certain circumstances chromium. Of these only vanadium is likely to give rise to any interference with the determination of titanium in rocks and minerals, and this only rarely. The maximum absorption of solutions containing the vanadium complex occurs at a wavelength of 460 nm, and it is therefore possible to determine both titanium and vanadium in the same solution by measuring optical densities at both 400 and 460 nm. The concentrations of the two elements can then be calculated from the simultaneous equations:

$$\log I_o/I_{(400\ nm)} = a(Ti) + b(V)$$

and

$$\log I_o/I_{(460\ nm)} = a'(Ti) + b'(V)$$

where a, a', b and b' are the slopes of the calibration graphs for titanium and vanadium at the two wavelengths.

Iron, chromium and nickel, elements that form coloured ions in solution, can also interfere. In silicate rocks, however, only iron is likely to be present in amounts sufficient to give rise to serious error, and this can be avoided by adding measured amounts of phosphoric acid to the rock solution and to the standards used for the calibration. Alternatively the optical density of the solution to which hydrogen peroxide has been added can be measured relative to the solution without it.

Fluoride ions interfere with the titanium determination by bleaching the yellow colour. A slight bleaching effect has also been observed in the presence of alkali salts, citric acid and phosphoric acid, but these do not interfere seriously with the determination in silicate rocks, even after a potassium pyrosulphate fusion or the addition of phosphoric acid to complex iron. High acid concentrations also cause some reduction in colour, and for most purposes a concentration of between 1.5 and 3 N sulphuric acid should be used. Perchloric acid concentrations of up to 3.5 M can be used, but hydrochloric acid should be avoided because of the strong colour given by ferric iron in this medium.

Even the very small quantities of platinum removed from old crucibles in the course of a pyrosulphate fusion can catalytically decompose hydrogen peroxide, and result in a slow fading of the yellow titanium-peroxide colour. A similar decomposition of hydrogen peroxide has been observed with a number of samples containing appreciable amounts of cerium.

Method

Reagents: Hydrogen peroxide, 20 or 30 volume.*

Potassium pyrosulphate solution, dissolve 20 g in water and dilute to 100 ml.

Standard titanium stock solution, 0.5 mg TiO_2 per ml; this can be prepared from either potassium titanyl oxalate or potassium fluotitanate by direct weighing, evaporation to fumes with sulphuric acid and dilute to volume with water. However, both these salts are hydrated, and the solutions obtained should be standardised by precipitation of titanium with ammonia, cupferron or N-benzoylphenylhydroxylamine and ignition to oxide. Alternatively, pure titanium metal foil or pure titanium dioxide may be used (Note 1).

Procedure. Accurately weigh approximately 0.5 g (Note 2) of the finely powdered rock material into a 25-ml platinum crucible, add 0.5 ml of concentrated nitric acid (Note 3), 1 ml of 20 N sulphuric acid and 10 ml of concentrated hydrofluoric acid. Place the crucible on a hot plate and evaporate to fumes of sulphuric acid. Allow to cool, rinse down the sides of the crucible with a little water and again evaporate, this time to dryness. Add 2 g of potassium pyrosulphate to the dry residue and fuse in the covered crucible over a low flame for the minimum time required to give a clear melt. Allow to cool.

Extract the melt with water containing 10 ml of 20 N sulphuric acid, transfer the solution to a 100-ml volumetric flask, dilute to volume with water and mix well (Note 4). Fill two matched 1-cm spectrophotometer cells with the solution. To one cell add 1 drop only of hydrogen peroxide solution and mix with a small polyethylene rod. Measure the optical density of this coloured solution relative to the solution in the other cell to which hydrogen peroxide has not been added. A wavelength in the range 400-410 nm should be used. If the recorded optical density at this wavelength is much less than 0.1, the measurement should be repeated in 4-cm cells.

* Unstable. Prepare by dilution of 100 volume when required.

Calibration. Although this method is not very sensitive it is very precise and there is a linear relation between optical density and titanium concentration.

Transfer aliquots of 4-20 ml of the standard titanium solution containing 2-10 mg TiO_2, to separate 100-ml volumetric flasks and add to each 10 ml of 20 N sulphuric acid and 10 ml of the potassium pyrosulphate solution. Dilute to volume, mix well and determine the optical densities of each solution by filling two 1-cm spectrophotometer cells with each solution, adding 1 drop of hydrogen peroxide to one of each pair and measuring the coloured solution against the solution to which hydrogen peroxide has not been added, as described for the sample solution above. Plot the relation of optical density to titanium concentration.

Notes: 1. For the preparation of the standard titanium solution from metal foil proceed as follows. Weigh 0.149 g into a 150-ml beaker, add 25 ml of 20 N sulphuric acid and 50 ml of water and gently boil on a hot plate until all the titanium has passed into solution, this usually takes 3 to 4 hours. Now add dilute hydrogen peroxide solution drop by drop until the violet colour of the solution just disappears and a very slight yellow colour is apparent, but avoiding an excess. Add a piece of platinum (a crucible lid will do) and gently boil the solution to decompose the titanium-peroxide complex, giving a completely colourless solution. Allow to cool, transfer to a 500-ml volumetric flask, add a further 25 ml of 20 N sulphuric acid and dilute to volume with water. Mix well. This solution contains 0.5 mg TiO_2 per ml. For the preparation from pure titanium dioxide, fuse 0.25 g of the oxide with 2.5 g of powdered potassium pyrosulphate, dissolve in water containing 50 ml of 20 N sulphuric acid and dilute to 500 ml.

2. A 0.5-g sample is adequate for most silicate rocks. If material is limited a much smaller portion can be taken and the quantities of reagents and final volumes reduced accordingly.

3. The volume of nitric acid added should be increased for those samples containing much ferrous iron or organic matter.

4. If the rock material contains more than a trace of barium, a white residue will be obtained on dissolution of the melt. Once the solution has been diluted to volume, this precipitate can be allowed to settle and may then be ignored, providing none of it is transferred to the spectrophotometer cells.

Spectrophotometric Method using Diantipyrylmethane

The reaction of the titanium ion Ti^{4+} with diantipyrylmethane to give an intensely yellow-coloured solution was originally described by Minin[4] for the photometric determination of titanium in the presence of iron, vanadium, fluorides and phosphates. Polyak[1] made a more extensive study of the use of this reagent, whilst Jeffery and Gregory[5] used it in the analysis of ores, rocks and minerals.

Solutions of diantipyrylmethane deteriorate slowly, particularly when exposed to direct sunlight. This deterioration is evidenced by gradual yellowing of the solution, and can be considerably lessened by adding ascorbic acid and by storing in the dark. The colour given with titanium forms in dilute hydrochloric or sulphuric acid solution, some reduction in intensity being noted at acid concentrations greater than 4 N hydrochloric. Perchloric acid precipitates the reagent. The colour develops rapidly, reaching a maximum after 3 hours, and is stable for several months. The maximum absorption occurs at a wavelength of 380 nm, and the Beer-Lambert Law is valid up to 400 µg TiO_2 per 100 ml.

Very few metallic ions interfere with the reaction. Niobium and tantalum precipitate with the reagent giving low recoveries of titanium, but this can be avoided by adding tartaric acid. Iron and vanadium do not interfere in their low valency states. Slight interference is encountered from uranium, molybdenum and tin when present in the solution, although 3-4 mg of each can be tolerated in 100 ml. There is no interference from other elements present in normal silicate or carbonate rocks.

Method

Reagents:

Ascorbic acid solution, dissolve 10 g of the reagent in 100 ml of water.

Diantipyrylmethane solution, dissolve 5 g of the reagent and 5 g of ascorbic acid in 150 ml of 2 N sulphuric acid and dilute to 500 ml with water. Transfer to a dark glass bottle and store in the dark.

Standard titanium working solution, pipette 10 ml of the stock solution containing 0.5 mg TiO_2 per ml into a 500-ml volumetric flask, add 83 ml of concentrated hydrochloric acid and dilute to volume with water. This solution contains 10 µg TiO_2 per ml.

Procedure. Accurately weigh approximately 0.1 g (Note 1) of the finely powdered silicate rock into a 10-ml platinum crucible and evaporate to fumes of sulphuric acid with 0.5 ml of concentrated nitric acid, 1 ml of 20 N sulphuric acid and 4 ml of hydrofluoric acid. Allow to cool, rinse down the sides of the crucible and again evaporate, this time to dryness. Fuse the residue with a small quantity of potassium pyrosulphate - not more than 2 g should be required - to give a completely clear melt. Dissolve this melt in 2 N hydrochloric acid, transfer to a 100-ml volumetric flask and dilute to volume with 2 N hydrochloric acid.

Pipette an aliquot of this solution containing not more than 500 μg TiO_2 to a 100-ml volumetric flask and add sufficient 2 N hydrochloric acid to bring the total volume of this acid up to 50 ml. Add 5 ml of ascorbic acid solution, mix by gently swirling the flask, and allow to stand for 30 minutes. Add 25 ml of the diantipyrylmethane solution, dilute to volume with water, mix well and allow to stand for 3 hours or overnight. Measure the optical density relative to a reference solution containing the same quantity of rock solution, 50 ml of 2 N hydrochloric acid, 5 ml of ascorbic acid and diluted to volume in a 100-ml volumetric flask without the addition of diantipyrylmethane reagent. The spectrophotometer should be set at a wavelength of 380 nm, and either 1-cm or 4-cm cells used.

Calibration. Prepare the calibration curve by taking aliquots of the standard titanium solution containing 10 μg/ml TiO_2 and diluting to volume with hydrochloric acid, ascorbic acid and diantipyrylmethane solution as described above. Aliquots of 2-8 ml containing 20-80 μg TiO_2 can be used for the calibration of 4-cm cells, and 10-50 ml containing 100-500 μg TiO_2 for 1-cm cells.

Note: 1. For granitic and other acidic rocks containing little titanium a sample weight of 0.25 g should be taken, for most other silicate rocks a 0.1 g portion will be sufficient.

Determination of Titanium by Atomic Absorption Spectroscopy

The method given below is based upon that described by Walsh,(2) using the decomposition procedure of Riley.(6) It is suitable for rocks containing normal amounts (12% or more) of alumina and both iron and alkali metals. If the aluminium content of the rock is low, or the iron and alkali metals absent, as in certain silicate minerals, these elements can be added to the rock solution. The method as described is not suitable for carbonate rocks or silicates with high (more than 30%) calcium oxide.

Method

Accurately weigh approximately 0.5 g of the finely powdered rock material into a platinum crucible and evaporate to dryness in the usual way with 5 ml of concentrated

(60% m/m) perchloric acid and 15 ml of hydrofluoric acid. Dissolve the moist perchlorates in water and dilute to 500 ml. Any insoluble material should be collected, the titanium recovered in the usual way and added to the solution before dilution to volume. Alternatively a PTFE pressure decomposition vessel may be used to effect solution.

Set the atomic absorption spectrometer at 365.3 nm and using the fuel-rich nitrous oxide-acetylene flame with the acetylene level set just short of that required for a luminous flame, and a titanium hollow cathode lamp, aspirate the rock solution and measure the absorption due to titanium at a wavelength of 364.3 nm or 365.3 nm.

Measure also the absorption of a series of standard solutions containing 30, 20, 10 and 5 µg Ti per ml together with 100 µg per ml of aluminium, 50 µg per ml of iron and 35 µg per ml of sodium in diluted perchloric acid (1% v/v).

References

1. POLYAK L Ya., Zhur. Anal. Khim. (1962) 17, 206
2. WALSH J N., Analyst (1977) 102, 972
3. VAN LOON J C and PARISSIS C M., Anal. Lett. (1968) 1, 249
4. MININ A A., Uch. Zap. Permsk. Univ. (1955) 9, 177
5. JEFFERY P G and GREGORY G R E C., Analyst (1965) 90, 177
6. RILEY J P., Anal. Chim. Acta (1958) 19, 413

CHAPTER 45

Uranium

One of the most sensitive of methods in general use is based upon measurement of the fluorescence produced by uranium compounds in fused media.[1] The fluorescence intensity depends upon the medium used, as well as on the conditions used for the fusion. A number of elements have enhancing or quenching effects and for this reason prior separation of the uranium is necessary before the method can be applied to silicate rocks. Polarographic methods have been applied to the determination of uranium, but these are generally less sensitive than recent photometric methods, and no longer find wide acceptance - possibly because polarographs are not standard items of equipment in many laboratories.

There is no lack of general photometric reagents. Not all of these are sufficiently sensitive for application to the analysis of silicate rocks, although some of them were formerly used for this purpose. These include thiocyanate[2] and peroxide[3] which give yellow colours with uranium.

More sensitive reagents that have been suggested for the determination of uranium include dibenzoylmethane,[4,5] arsenazo I,[6] 1-(2-pyridylazo)-2-naphthol (PAN)[7] and 4-(2-pyridylazo)-resorcinol (PAR).[8] The molar absorptivities of the uranium complexes with these reagents range up to about 35,000 but interference occurs from the presence of many other elements, making prior separation essential.

Arsenazo III[9,10] has been used without extensive separation from other elements, but interference does occur and a separation procedure is recommended. The molar absorptivity of the uranium IV complex has a value of about 100,000. The absorption spectrum shows maximum absorption at a wavelength of 662.5 nm but this is a very sharp peak and when a spectrophotometer is being used, care must be taken to ensure that the optical densities are measured at this peak value. It is advisable to prepare a calibration solution and measure this at the same time as the sample solution, as slight variations in the slope of the calibration graph have been found to occur from day to day. The reagent itself has considerable absorption at wavelengths below 600 nm, and for this reason all measurements of optical density are made against a "blank solution" containing the same quantity of arsenazo III reagent.

Methods for the determination of uranium by atomic absorption spectroscopy are of low sensitivity, probably due to the thermal stability of uranium oxide. Interference from CN-band emission gives rise to high background noise levels, and enhancement effects have been recorded in the presence of alkali metals and other elements. It is perhaps not surprising therefore that the method has not been widely reported for silicate rocks. An indirect method by Alder and Das[11] is based upon the reduction of copper II to copper I by uranium IV, extraction of copper I as its neocuproine complex and determination of the extracted copper by atomic absorption using an air-acetylene flame. It is doubtful if this procedure can be used for normal silicate rocks containing only a few ppm uranium or less.

Uranyl nitrate is appreciably soluble in a number of organic solvents, and solvent extraction procedures have often been used to recover uranium from aqueous solutions. The efficiency of the separation depends very much upon the nature and amount of other constituents present in the solution, and small quantities of other metals are usually extracted with the uranium into the organic phase. A number of organic solvents can be used, including diethyl ether, ethyl acetate and trioctylphosphine oxide,[12] which have all been used for the determination of uranium in silicate rocks.

Korkisch _et al_[13,14] used strongly basic anion exchange resins (Dowex 1x8) to separate uranium, thorium, rare earths, cadmium, bismuth and lead from all other elements present in a deep-sea sediment. Somewhat similar separations were described by Strelow _et al_,[15] and by Kiriyama and Kuroda.[16]

One of the procedures commonly used for the recovery of uranium from rocks, ores and minerals involves elution with an ether-nitric acid mixture from a column packed with cellulose and alumina.[17,18,19] Small amounts of both thorium and zirconium tend to bleed through the cellulose-alumina column and give rise to high values for uranium where the determination is completed photometrically with arsenazo III. This error can be avoided by dividing the solution into two parts in only one of which is the unreactive uranium (VI) reduced to the reactive uranium (IV). After the addition of arsenazo III reagent, the unreduced solution is used as the reagent blank for the reduced sample solution.

Spectrophotometric Determination Following Anion Exchange Separation

In the procedure described here, based upon that given by Kiriyama and Kuroda[16] the silicate rock material is decomposed by evaporation with perchloric and hydrofluoric acids. After evaporation with sulphuric acid, the sulphate solution is transferred to an anion exchange column. Thorium, zirconium and uranium are all recovered from the resin by washing with the appropriate eluates, and the separate determination of all three elements completed spectrophotometrically with arsenazo III. For thorium a second anion exchange separation is recommended.

Reagents: Arsenazo III solution, 0.1% (w/v) in water.

Hydrogen peroxide, 30 volume solution.

Ion-exchange resin, Dowex 1x8, 100-200 mesh. Slurry 10 g with water and transfer to a glass column approximately 1 cm internal diameter to give a bed depth of about 25 cm. Convert to the sulphate form by washing with a solution 0.025 M in sulphuric acid, 0.1 M in ammonium sulphate until the eluate is free from chloride ions.

Standard uranium stock solution, dissolve 0.118 g of uranium oxide in a few ml of concentrated nitric acid and evaporate just to dryness. Moisten with water, add hydrochloric acid and again evaporate to dryness. Dissolve the chloride residue in water and dilute to 1 litre. This solution contains 100 µg U per ml.

Standard uranium working solution, dilute 5 ml of the stock solution with water to 250 ml in a volumetric flask. This solution contains 2 µg U per ml.

Procedure. Accurately weigh approximately 1 g of the finely powdered rock material into a platinum or PTFE dish, moisten with water, add 2 ml of concentrated perchloric acid and 5 ml of concentrated hydrofluoric acid and evaporate almost to dryness on a steam bath. Evaporate twice more with 1-ml portions of hydrofluoric acid and a few drops of perchloric acid, finally to complete dryness. Dissolve the residue in 10 ml of 6 M hydrochloric acid, dilute with water to a volume of about 100 ml and precipitate metal hydroxides from the hot solution by cautious addition of diluted (1+1) aqueous ammonia until a slight excess is present. Allow to digest on a steam bath for about an hour.

Collect the precipitate on an open textured filter paper such as a Whatman No. 41 and wash with a warm 2% ammonium chloride solution. Dissolve the residue from the paper using 30 ml of 6 M hydrochloric acid, add 5 ml of 6 M sulphuric acid and evaporate to dryness. To the dry residue add 0.5 ml of 5 M sulphuric acid and 10 ml of water, and dissolve by warming (Note 1). Dilute to a volume of 200 ml with water and transfer to the ion-exchange column. Rinse the column with 100 ml of the 0.025 M sulphuric acid - 0.1 M ammonium sulphate solution containing 5 ml of hydrogen peroxide and discard the eluate. Elute the thorium with 100 ml of 0.025 M sulphuric acid - 0.1 M ammonium sulphate solution, the zirconium with about 50 ml of 4 M hydrochloric acid and the uranium with 30 ml of 1 M perchloric acid. Complete the determination spectrophotometrically as described below.

Determination of Thorium. Add a solution of ferric chloride containing about 6 mg Fe to the eluate, heat, precipitate the iron hydroxide with diluted ammonia. Collect the precipitate on an open textured filter paper, wash with 2% ammonium chloride solution and dissolve in a small volume of 6 M hydrochloric acid. Transfer the solution to a short column of Dowex 1x8 anion exchange resin previously washed with 6 M hydrochloric acid and elute with 6 M hydrochloric acid. Determine the thorium content with arsenazo III solution as described in chapter 42.

Determination of Zirconium. Transfer the zirconium eluate or a suitable aliquot of it to a 25-ml measuring flask, add 18.5 ml of concentrated hydrochloric acid, 1 ml of a freshly prepared 1% (w/v) gelatin solution and 1 ml of arsenazo III solution. Dilute to volume with water, measure the optical density at 665 nm and complete the determination of zirconium by reference to a calibration graph prepared from standard zirconium solutions.

Determination of Uranium. Evaporate the uranium eluate to dryness. Add 1 ml of perchloric acid and 3-4 ml of nitric acid, and again evaporate to dryness. Dissolve the residue in 5 ml of 8 M hydrochloric acid, add 0.7 g of metallic zinc and allow to stand for 15 minutes. Using 14 ml of a solution 7 M in hydrochloric acid and saturated with oxalic acid, transfer the solution to a 25-ml volumetric flask, add 1 ml of arsenazo III solution and dilute to volume with water. Measure the optical density of the solution at a wavelength of 662.5 (Note 2) against a reagent blank solution, and determine the uranium content by reference to a calibration curve prepared from solutions containing 0 to 10 μg uranium in 25 ml. The procedure given above, beginning with evaporation to dryness, should be followed for each calibration solution.

Notes: 1. The original description of the method suggests that any residue should be collected at this stage. It should be ignited and fused with 200 mg of sodium carbonate. Dissolve the melt with 2 ml of 6 M sulphuric acid, add 2 ml of hydrofluoric acid and evaporate to dryness. Add 0.5 ml of 5 M sulphuric acid and 10 ml of water and heat to dissolve the bulk of the residue. Add 100 ml of water and warm to give a clear solution. Cool, add 10 ml of hydrogen peroxide, combine with the filtrate obtained earlier and transfer to the ion-exchange column. In normal practice, the residue should be collected and decomposed following the rock decomposition with hydrofluoric and perchloric acids.

2. See earlier comment on the need to ensure that optical densities are measured at their peak value.

References

1. STRASHEIM A., Spectrochim. Acta (1950) 4, 200
2. NEITZEL O A and DESESA M A., Analyt. Chem. (1957) 29, 756
3. HARVEY C O., Bull. Geol. Surv. Gt. Brit. (1951) (3), 43
4. YOE J H, WILL F and BLACK R., Analyt. Chem. (1953) 25, 1200
5. UMEZAKI Y., Bull. Chem. Soc. Japan (1963) 36, 769
6. HOLCOMB H P and YOE J H., Analyt. Chem. (1960) 32, 616
7. CHENG K L., Analyt. Chem. (1958) 30, 1027
8. FLORENCE T M and FARRAR Y., Analyt. Chem. (1963) 35, 1613
9. SINGER E and MATUCHA M., Zeit. Anal. Chem. (1962) 191, 248
10. NEMODRUK A A and GLUKHOVA L P., Zhur. Anal. Khim. (1966) 21, 688
11. ALDER J F and DAS B C., Anal. Chim. Acta (1977) 94, 193
12. KONSTANTINOVA M H, MARETA S and IORDANOV N., Dokl. Bolg. Akad. Nauk. (1977) 30, 1293
13. KORKISCH J and ARRHENIUS G., Analyt. Chem. (1964) 36, 850
14. KORKISCH J, STEFFAN I and GROSS H., Mikrochim. Acta (1976) I, 503
15. STRELOW F W E, KOKOT M L, VAN DER WALT T N and BHAGA B., J S Afr. Chem. Inst. (1976) 29, 97
16. KIRIYAMA T and KURODA R., Anal. Chim. Acta (1974) 71, 375
17. The Determination of Uranium and Thorium, Nat. Chem. Lab., D.S.I.R., HMSO, 1963 (2nd ed.)
18. BURSTALL F H and WELLS R A., Analyst (1951) 76, 396
19. ADAMS J A S and MAECK W J., Analyt. Chem. (1954) 26, 1635

CHAPTER 46

Vanadium

Titrimetric methods were at one time used for rocks rich in vanadium. These methods have now been replaced by spectrographic, spectrophotometric and atomic absorption procedures.

There is no shortage of colour-forming reagents for vanadium; these have been reviewed by Svehla and Tolg.[1] They give rise to a multiplicity of photometric methods in a wide variety of matrices. Two methods that are applicable to silicate rocks and minerals are given in detail below. These are based respectively upon the yellow colour given by vanadium with tungstate in phosphoric acid solution, and the purple colour of the complex formed with N-benzoyl-o-tolyl-hydroxylamine in 6 N hydrochloric acid solution.

For determination by atomic absorption spectroscopy, a fuel rich nitrous oxide-acetylene flame is now commonly used. The vanadium spectrum contains a large number of lines, 318.5 is the most sensitive, 318.4 is the most intense: these lines are not resolved on most analytical instruments. For silicate rocks, where alkali metals, aluminium and iron may all be major constituents, it is necessary to add an ionisation buffer to samples and standards to overcome the varying effect of ranging concentrations of these ionisation suppressants on the vanadium absorption. The determination of vanadium by this technique may be combined with the determination of other elements.

Spectrophotometric Determination as Phosphotungstate

Vanadium forms a yellow-coloured complex with alkali tungstate in solutions containing phosphoric acid. The yellow solution has a maximum optical density in the ultraviolet region of the spectrum, but measurements can conveniently be made at 400 nm within the range of even the simplest spectrophotometer. The Beer-Lambert Law is valid over the concentration range normally encountered in silicate rocks. The reaction is not sufficiently sensitive for direct application to silicate rocks and it is necessary to use a procedure for concentrating the vanadium. In the method described below, adapted from Bennett and Pickup,[2] a solvent extraction stage is incorporated removing vanadium from dilute acetic acid solution as its complex with 8-hydroxyquinoline.

Method

Reagents: 8-Hydroxyquinoline solution, dissolve 1 g of the reagent in 100 ml of 2 N acetic acid.

Chloroform.

Sulphuric acid, 6 N.

Phosphoric acid, 5 N.

Sodium tungstate solution, dissolve 8.25 g of the dihydrate in 50 ml of water.

Standard vanadium stock solution, dissolve 0.230 g of dried ammonium metavanadate in water and dilute to 500 ml with water. This solution contains 200 μg V/ml.

Standard vanadium working solution, dilute the stock solution with water to give working standard containing 10 μg V/ml.

Procedure. Accurately weigh approximately 0.5 g of the finely ground rock material into a platinum crucible and add 3 g of anhydrous sodium carbonate and 0.1 g of potassium nitrate (Note 1). Fuse the mixture over a Bunsen burner for 30 minutes, or longer if refractory minerals are present, and allow to cool. Extract the melt with hot water, filter using a medium or close-textured paper, and wash the residue well with hot 2 per cent (w/v) sodium carbonate solution. Discard the residue and combine the filtrate and washings.

Transfer this solution, or a suitable aliquot of it containing not more than 60 μg vanadium, to a 100-ml separating funnel, add 2 drops of methyl orange indicator solution and titrate with 6 N sulphuric acid until the indicator turns red. Swirl the solution to remove as much as possible of the liberated carbon dioxide, and add 1 ml of 8-hydroxyquinoline solution and 3 ml of chloroform. Shake the solution for 1 minute to extract the dark-coloured vanadium complex, allow the phases to separate and remove the chloroform layer. Rinse the funnel with a little chloroform. Add a further 0.5 ml of 8-hydroxyquinoline solution and 3 ml of chloroform and again extract by shaking for 1 minute. If the extract shows an appreciable dark colour, repeat the extraction for a third time. Discard the aqueous solution.

Collect all the chloroform extracts in a small platinum crucible, add 0.1 g of sodium carbonate and allow the chloroform to evaporate. Burn off the organic matter and fuse the residue to convert all vanadium to sodium vanadate. Dissolve the melt by warming with 2-3 ml of water, add 1 ml of 6 N sulphuric acid, 1 ml of 5 N phosphoric acid and 0.5 ml of sodium tungstate solution. Heat just to boiling, cool, transfer to a 10-ml volumetric flask and dilute to the mark with water. Measure the optical density of the solution relative to water in 2-cm cells using the spectrophotometer set at a wavelength of 400 nm. Measure also the optical density of a reagent blank solution, prepared in the same way as the sample solution but omitting the rock powder.

Calibration. Transfer aliquots of 1-6 ml of the standard vanadium working solution containing 10-60 μg V, to separate 10-ml volumetric flasks and add to each sulphuric acid, phosphoric acid and sodium tungstate solution as described for the sample solution above. Plot the relation of optical density to vanadium concentration.

Note: 1. The exclusion of potassium nitrate leads to low recoveries of vanadium, particularly from rocks rich in ferrous iron. The prolonged heating of rock samples with a flux containing excessive potassium nitrate or other oxidising agent should, however, be avoided as this will result in an appreciable attack of the platinum crucible.

Spectrophotometric Determination using N-Benzoyl-o-Tolylhydroxylamine

A number of hydroxylamine derivatives have been suggested as reagents for vanadium, but very few combine a sufficient sensitivity with the necessary specificity to enable vanadium to be determined in silicate rocks without a prior separation from interfering elements. One reagent that has been used for this is N-benzoyl-o-tolylhydroxylamine which reacts with vanadium in strongly acid solution to give a purple-coloured complex,[3] readily soluble in organic solvents. The molar absorptivity is 5250, and the Beer-Lambert Law is valid in the range 0-60 μg vanadium in 10 ml of extract. The maximum optical density occurs at a wavelength of 510 nm. The reaction occurs only with vanadium(V), and somewhat lower calibration curves are obtained unless the aliquots of the standard vanadium solution are reoxidised prior to colour formation. For this oxidation an aqueous potassium permanganate solution is used, added drop by drop until an excess is present. Any chlorine liberated on adding hydrochloric acid is removed with sulphuric acid.

N-benzoyl-o-tolylhydroxylamine has been used for the determination of vanadium in silicate rocks and minerals by Jeffery and Kerr,[4] whose procedure is given in detail below. The most serious interference is from titanium which gives an intense yellow colour with the reagent. This colour can be completely suppressed by adding a small quantity of sodium fluoride.

This reagent also gives yellow colours with gold and platinum. Whilst the small amounts of gold and platinum that occur in silicate rocks are insufficient to interfere with the determination of vanadium, any platinum introduced from the platinum apparatus may completely obscure the violet colour from rocks containing only a few ppm of vanadium. For this reason prolonged fusion in platinum should be avoided and the alternative decomposition in PTFE and silica used - particularly for samples low in vanadium.

The complex formed between vanadium and N-benzoyl-o-tolylhydroxylamine is soluble in a number of organic solvents. Solutions in carbon tetrachloride, chloroform, iso-butylmethyl ketone and toluene all give similar calibration curves. Chloroform should be avoided because of possible interference from the presence of traces of ethanol which give rise to a yellow-coloured complex of vanadium with the reagent. Carbon tetrachloride is a convenient solvent to use.

Method

Reagents: N-benzoyl-o-tolylhydroxylamine solution, dissolve 0.02 g of the recrystallised reagent in 100 ml of carbon tetrachloride. (The preparation of this reagent is described by Majundar and Das.)[3]

Potassium permanganate solution, approximately 0.02 M.

Sulphamic acid solution, approximately 0.05 M.

Sodium fluoride solution, saturated aqueous, store in a polythene bottle.

Standard vanadium stock solution, prepare as described above.

Procedure. Decompose a 100-mg portion of the finely powdered silicate rock or mineral (Note 1) by evaporation in platinum apparatus with sulphuric, nitric and hydrofluoric acids in the usual way, removing excess sulphuric acid by heating on a hot plate. Fuse the dry residue in platinum with potassium pyrosulphate, and extract the fused melt with 10 ml of water containing 2 drops of 20 N sulphuric acid. Transfer the solution to a separating funnel and add potassium permanganate solution drop by drop until an excess is present, giving a pink colour that persists for 5 minutes.

The volume of the solution at this stage should be about 20 ml. Add 2 ml of sulphamic acid solution, 2 ml of sodium fluoride solution and 20 ml of concentrated hydrochloric acid. Now add by pipette 10 ml of the N-benzoyl-o-tolylhydroxylamine reagent solution, stopper and shake for 30 seconds. Allow the two layers to separate and filter the lower organic layer through a small wad of cotton wool into a 2-cm spectrophotometer cell. Measure the optical density relative to carbon tetrachloride at a wavelength of 510 nm. Determine also the optical density of a reagent blank solution similarly prepared but omitting the sample material.

Calibration. Transfer aliquots of 1-6 ml of the standard vanadium solution containing 10-60 μg vanadium to a series of separating funnels and dilute each with water to a volume of 20 ml. Add potassium permanganate solution drop by drop until a slight excess is present, then add sulphamic acid solution, sodium fluoride solution and hydrochloric acid, and continue as described above for the sample solution. Plot the relation of optical density to vanadium concentration.

Note : 1. This sample weight is sufficient for rock samples containing 50-500 ppm V. For rocks containing less vanadium than this, use a larger sample weight, for rocks containing more, dilute the rock solution to volume and use an aliquot for the determination.

Determination of Vanadium by Atomic Absorption Spectroscopy

As noted earlier, vanadium can readily be determined by atomic absorption spectroscopy. Interference has been recorded from a number of elements. In an early procedure Terashima[5] noted that aluminium present in silicate rocks suppressed this interference, and added aluminium, calcium and potassium to the standard solutions used to calibrate the spectrometer.

A somewhat larger range of additions to the standards was used by Warren and Carter[6] including a further quantity of potassium to serve as an ionisation buffer for both standards and the rock solution. The rock sample was decomposed by a mixture of nitric, perchloric and hydrofluoric acids in a PTFE bomb at a temperature of 100° for one hour. After evaporation to remove nitric and hydrofluoric acids, the precipitated fluorides were dissolved by the addition of boric acid. The determination was completed by absorption measurement at 318.4 nm in a nitrous oxide-acetylene flame in the usual way. As described the limit of determination is 10 ppm, which is adequate for most silicate rocks. The procedure used is described in detail in the chapter on nickel.

A more elaborate procedure involving an ion-exchange separation using a strongly basic resin was described by Korkisch and Gross.[7] The procedure was used for the analysis of uranium minerals and yellow cake, and its application to silicate rocks and minerals does not appear to have been extensively studied.

The possibility of using a graphite furnace for the determination of vanadium by flameless atomic absorption was indicated by Schweizer.[8] Interference from other elements is avoided by the technique of 'standard additions'. The method was devised particularly for carbonate rocks - although reported as applicable to silicate rocks, further work is clearly necessary before it can be described as a standard or routine method for this application.

References

1. SVEHLA G and TOLG G., Talanta (1976) 23, 755
2. BENNETT W H and PICKUP R., Colon. Geol. Min. Res. (1952) 3, 171
3. MAJUMDAR A K and DAS G., Anal. Chim. Acta (1964) 31, 147
4. JEFFERY P G and KERR G O., Analyst (1967) 92, 763
5. TERASHIMA S., Japan Analyst (1973) 22, 1317
6. WARREN J and CARTER D., Can. J. Spectrosc. (1975) 20, 1
7. KORKISCH J and GROSS H., Talanta (1973) 20, 1053
8. SCHWEIZER V B., Atom. Absorp. Newsl. (1975) 14, 137

CHAPTER 47

Zinc

Dithizone extraction procedures have been used[1] for the determination of zinc in silicate rocks, but Carmichael and McDonald[2] have shown that such procedures are subject to interference from other metals, particularly copper, cobalt and nickel. This interference leads to high results - in some analyses twice the accepted values were obtained. This conclusion is in agreement with the work of Greenland,[3] who particularly implicated nickel as the cause of high zinc values when a dithizone extraction was followed by photometric determination.

A number of spectrophotometric procedures for small amounts of zinc were investigated by Margerum and Santacana,[4] who preferred a single colour dithizone procedure with bis(2-hydroxyethyl)dithio-carbamate as a masking agent. The use of zincon was not recommended when certain impurities were present. When zinc was isolated by the use of a column of anion exchange resin, either dithizone or zincon could be used. Such procedures have been described by Rader et al[5] and Huffman et al.[6]

The ease and simplicity and freedom from interference which characterises the determination of zinc by atomic absorption spectroscopy is in direct contrast to the lengthy, difficult determination by spectrophotometry, for which a careful separation from interfering elements is required. The zinc line at 213.9 nm is of more than adequate sensitivity in an air-acetylene flame. The conditions are not critical and perchlorate, chloride, sulphate and nitrate solutions have all been recommended.

Spectrophotometric Determination Following Ion-Exchange Separation

Zincon is the trivial name for the red organic reagent 5-(2-carboxy-phenyl)-1-(2-hydroxy-5-sulphophenyl)-3-phenylformazan which forms a blue complex with zinc in alkaline solution.[7,8] The maximum absorption occurs at a wavelength of 620 nm, at which the reagent has only very slight absorption. The Beer-Lambert Law is followed in the range 0-120 µg zinc in 50 ml. The reagent solution is stable for at least a week, but the zinc complex for only a few hours.

Many ions interfere with the determination of zinc, including aluminium, beryllium, bismuth, cadmium, cobalt, chromium (III), copper, iron (III), manganese, mercury, molybdenum (VI), nickel and titanium, which all react with the reagent, and must be separated before the zinc complex is formed. Anion-exchange separation[7,9] can conveniently be used for this.

The procedure given below is essentially that of Huffman <u>et al</u>,[6] designed to determine zinc in rocks containing as much as 20 per cent iron. The rock material is decomposed by evaporation with hydrofluoric, nitric, perchloric and sulphuric acids, and the residue dissolved in dilute hydrochloric acid. On passing the hydrochloric acid solution through a column of anion exchange resin, zinc, lead, molybdenum, iron, uranium and cadmium are absorbed on the resin, and the remaining constituents eluted. Iron is removed by washing the column with 1.2 M hydrochloric acid, and the zinc collected by elution with 0.01 M hydrochloric acid. Zinc is precipitated from this eluate with diethyl-dithiocarbamate at a pH of 8.5, and the precipitate extracted into chloroform. Dilute hydrochloric acid is used to strip the zinc from the organic solution, and the determination is then completed spectrophotometrically with zincon.

Method

<u>Apparatus</u>. Ion-exchange column, set up a small ion-exchange column 8-cm in length and 0.8-cm in diameter, of a strongly basic ion exchange resin, such as Dowex 1 x 8, and wash thoroughly with 1.2 M hydrochloric acid, followed by 0.01 M hydrochloric acid and then water. Leave the column 1.2 M in hydrochloric acid ready for use.

<u>Reagents</u>:

Ammonium citrate buffer solution, dissolve 25 g of citric acid in 300 ml of water, bring pH of the solution to 8.5 by adding dilute aqueous ammonia, and dilute to 500 ml with water.

Sodium borate buffer solution, dissolve 14.4 g of sodium hydroxide, 24.7 g of boric acid and 29.8 g of potassium chloride in about 950 ml of water, cool to room temperature and dilute to 1 litre. The pH should be 10.2.

Zincon solution, transfer 0.130 g of the solid zincon reagent to a 100-ml volumetric flask, add 2 ml of 1 M sodium hydroxide solution and dilute to volume with water after the zincon is in solution. The solution is stable for about 10 days.

Standard zinc stock solution, ignite zinc sulphate heptahydrate to constant weight in an electric muffle furnace set at a temperature of 450°. Allow to cool, and weigh 0.247 g of the anhydrous salt into a 200-ml volumetric flask and dilute to volume with water. This solution contains 500 µg Zn per ml.

Standard zinc working solution, dilute 5 ml of the stock solution to 250 ml with 1.2 M hydrochloric acid. This solution contains 10 µg Zn per ml.

Diethyldithiocarbamate solution, dissolve 0.5 g of the sodium salt in 100 ml water

<u>Procedure</u>. Weigh approximately 0.5 g of the finely ground sample material into a small platinum dish or crucible, moisten with water and add 10 ml of concentrated nitric acid and 10 ml of hydrofluoric acid. Cover the vessel, and set it aside for several hours, preferably overnight.

Add to the dish 5 ml of perchloric acid and 10 ml of 20 N sulphuric acid. Transfer the dish to a hot plate and evaporate to a volume of about 3 ml. Cool the dish, wash down the sides with a little water, add a further 5 ml of perchloric acid and again evaporate, this time just to dryness - but avoiding baking the residue. Add 10 ml of concentrated hydrochloric acid and 25 ml of water to the residue, rinse the solution into a beaker and digest on a steam bath for 30 minutes (Note 1). Cool the solution and transfer to a 100-ml volumetric flask.

Transfer the entire 100 ml of solution (or a suitable portion of it containing from 20 to 100 µg zinc and diluted to a volume of 100 ml with 1.2 M hydrochloric acid) to the resin column. Regulate the flow rate of the solution through the column to about 1 ml per minute by adjusting the stopcock at the bottom of the column. When the flow ceases, discard the solution that has passed through, wash the column with 50 ml of 1.2 M hydrochloric acid and discard the wash solution. Place a clean 150-ml beaker under the column and elute the zinc by passing 45 ml of 0.01 M hydrochloric acid through the resin also at a flow rate of about 1 ml per minute.

Add 3 drops of phenolphthalein indicator solution to the beaker containing the zinc solution, and adjust the pH to 8.5 ± 0.5 by adding dilute ammonia solution dropwise until the pink indicator colour just but only just forms. Quantitatively transfer the solution to a 125 ml separating funnel, add 2 ml of the diethyldithiocarbamate solution, stopper the funnel and shake the solution. Add 10 ml of chloroform to the funnel and shake to extract the zinc-carbamate complex, draining the chloroform solution into a clean separating funnel. Rinse the stem of the extraction funnel with about 2 ml of chloroform and add this chloroform to the organic extract. Repeat the extraction and rinsing operations once more using 5 and 2 ml volumes of chloroform respectively, and then discard the aqueous solution.

Add 10 ml of water to the chloroform solution and wash by shaking the funnel for about 30 seconds. Drain the chloroform into a clean 125-ml separating funnel, rinse the stem with about 2 ml of chloroform and add the chloroform wash liquor to the main portion of chloroform. Add 10 ml of 0.16 M hydrochloric acid to the combined chloroform solution and strip the zinc from the organic layer by shaking for at least 1 minute. Remove and discard the lower chloroform layer. Wash the aqueous phase with

10 ml of chloroform by shaking for 30 seconds and again drain, remove and discard the organic layer.

Filter the aqueous solution through a 5.5-cm filter paper previously washed with 0.16 M hydrochloric acid, and collect the filtrate in a 50-ml volumetric flask. Wash the separating funnel twice with about 3 ml of water, passing the washings through the paper into the flask. Add 10 ml of the sodium borate buffer solution to the flask and mix with the solution. A pH of 9 $\pm$ 0.5 should be obtained. Add 3 ml of the zincon reagent solution to the flask to give a brownish mixed colour, with a transition to blue as the zinc concentration increases. Dilute the solution to 50 ml with water. The intensity of the colour complex reaches a maximum very quickly and is stable for a few hours.

Measure the optical density of this solution in 1-cm cells, against the reagent blank as the reference solution, using the spectrophotometer set at a wavelength of 620 nm.

Calibration. Transfer aliquots of 0-10 ml of the standard solution containing 0-100 µg Zn to separate 50-ml volumetric flasks, add 10 ml of 0.16 M hydrochloric acid, and 10 ml of the sodium borate buffer solution and mix well. Add 3 ml of the zincon reagent solution, dilute to volume with water and mix well. Measure the optical densities in 1-cm cells at a wavelength of 620 nm and plot these values against zinc concentrations to obtain a standard working curve.

Note: 1. If any insoluble residue remains, collect it on a small filter, wash with a little water, dry, ignite and fuse with the minimum quantity of sodium carbonate. Dissolve the melt in the minimum amount of hydrochloric acid and combine with the solution in the 100-ml volumetric flask before dilution to volume.

Determination of Zinc by Atomic Absorption Spectroscopy

The method below is based upon that described by Sanzolone and Chao.[10] The determination of zinc by this method can be combined with the determination of a number of other elements of adequate sensitivity by atomic absorption spectroscopy. The calibration curve for zinc is slightly convex towards the concentration axis.[11] A procedure for determining zinc and copper in the same solution has been given by Belt[12] and the technique has been used by Burrell[13] for determining zinc in amphibolites.

Procedure. Accurately weigh approximately 0.5 g of the finely powdered silicate rock material into a small platinum dish or crucible, and evaporate to dryness with 1 ml of concentrated perchloric acid and 5 ml of concentrated hydrofluoric acid. Moisten the dry residue with a further 1 ml of perchloric acid and again evaporate to dryness. Allow to cool and add 0.3 ml of perchloric acid and rinse the residue into a small beaker. Warm until dissolution is complete, then cool, and dilute to 25 ml with water in a volumetric flask.

Using an atomic absorption spectrophotometer fitted with a zinc lamp and operating according to the manufacturer's instructions, measure the absorption at a wavelength of 213.9 nm. Use a 10 ppm zinc standard solution to prepare a series of working standards covering the concentration range 0 to 1 ppm Zn. The calibration line is appreciably curved.

References

1. SANDELL E B., Ind. Eng. Chem. Anal. Ed. (1937) 9, 464
2. CARMICHAEL I and McDONALD A J., Geochim. Cosmochim. Acta (1961) 22, 87
3. GREENLAND L P., Geochim. Cosmochim. Acta (1963) 27, 269
4. MARGERUM D W and SANTACANA F., Analyt. Chem. (1960) 32, 1157
5. RADER L F, SWADLEY W C, LIPP H H and HUFFMANN C Jr., U S Geol. Surv. Prof. Paper 400-B, p.B437, 1960
6. HUFFMAN C Jr, LIPP H H and RADER L F., Geochim. Cosmochim. Acta (1963) 27, 209
7. YOE J H and RUSH R M., Anal. Chim. Acta (1952) 6, 526
8. YOE J H and RUSH R M., Analyt. Chem. (1954) 26, 1345
9. JACKSON R K and BROWN J G., Proc. Amer. Soc. Hort. Sci. (1956) 68, 1
10. SANZOLONE R F and CHAO T T., Anal. Chim. Acta (1976) 86, 163
11. ERDEY L, SVEHLA G and KOLTAI L., Talanta (1963) 10, 531
12. BELT C B Jr., Econ. Geol. (1964) 59, 240
13. BURRELL D C., Norsk. Geol. Tidsskr. (1965) 45, 21

CHAPTER 48

Zirconium and Hafnium

One of the oldest procedures for the determination of zirconium (+ hafnium) is based upon precipitation as phosphate from dilute sulphuric acid solution. This determination can readily be combined with those of a number of other minor constituents of silicate rocks, such as chromium, vanadium, sulphur and chlorine in an initial alkaline filtrate, and the rare earths and barium with the zirconium in the residue. This method is given in textbooks of rock analysis, but the published procedures do not stress the difficulties in making accurate determinations of these small amounts of zirconium.[1] It was given in detail in the earlier editions of this book, but now appears to be very little used. Other gravimetric methods for zirconium[2,3] have not been widely adopted for silicate rocks.

Although a large number of reagents have been suggested for the photometric determination of zirconium, none of the reactions is completely specific and few are even selective. In their application to silicate rocks it is usually necessary to make a separation from interfering elements. Babko and Vasilenko[4] studied a total of eighteen reagents for zirconium and considered xylenol orange and methylthymol blue to be the best. Arsenazo I[2], arsenazo III[5] and quinalizarin sulphonic acid[6] have all been used to determine zirconium in rocks and minerals. A procedure involving an ion exchange separation of zirconium, uranium and thorium, followed by spectrophotometric determination of all three with arsenazo III, based upon the work of Kiriyama and Kuroda[7] is described in chapter 45.

Both zirconium and hafnium are difficult to determine by atomic absorption spectroscopy - the sensitivity is poor, matrix effects considerable and interelement effects serious. This technique cannot as yet be recommended for application to silicate rocks.

Spectrophotometric Determination of Total Zirconium and Hafnium

The procedure described in detail below is based upon the use of xylenol orange. This reagent, which is yellow in colour, forms a red-coloured complex with zirconium, with a maximum absorption at a wavelength of 535 nm. In a dilute acid medium the only other elements to form complexes with the reagent are hafnium, bismuth, tin, molybdenum and iron (III).[8] In the determination of zirconium (+ hafnium), ferric iron is reduced with ascorbic acid. None of the remaining elements is likely to be present in amounts sufficient to interfere. The separation of zirconium from other elements, other than from silicon, is therefore not required with this reagent.

Method

Reagents: Ascorbic acid solution, dissolve 2.5 g of the reagent in 50 ml of water. Prepare freshly as required.

Xylenol orange solution, dissolve 50 mg of the reagent in water and dilute to 100 ml.

Standard zirconium stock solution, dissolve 3.9 g of zirconium sulphate $Zr(SO_4)_2.4H_2O$ in dilute sulphuric acid and dilute to 1 litre with a final acid concentration of 1 M. Standardise by back titration with bismuth (after adding an excess of EDTA) using xylenol orange as indicator. This solution contains 1 mg Zr per ml.

Standard zirconium working solution, transfer 5 ml of the stock zirconium solution to a 1 litre volumetric flask, add 200 ml of 6 N hydrochloric acid, dilute to volume with water and mix well. This solution contains 5 µg Zr per ml.

Procedure. Accurately weigh approximately 0.5 g of the finely powdered silicate rock into a platinum dish and add 5 ml of 20 N sulphuric acid, 1 ml of concentrated nitric acid and 10 ml of concentrated hydrofluoric acid. Transfer the dish to a hot plate and evaporate to fumes of sulphuric acid. Allow to cool, dilute with a few ml of water, add 5 ml of concentrated hydrofluoric acid and evaporate to fumes. Allow to cool, dilute with water and again evaporate to fumes. Repeat the evaporation once more, this time to dryness (Note 1).

Allow to cool, moisten the residue with 5 ml of water, add 10 ml of 6 N hydrochloric acid, warm to dissolve most of the residue and then rinse into a 150-ml beaker containing 50 ml of water. Warm the solution on a hot plate until all soluble material has dissolved.

Transfer the solution to a 100-ml volumetric flask, add a further 10 ml of 6 N hydrochloric acid and dilute to volume with water. Mix well and transfer an aliquot of the solution containing not more than 40 µg of zirconium to a 50-ml volumetric flask. Add sufficient hydrochloric acid to bring the final concentration to 0.8 N, followed by 5 ml of the ascorbic acid solution to reduce ferric iron and 2 ml of the xylenol orange solution. Dilute to volume with water, mix well and measure the optical density using the spectrophotometer set at a wavelength of 525 nm, against a reagent blank solution prepared in the same way as the sample solution but omitting the sample material.

Calibration. For the calibration graph, pipette 0-8 ml aliquots of the standard zirconium working solution containing 0-40 μg zirconium to 50-ml volumetric flasks and proceed as described above.

Note: 1. A major part of the zirconium may be present in the form of the mineral zircon, which is decomposed by acid only with some difficulty. If any residue does remain undecomposed, collect it on a small filter, wash with a little water, dry and ignite in a small platinum crucible. Fuse with a little sodium carbonate, extract with water, filter, dissolve the residue in a little hydrochloric acid and add to the main rock solution.

Determination of Hafnium

Although as noted above, hafnium is present in silicate rocks to a smaller extent than zirconium, it is by no means as small as the amounts of some other elements that are readily determined by photometric methods. Chemical methods, including spectrophotometry, have not been successfully applied to the determination of hafnium in silicate rocks. For this purpose emission spectrography,[9] X-ray spectrography[3] and neutron activation analysis[9,10,11] are commonly employed.

References

1. BENNETT W H and PICKUP R., Colon. Geol. Min. Res. (1952) 3, 171
2. TSERKOVNITSKAYA I A and BOROVAYA N S., Vestn. Leningr. Univ. (1962) No. 16, Ser. Fiz. i. Khim. (3), 148
3. TUZOVA A M and NEMODRUK A M., Zhur. Anal. Khim. (1958) 13, 674
4. BABKO A K and VASILENKO V T., Zavod. Lab. (1961) 27, 640
5. GORYUSHINA V G and ROMANOV E V., Zavod. Lab. (1960) 26, 415
6. CULKIN F and RILEY J P., Anal. Chim. Acta (1965) 32, 197
7. KIRIYAMA T and KURODA R., Anal. Chim. Acta (1974) 71, 375
8. CHENG K L., Talanta (1959) 2, 61
9. SETSER J L and EHMANN W D., Geochim. Cosmochim. Acta (1964) 28, 769
10. BUTLER J R and THOMPSON A J., Geochim. Cosmochim. Acta (1965) 29, 167
11. MORRIS D F C and SLATER D N., Geochim. Cosmochim. Acta (1963) 27, 285

Author Index

ABBEY S. 6, 16, 51, 52, 53, 64, 65, 68, 224; 297, 314, 315, 330, 331
ABDULLA M.I. 120, 232, 236
ADAMS S.J. 17, 196
AHRENS L.H. 275, 290
AKAIWA H. 138
ALDER J.F. 353
ALKHIMENKOVA G.I. 312
ALLEN W.J.F. 61
ALLMANN R. 306
ALONSO S.J. 110
ANDERSON W.L. 288
ANDO A. 7
ANTWEILER J.C. 20
ARMANNSSON H. 161
ASLIN G.E.H. 56
ASTON S.R. 240
ATKINS D.H.F. 264

BAADSGAARD H. 269
BABKO A.K. 368
BAFFI F. 56
BAKHMATOVA T.K. 275
BAREEV V.V. 330
BARREDO F.B. 55,115,312
BARTEL A.J. 339
BASKOVA Z.A. 213, 214
BAUR W.H. 265
BEHNE W. 142
BELCHER R. 133
BELOPOL'SKI M.P. 274
BELT C.B.Jr. 50, 366
BENNETT H. 27, 290, 295
BENNETT W.H. 96, 140, 149, 269, 357
BERGMANN J.G. 141
BERNAS B. 50
BERZELIUS J.J. 59, 165
BETTELEY I.G. v
BEYERS H.P. 7
BISKUPSKY V.S. 27
BISQUE R.E. 239
BLANCHET M.L. 45
BLANKLEY M. 284
BLYUM I.A. 264
BOCK R. 27
BODART D.E. 160
BODKIN J.B. 271
BOL'SHAKOVA L.I. 175
BORGSTROM L.H. 127
BRANNOCK W.W. 43, 45, 48, 50, 124, 125, 186, 269, 297
BRECKENRIDGE J.G. 160
BROOKES C.J. v
BROOKS R.R. 284, 285, 288
BRUDZ V.G. 274
BURRELL D.C. 366
BURTON J.D. 175, 179
BUSH P.R. 133

CAMPBELL E.Y. 102, 261, 262, 289
CAMPBELL W.C. 217
CARMICHAEL I. 155, 363
CARPENTER F.G. 134
CARTER D. 156, 159, 161, 163, 253, 258, 361
CHALMERS R.A. 13, 208
CHAN K.M. 245, 247
CHAO T.T. 84, 217, 233, 304, 366
CHAPMAN J.E. 211
CHAU Y.K. 284, 285
CHENG K.L. 284
CHESHER S.E. 65

CHILDRESS A.E. 261
CHIRNSIDE R.C. 32, 43
CHOW A 177, 306
CHRISTIE O.H.J. 8
CHUECAS L. 152
CIONI R. 99
CLAEYS A. 179
CLARK A.H. 138
CLULEY H.J. 119, 179
CORBETT J.A. 127, 135
COREY R.B. 45
COURVILLE S. 186
CRESSHAW G.L. 284
CROCKET J.H. 306
CROUCH W.H.Jr. 143
CRUZ R.B. 217
CULKIN F. 175, 331

DADONE A. 56
DANIELSSON L. 110
DAS B.C. 353
DEAN J.A. 99
DE LAETER J.R. 339
DEWEY D.W. 156
DEUTSCH S. 61
DIEZ L.P. 55
DIMITRIADIS D. 331, 332
DOCEKAL B. 95
DOLEZAL J. 27
DONALDSON E.M 196
DURST R.A. 174

EARDLEY R.P. 61
EASTON A.J. 192
EDGE R.A. 275
ELINSON M.M. 182
ENGELS J.C. 5
ESSON J. 228, 260
EVANS W.H. 71, 166, 223

FABROKOVA E.A. 312
FAHEY J.J. 125, 195
FAIRBAIRN H.W. 4
FARZANEH A. 110, 142, 172
FEIGL F. 255
FEIK F. 213
FERRIS A.P. 135
FICKLIN W.H. 143
FILBY R.H. 143
FINN A.N. 23
FISHKOVA N.L. 308
FITTON J.G. 194
FLANAGAN F.J. 6
FLASCHKA H. 24
FLEET M.E. 109, 110
FLOYD M. 239
FORNASERI M. 312
FRACHE R. 56, 99
FRANTSUZOVA T.A. 182
FRATTA M. 325
FRIEDHEIM C. 192
FRENCH W.J. 17, 196
FROHLICH F. 151
FRYER B.J. 309
FUGE R. 152

GAGE J.C. 214
GALKINA L.L. 275
GIBSON E.K.Jr. 266
GILBERT D.D. 340
GILL R.C.O. 194
GILLIS J. 179, 284
GOLEMBESKI T. 284, 285
GONG H. 115
GOOCH F.A. 190
GORYUSHINA V.G. 277
GOVINDARAJU K. 6, 7, 55, 67
GRAFF P.R. 48
GRANDI L. 312

GREENLAND L.P. 261, 262, 289, 363
GREGORY G.R.E.C. 349
GRIMALDI F.S. 86, 88, 143, 149, 155, 255, 260, 308
GROGAN C.H. 152
GROSS H. 214, 217, 361
GROVES A.W. 188, 196, 311
GUIMONT J. 344

HAMILTON W.B. 12
HANNAKER P. 325
HANSON C.K. 288
HARRIS F.R. 195
HARRISON A. 122
HARRISON M.P. 178
HARVEY C.O. 186, 187
HARWOOD H.F. 36, 252
HATCHER J.T. 110
HAWLEY W.G. 27
HAYNES S.J. 138
HEAD P.C. 240
HEINRICHS H. 107, 115, 325
HEISIG G.B. 200
HESS D.C. 213, 218
HEY M.H. 200
HILDON M.A. 61
HILLEBRAND W.F. 1, 16, 36, 183
HOFFMAN J.I. 23, 31, 32
HOLT E.V. 36
HOOREMAN M. 255
HORSTMAN E.L. 66
HOSTE J. 160, 179, 284
HUANG W.H. 141, 169
HUBERT A.E. 288
HUFFMAN C.Jr. 339, 363, 364
HUGHES T.C. 325
HUME D.S. 208
HUSLER J. 262
HUTCHISON D. 172, 246
HUTTON R.C. 99
HYBBINETTE A.G. 179

INGAMELLS C.O. 5, 27, 49, 50, 302
IORDANOV N. 213
ISAEVA A.G. 312
ISKANDAR I.K. 238
IVANOV V.V. 191

JACKSON M.L. 45
JEANROY E. 56
JECKO G. 8
JEFFERY P.G. 123, 127, 133, 134, 184, 188, 247, 300, 302, 349, 359
JEFFERY P.M. 339
JEMANEANU M. 239
JENKINS L.B. 331, 332
JENKINS N. 261
JEPSON W.B. 135
JOHNS W.D. 141, 169
JOHNSON D.J. 181
JONES P.D. 340

KANE J.S. 107
KARANOVICH G.G. 102, 103
KATZ A. 65, 312
KAWABUCHI K. 247
KERR G.O. 359
KERRICH R. 309
KHALIZOVA V.A. 291
KING H.G.C. 223
KIPPING P.J. 123, 133, 134
KIRILLOV A.I. 312
KIRIYAMA T. 331, 353, 368
KISS E. 155, 203
KLEKOTKA J.F. 23
KOCHEVA L. 213
KOLTHOFF I.M. 208
KORKISCH J. 102, 213, 214, 217, 331, 332, 353, 361

KRAMER H. 112, 122
KUKHARENKO A.A. 8
KURODA P.K. 140
KURODA R. 247, 331, 353, 368
KUZNETZOV V.I. 175
KUZNETZOV V.K. 175

LAKIN H.W. 81, 284, 288
LANGMYHR F.J. 19, 20, 48, 51, 56, 67, 115, 177, 217, 231, 325
LEBEDEV V.I. 67
LEONARD M.A. 121
LIEBENBERG C.J. 56
LINDNER J. 133
LIPINSKY W. 177
LITEAUNU C. v
LO-SUN JEN. 195
LOVERING J.F. 192
LOVERING T.G. 288
LOXTON S.M. v
LUECKE C. 61
LYPKA G.N. 177

McCARTHY J.H. 288
McDONALD A.J. 155, 284, 363
McDONALD C.W. 115
McHUGH J.B. 84
MALAPRADE L. 45
MARGERUM D.W. 363
MARSHALL N.J. 94
MARSHALL R.R. 213, 218
MARSTON H.R. 156
MATTHEWS A.D. 327
MAXWELL J.A. 16, 64, 224
MAY I. 19, 331, 332
MAZZUCOTELLI A. 56, 99, 261, 262
MEGREGIAN S. 166, 169
MERCY E.L.P. 75, 76, 203
MERWIN H.E. 165
MEYROWITZ R. 196
MIHALKA S. 306
MILLER A.D. 143
MILLER C.C. 208
MININ A.A. 348
MIZON K.J. 127, 135
MOLDAN B. 344
MOORE C.E. 266
MOORE F.L. 115
MORGAN G.T. 127
MORRIS D.F.C. 264
MURPHY J.M. 201, 321
MURRAY-SMITH R. 306, 308
MYERS G. 24

NEUBAUER W. 110
NEUERBURG C.J. 288
NEWMAN E.J. 340
NICHOLLS G.D. 200
NICHOLSON R.A. 239, 240
NYDAHL F. 231

O'GORMAN J.V. 65
OLYA A. 264
OMANG S.E. 240
ONISHI H. 81, 91, 93, 175, 340
OTTAWAY J.M. 122, 217
OZAWA T. 317

PAGE E.S. 13
PAKALNS P. 331
PARISSIS C.M. 52, 302, 345
PARKES A. 306, 308
PATROVSKY V. 264
PAUS P.E. 51, 240
PAVLOVA N.N. 264
PECK L.C. 36
PENFIELD S.L. 186
PETERSON H.E. 288
PICKUP R. 96, 140, 149, 269, 357
POLYAK L.Ya. 349

POOLE A.B. 8
POPEA F. 239
POPOV M.A. 338
POPOV N.P. 274
PORTMANN J.E. 86, 87
PRATT J.H. 195
PRIBIL R. 211, 228
PRICE W.J. 55, 233
PRUDEN G. 223
PYRIH R.Z. 239

RADER L.F. 86, 88, 149, 155, 160, 255, 363
RAFTER R.A. 26
RASMUSSEN S. 177
READ J.I. 133, 135
REED R.A. 61, 295
REICHEN L.E. 195
RESMANN A. 306
RICA I v
RICE T.D. 64
RICHARDSON W.A. 290
RIDDLE C. 268
RIDSDALE P.D. 7
RILEY J.P. 45, 48, 72, 77, 86, 87, 120, 130, 152, 160, 161, 175, 189, 232, 240, 245, 247, 268, 284, 285, 327, 331, 350
ROMANOVA E.V. 277
ROSS W.J. 340
ROWE J.J. 19
ROWLEDGE H.P. 196
RUBESKA I. 306
RYABCHIKOV D.I. 276

SANCHEZ G.A. 110
SANDELL E.B. 81, 91, 93, 140, 151, 156, 175, 179, 186, 245, 269, 340
SANIK J.Jr. 141
SANTACANA F. 363
SANTOS A.M. 115
SANZOLONE R.F. 86, 95, 217, 233, 366
SARMA P.L. 166, 169
SAUNDERS M.J. 75, 76, 203
SAVVIN S.B. 277, 330
SAXBY J.D. 136
SCHMIDT K. 8
SCHNEIDER L.A. 143
SCHNEIDER W.A. 179
SCHNEPFE M.M. 81, 82, 143, 284, 308
SCHROLL E. 8
SCHWEIZER V.B. 146, 161, 362
SEELYE F.T. 26
SEN GUPTA J.G. 133, 135, 277, 282
SERGEANT G.A. 72, 78, 166, 190, 321
SEVERNE B.C. 284, 285, 288
SHAPIRO L. 43, 45, 48, 49, 50, 124, 125, 186, 196, 198, 269, 297
SHAW D.M. 12
SHIMIZU T. 274
SHOROKHOV N.R. 182
SIGHINOLFI G.P. 102, 115, 325
SIMON F.O. 102
SIMONSEN A. 309
SINHA B.C. 291
SINHASENI P. 160, 161
SKREBKOVA L.M. 175
SLOVAK Z. 95
SMALES A.A. 264
SMERAL J. 136
SMITH A.E. 181
SMITH J.D. 240, 340
SMITH J.L. 59
SMITH V.C. 36
SNEESBY G. 290
SOMMERS L.E. 239
SORIO A. 102
SPANGENBERG J.D. 152
STANTON R.E. 284
STEELE T.W. 5

STEIGER G. 165
STEPANOVA N.A. 247
STRELOW F.W.E. 56, 64, 353
STREL'TSOVA S.A. 275
STEVENSON F.J. 265
STONE M. 65
SUHR N.H. 8, 65, 115
SUNDBERG L.L. 253
SVEHLA G. 357
SVEEN S. 19

TANAKA K. 340
TANANAEV N.A. 175
TARTEL N. 65
TAUSON L.V. 7
TERADA K. 284
TERASHIMA S. 84, 86, 94, 361
THEOBALD L.S. 252
THOMASSEN Y. 67
THOMPSON K.C. 204
THOMPSON M. 156
THORNTON W.M.Jr. 211
TOKHTUYEV G.V. 182
TOLG G. 357
TREADWELL F.P. 195
TROLL G. 110, 142, 172, 173, 174
TSUBAKI I. 293
TUREK A. 268

URE A.M. 240

VALCHA Z. 5
VAN DER MEULEN J.H. 142
VAN LOON J.C. 52, 217, 302, 345
VARSHAL G.M. 276
VASILENKO V.T. 368
VESELY V. 211, 228
VICTOR A.H. 56
VINCENT E.A. 196
VINOGRADOV A.P. 265
VOLKOV I.I. 317
VOSKRESENSKAYA N.T. 325
VOSTERS M. 61

WAGSTAFF K. 204
WALSH J.N. 345, 350
WARD F.N. 81
WARREN H.V. 240
WARREN J. 156, 159, 161, 163, 178, 253, 258, 361
WASHINGTON H.S. 1
WATTERSON J.R. 288
WELSCH E.P. 84
WHITE J.C. 340
WHITESIDE P.J. 55, 233
WILCOX L.V. 110
WILLARD H.H. 165
WILLIAMS H.P. 45, 232
WILSON A.D. 72, 78, 127, 184, 187, 188, 195, 196, 300, 302, 319
WINTER O.B. 165
WLOTZKA F. 265, 266

YAKUMINA G.A. 247
YOSHIDA S. 224

ZELYUKOVA Yu.V. 65

Subject Index

Entries in capitals indicate complete chapters or selections of chapters

Accuracy 2, 4

ALKALI METALS 2, 3, 48, 49, 59-69
  See also Caesium, Lithium, Potassium, Rubidium and Sodium

Alkaline Earths
  See Calcium and Magnesium

ALUMINIUM 46-52, 54, 56, 70-80
  determination by difference 38, 70
  "Aluminium traces" 37

Ammonia 265-266

Andalusite 21, 70

Ankerite 124, 221

ANTIMONY 81-85, 86

ARSENIC 81, 86-95

BARIUM 3, 48, 49, 52, 54, 96-101

Beryl 21, 102

BERYLLIUM 102-106

BISMUTH 107-108

BORON 109-114
  removal of 26-27

BROMINE 142-143

Bureau of Analysed Samples 3

CADMIUM 115-116

CAESIUM 49, 52, 59, 60, 65-69

Calcite 123, 124

CALCIUM 3, 39, 47, 50, 51, 52, 54, 117-122, 311

CARBON 123-137

Carbonate Rocks and Minerals 18, 123, 125, 126, 221

Carbonatites 10, 18, 123

Cassiterite 21, 22, 338-342

CHLORINE 138-142
  acid-soluble 138-139
  total 139-140
  water-soluble 138

Chromite 21, 22, 46

CHROMIUM 48, 49, 52, 54, 70, 76, 145-154

Classical scheme of analysis 3, 29-42

COBALT 52, 155-159

Complete analysis 2, 29-42

Consensus value 6

Contamination 16-17

COPPER 52, 160-164

Corundum 46

Crushing and grinding 14-16

"Dean and Stark" 185

Dolomite 123, 124, 221

Elements determined 9-12

Errors 2

FLUORINE 165-174
  removal of 20

GALLIUM 81, 175-178

Geochemical reference material 3-8

GERMANIUM 179-181

GOLD 81, 305-308

Graphite crucibles 50

Grinding see Crushing and grinding

HAFNIUM 368-370

HYDROGEN 182-190
total water 47, 186-190
water evolved at 105° 182-185

Ilmenite 21
INDIUM 191
IODINE 143
IRON 47, 48, 49, 50, 52, 54, 70, 73, 81, 192-212
ferric 201-203
ferrous 47, 194-201
metallic 192-194
total 203-211

Kyanite 21, 26, 27, 70

Lanthanum see Rare Earths
LEAD 52, 213-220
Limestone 10, 123, 120-121
See also Carbonate Rocks and Minerals
LITHIUM 52, 59, 64-65
Lithium metaborate fusion 49, 50, 52, 53

Magnesite 123, 221
MAGNESIUM 3, 40, 47, 49, 50, 51, 52, 54, 120, 221-230
MANGANESE 42, 47, 48, 50, 51, 52, 54, 70, 71, 231-237
MERCURY 238-244
"Mixed Oxides" 70
determination of 34
ignition of 37
"Moisture" 47
See also Hydrogen, water evolved at 105°
MOLYBDENUM 245-251
Monazite 21

National Bureau of Standards 3
NICKEL 49, 52, 54, 70, 71, 252-259
Nickel crucibles 25
NIOBIUM 260-263
NITROGEN 265-266

Organic matter 135-136
OSMIUM 308-309

PALLADIUM 308-309
Perovskite 345
PHOSPHORUS 47, 48, 49, 50, 54, 70, 267-273
Platinum crucibles 19, 22, 24
PLATINUM METALS 308-309
POTASSIUM 49, 50, 51, 52, 54, 59, 61-64
Precision 4
PTFE decomposition vessels 19, 51, 105

Quartzites 9

Rapid analysis 43-58
RARE EARTHS 274-283
Reporting analyses 12-13
RUBIDIUM 49, 52, 59, 60, 65-69
RUTHENIUM 308-309
Rutile 21, 27, 46

Sample decomposition 18-28
Sample preparation 14-17
Sample size 14
Sampling 13-17
Sandstones 9
SCANDIUM 274-275
Scapolite 18, 130
Selection of material 13-14
SELENIUM 86, 284-289

Siderite 124
SILICA (silicon) 3, 9-10, 32, 46, 48, 49, 50, 51, 52, 54, 56, 290-303
ignition and volatilisation of 34
"Silica traces" 38
Sillimanite 21, 26, 27, 70
SILVER 304-305
SODIUM 49, 50, 51, 52, 54, 59, 61-63
Standards, see Geochemical Reference Material
STRONTIUM 3, 39, 48, 52, 54, 311-316
Strontium metaborate fusion 56
SULPHUR 317-324
acid soluble 319
free 317
sulphide 317
total 320-324
water-soluble 319

TANTALUM 264
TELLURIUM 284-289
THALLIUM 81, 325-329
THORIUM 330-337, 355
TIN 81, 86, 338-344
TITANIUM 38, 47, 48, 49, 51, 52, 54, 70, 73, 78, 345-351
Topaz 27
Tourmaline 109, 112-113
TUNGSTEN 81, 245-251

URANIUM 352-356

VANADIUM 38, 48, 51, 70, 71, 73, 357-362

Water see Hydrogen

YTTRIUM 275

ZINC 48, 52, 363-367
Zircon 21, 22, 27, 370
ZIRCONIUM 355, 368-370
Zirconium crucibles 25